TRITIUM IN
THE ENVIRONMENT

The following States are Members of the International Atomic Energy Agency:

AFGHANISTAN
ALBANIA
ALGERIA
ANGOLA
ANTIGUA AND BARBUDA
ARGENTINA
ARMENIA
AUSTRALIA
AUSTRIA
AZERBAIJAN
BAHAMAS, THE
BAHRAIN
BANGLADESH
BARBADOS
BELARUS
BELGIUM
BELIZE
BENIN
BOLIVIA, PLURINATIONAL STATE OF
BOSNIA AND HERZEGOVINA
BOTSWANA
BRAZIL
BRUNEI DARUSSALAM
BULGARIA
BURKINA FASO
BURUNDI
CABO VERDE
CAMBODIA
CAMEROON
CANADA
CENTRAL AFRICAN REPUBLIC
CHAD
CHILE
CHINA
COLOMBIA
COMOROS
CONGO
COOK ISLANDS
COSTA RICA
CÔTE D'IVOIRE
CROATIA
CUBA
CYPRUS
CZECH REPUBLIC
DEMOCRATIC REPUBLIC OF THE CONGO
DENMARK
DJIBOUTI
DOMINICA
DOMINICAN REPUBLIC
ECUADOR
EGYPT
EL SALVADOR
ERITREA
ESTONIA
ESWATINI
ETHIOPIA
FIJI
FINLAND
FRANCE
GABON
GAMBIA, THE
GEORGIA
GERMANY
GHANA
GREECE
GRENADA
GUATEMALA
GUINEA
GUYANA
HAITI
HOLY SEE
HONDURAS
HUNGARY
ICELAND
INDIA
INDONESIA
IRAN, ISLAMIC REPUBLIC OF
IRAQ
IRELAND
ISRAEL
ITALY
JAMAICA
JAPAN
JORDAN
KAZAKHSTAN
KENYA
KOREA, REPUBLIC OF
KUWAIT
KYRGYZSTAN
LAO PEOPLE'S DEMOCRATIC REPUBLIC
LATVIA
LEBANON
LESOTHO
LIBERIA
LIBYA
LIECHTENSTEIN
LITHUANIA
LUXEMBOURG
MADAGASCAR
MALAWI
MALAYSIA
MALI
MALTA
MARSHALL ISLANDS
MAURITANIA
MAURITIUS
MEXICO
MONACO
MONGOLIA
MONTENEGRO
MOROCCO
MOZAMBIQUE
MYANMAR
NAMIBIA
NEPAL
NETHERLANDS, KINGDOM OF THE
NEW ZEALAND
NICARAGUA
NIGER
NIGERIA
NORTH MACEDONIA
NORWAY
OMAN
PAKISTAN
PALAU
PANAMA
PAPUA NEW GUINEA
PARAGUAY
PERU
PHILIPPINES
POLAND
PORTUGAL
QATAR
REPUBLIC OF MOLDOVA
ROMANIA
RUSSIAN FEDERATION
RWANDA
SAINT KITTS AND NEVIS
SAINT LUCIA
SAINT VINCENT AND THE GRENADINES
SAMOA
SAN MARINO
SAUDI ARABIA
SENEGAL
SERBIA
SEYCHELLES
SIERRA LEONE
SINGAPORE
SLOVAKIA
SLOVENIA
SOMALIA
SOUTH AFRICA
SPAIN
SRI LANKA
SUDAN
SWEDEN
SWITZERLAND
SYRIAN ARAB REPUBLIC
TAJIKISTAN
THAILAND
TOGO
TONGA
TRINIDAD AND TOBAGO
TUNISIA
TÜRKİYE
TURKMENISTAN
UGANDA
UKRAINE
UNITED ARAB EMIRATES
UNITED KINGDOM OF GREAT BRITAIN AND NORTHERN IRELAND
UNITED REPUBLIC OF TANZANIA
UNITED STATES OF AMERICA
URUGUAY
UZBEKISTAN
VANUATU
VENEZUELA, BOLIVARIAN REPUBLIC OF
VIET NAM
YEMEN
ZAMBIA
ZIMBABWE

The Agency's Statute was approved on 23 October 1956 by the Conference on the Statute of the IAEA held at United Nations Headquarters, New York; it entered into force on 29 July 1957. The Headquarters of the Agency are situated in Vienna. Its principal objective is "to accelerate and enlarge the contribution of atomic energy to peace, health and prosperity throughout the world".

TECHNICAL REPORTS SERIES No. 495

TRITIUM IN THE ENVIRONMENT

I.W. CROUDACE, A. ULANOWSKI

INTERNATIONAL ATOMIC ENERGY AGENCY
VIENNA, 2025

Publishing Section
International Atomic Energy Agency
Vienna International Centre
PO Box 100
1400 Vienna, Austria
tel.: +43 1 2600 22529 or 22530
email: sales.publications@iaea.org
www.iaea.org/publications

Printed by the IAEA in Austria
July 2025
STI/DOC/010/495
https://doi.org/10.61092/iaea.0qjt-xo5o

IAEA Library Cataloguing in Publication Data

Names: International Atomic Energy Agency.
Title: Tritium in the environment / International Atomic Energy Agency.
Description: Vienna : International Atomic Energy Agency, 2025. | Series: Technical reports series (International Atomic Energy Agency), ISSN 0074–1914 ; no. 495 | Includes bibliographical references.
Identifiers: IAEAL 25-01750 | ISBN 978-92-0-131824-4 (paperback : alk. paper) | ISBN 978-92-0-131924-1 (pdf) | ISBN 978-92-0-132024-7 (epub)
Subjects: LCSH: Tritium. | Tritium — Environmental aspects. | Radioisotopes. | Marine pollution. | Radioactive pollution.
Classification: UDC 546.1 | STI/DOC/010/495

FOREWORD

Tritium is both a natural and an anthropogenic radionuclide. Tritium attracted considerable attention following its substantial production and release to the environment from over 500 atmospheric nuclear weapon tests of varying magnitude conducted since the early 1950s, and continues to attract interest in its production, releases and impacts associated with current and planned nuclear energy developments. Much of the relevant scientific activity currently focuses on the sources and properties of tritium, its applications, its changing inventory and behaviour in the environment, and the associated radiological risks.

This publication incorporates contemporary information and scientific knowledge on the behaviour of tritium in various environmental compartments and media, with the aim of informing and addressing existing and future societal concerns relating to anthropogenic sources of tritium. This information serves to contextualize the possible impacts of power generation using nuclear fission and fusion technologies, nuclear fuel cycle activities, industrial and medical applications, accidental and deliberate releases, as well as military activities. The publication is also intended to serve as a technical reference on the radioecology of tritium, addressing current needs for updated and verified information, and to guide specialists and members of the public in understanding future environmental impacts.

This book is part of the series of IAEA Technical Reports on the environmental behaviour of radionuclides of radium, polonium and uranium, and adds to a growing body of IAEA books and proceedings on tritium and its applications in research, industry and the environment.

This publication was developed in the IAEA by an international group of experts and professionals, whose dedicated and dependable work is gratefully acknowledged. The IAEA is also grateful to N. Priest (United Kingdom) and K. Rozanski (Poland) for their valuable contributions.

The IAEA officer responsible for the publication was A. Ulanowski of the Division of Physical and Chemical Sciences.

CONTENTS

Chapter 1

INTRODUCTION

I.W. CROUDACE, M. GRÖNING, A. ULANOWSKI

1.1. BACKGROUND

Tritium, ^3H, a heavy isotope of hydrogen, belongs to the group of naturally occurring radionuclides. Some tritium is continuously produced mostly in the Earth's atmosphere via cosmic radiation induced nuclear reactions with nuclei of atmospheric nitrogen and oxygen (cosmogenic production). A greater amount of tritium was introduced to the environment in the middle of the 20th century with the testing of over 500 nuclear weapons and devices in the atmosphere (anthropogenic production). By the late 1950s, the scientific understanding of the impacts of these atmospheric tests was spearheaded by the United Nations Scientific Committee on Effects of Atomic Radiation (UNSCEAR)[1], which raised concerns about the increasing levels of globally distributed human made radioactive fallout, including tritium [1.1].

The concentration of environmental tritium generated from nuclear weapon testing reached a maximum in 1963 and was orders of magnitude greater than the natural fluxes and inventories of tritium. The greatest impact was identified in the Northern Hemisphere, where most testing and deposition had occurred. The broad international acceptance that testing in the atmosphere needed to cease led to the ratification of a series of international nuclear test ban treaties, beginning with the Partial Test Ban Treaty, signed by the Union of Soviet Socialist Republics (USSR), the United States of America (USA) and the United Kingdom (UK) in Moscow in August 1963. Besides the nuclear weapon tests, the significant international development of nuclear power production and related nuclear fuel cycle operations created another substantial source of anthropogenic tritium in the environment.

Scientific and societal concerns motivated the assessment of radiological impacts from all fallout radionuclides and of their potential to produce detrimental effects for human health and the environment. Since the middle of the 20th century, it has been generally understood, with some provisos, that tritium posed a low radiological hazard. This was based on the low energy output

---

[1] https://www.unscear.org

from the radioactive decay of tritium, its relatively short half-life and its tendency to become rapidly dispersed and diluted in the Earth's hydrosphere.

From the 1950s onwards, much international tritium research focused on its deposition, dispersion, bioaccumulation, biokinetics and dosimetry. Increasingly significant were its applications as an important hydrological radioactive tracer. There are many publications on tritium in the environment, with the IAEA making a substantive contribution. The first IAEA publication on the subject, entitled Tritium in the Physical and Biological Sciences, was published in 1962 [1.2, 1.3], following a symposium held in 1961, only a few years after the inception of the IAEA. This was soon followed by several Technical Reports Series publications providing an overview and discussion on scientific applications of tritium in the field of hydrology [1.4–1.7].

In the 1970s and 1980s, several reports on tritium measurement techniques and on tritium's presence in ecosystems were published by the IAEA [1.8–1.12]. In 1979, the IAEA published the conference proceedings Behaviour of Tritium in the Environment [1.13], which can be regarded as a predecessor to this publication.

In the early 1960s, the use of tritium as an ideal tool for tracing young groundwater became an important part of hydrological studies involving groundwater age estimation. This was enhanced and supplemented later by sophisticated groundwater age determinations combining the use of tritium and its noble gas decay product, ^{3}He. All these developments contributed to a distinct and innovative branch of hydrology supporting diverse marine and terrestrial investigations [1.14]. The Global Network of Isotopes in Precipitation (GNIP), jointly operated by the World Meteorological Organization (WMO) and the IAEA, established a key database of tritium in precipitation using worldwide monthly data collections for over 60 years [1.15]. This database was later complemented by a database on tritium in rivers and fresh water [1.14, 1.16–1.21].

Various industrial activities and past nuclear accidents have also contributed to the persisting presence of anthropogenic tritium in the environment. Recent achievements in the development of thermonuclear facilities such as ITER [1.22] and anticipation of future fusion power sources continue to stimulate interest in environmental tritium. The accident at the Fukushima Daiichi NPP [1.23] and the subsequent release of processed water with residual tritium [1.24] increased worldwide attention on environmental tritium. This highlighted aspects such as the technological difficulties in separating tritium from stable hydrogen isotopes and the potential radiological risks to humans and the environment.

Relative to other radionuclides, tritium dispersed in the environment, especially as tritiated water, is generally considered to pose a low radiological risk because of the low energy output of its emitted radiation. Because tritium has one of the lowest dose coefficients among all radioactive isotopes, it has higher authorized discharge limits and exemption levels [1.25]. However, experimental evidence collected during decades of radiobiological research suggests that the low energy beta radiation of tritium may have a higher potential for producing detrimental effects to biological tissues compared with the other sparsely ionizing radiations of photons and high energy electrons [1.26]. These findings support discussions on appropriate radiation weighting coefficients for tritium and justify the interest in doses and radiological risks from tritium incorporated in the body in the form of organically bound tritium (OBT). These risks are believed to be higher than those of water-incorporated tritium molecules ($^1H^3HO$ or HTO), owing to the longer residence of OBT in cells and its proximity to DNA and other radiosensitive targets [1.27]. It is recognized that direct studies of OBT-related biological effects are exceedingly difficult because of the low concentrations of tritium in biological tissues [1.26]. Correspondingly, the properties and interactions of tritium remain under continuing scrutiny using a variety of new methods to investigate potential unforeseen harmful effects.

It has become internationally recognized that there is a need to strengthen legal and regulatory frameworks and their implementation in relation to emergency preparedness and response [1.28–1.30]. This includes the need for a sound scientific basis for assessment and evaluation of the possible radiological impacts resulting from tritium releases under emergency conditions [1.31]. Through the IAEA's international model validation and data compilation programmes, standardized conceptual and mathematical models have been developed and tested to simulate tritium transfer in the environment following an emergency. These include the Environmental Modelling for Radiation Safety (EMRAS I and II) programmes and the Modelling and Data for Radiological Impact Assessments (MODARIA I and II) programmes. The validation of the models has required the consideration of a wide range of environmental and weather conditions and seasonal dynamics [1.31].

1.2. OBJECTIVE

This publication complements the Technical Reports Series books dedicated to the environmental transfer and behaviour of specific radionuclides, including radium [1.32], polonium [1.33], uranium [1.34] and thorium (in preparation). It also provides an update on environmental tritium science through a systematic and comprehensive review of properties, sources, distribution, dosimetry, radiobiology and measurement techniques.

Guidance and recommendations provided here in relation to identified good practices represent expert opinion but are not made on the basis of a consensus of all Member States.

1.3. SCOPE

In recent decades, new data and knowledge have substantially enhanced understanding of the behaviour, distribution, dispersion, accumulation, measurement and dosimetry of tritium. This publication draws on numerous sources of high quality data and compiles these advances in a single document. It aims to provide an up to date reference framework of lasting value for a broad spectrum of scientists, specialists, and decision and policy makers regarding the occurrence, behaviour and potential future impacts of tritium in the main compartments of the global environment. Examples include the releases of anthropogenic tritium trapped in permafrost and glaciers linked to global warming; the operation of experimental fusion reactors; the controlled discharges of tritiated aqueous wastes from nuclear fuel reprocessing plants; and decontamination and recovery works after a nuclear accident. Data from international peer reviewed publications and accessible databases are used to illustrate the origins, scale, temporal changes and impacts of tritium in the environment, and representative studies from different Member States are provided as examples.

The use of tritium as a tracer and age indicator in hydrology is only briefly discussed in this book, as there is already a wealth of IAEA publications and other material on this subject [1.14]. Other specialized topics (e.g. safety assessments or detailed studies of tritium related to industrial practices, nuclear facilities and occupational radiation protection) are also covered in other IAEA publications and are not discussed in detail here.

1.4. STRUCTURE

This publication consists of nine chapters, each of which systematically addresses current scientific knowledge on a specific topic relevant to tritium in the environment. As each chapter is intended to be an independent, detailed reference document, several chapters address similar information on tritium source terms, properties and its mobility in the environment, and no effort was made to avoid such repetition.

Chapter 2 describes the physical and chemical properties of tritium, including tritium radioactive decay, its chemical occurrence in the environment, and its transfer and exchange. The chapter concludes with an overview of relationships between tritium activity concentrations and atom ratios for various compounds.

Chapter 3 elaborates on the sources of tritium: its natural production by interaction of cosmic rays with the nuclei of atmospheric constituents, as well as its anthropogenic sources (i.e. nuclear weapon tests, nuclear power plants (NPPs), nuclear fuel reprocessing plants, tritium production sites, accidental releases and future fusion reactors).

Chapter 4 discusses different aspects of tritium in the atmosphere and deals with tritium source terms, tritium transport in the atmosphere, monitoring programmes of tritium content in precipitation (e.g. GNIP) and the impact of technogenic tritium releases to the atmosphere and tritium transfer at the soil–air interface.

Chapter 5 discusses tritium in marine and brackish waters, including tritium transport in the marine environment. It also presents examples of the powerful uses of tritium and its radiogenic progeny ^{3}He as a tracer for ocean circulation and dispersal processes, estimates the content of tritium in different oceans and addresses the marine dispersion of tritium from continental sources. Finally, it considers aspects of tritium sources and bioavailability.

Chapter 6 addresses the presence of tritium in freshwater environments, including relevant sources, transport and exchange processes in freshwater systems, and physical factors affecting tritium concentrations; discusses the transient concentration development with time during the nuclear era; and considers OBT in the terrestrial environment.

Chapter 7 discusses tritium in the terrestrial environment, including tritium sources and tritium incorporation into organic matter, its behaviour and fate in organic matter. It also provides a prediction of the long term fate of tritium in the terrestrial environment.

Chapter 8 deals with dosimetry and the radiological properties of tritium, covering different intake pathways of tritium, its behaviour in the human body, data and methods for assessment of radiation doses and radiological impacts for humans, animals and plants.

Chapter 9 is devoted to techniques of tritium assay, including sampling, speciation, possible contamination issues, sample preparation and tritium extraction, including electrolytic enrichment of tritium in water samples required to measure low activity concentrations. It also describes three measurement techniques: liquid scintillation spectrometry, gas flow proportional counting and ^{3}He ingrowth mass spectrometry.

REFERENCES TO CHAPTER 1

[1.1] UNITED NATIONS, Radioactive Contamination of the Environment by Nuclear Tests (Report to the General Assembly), Annex A, Supplement No. 14 (A/5814), Scientific Committee on the Effects of Atomic Radiation (UNSCEAR), UN, New York (1964).

[1.2] INTERNATIONAL ATOMIC ENERGY AGENCY, Tritium in the Physical and Biological Sciences, Vol. I, Proceedings Series, IAEA, Vienna (1962).
[1.3] INTERNATIONAL ATOMIC ENERGY AGENCY, ibid., Vol. II.
[1.4] INTERNATIONAL ATOMIC ENERGY AGENCY, Application of Isotope Techniques in Hydrology, Technical Reports Series No. 11, IAEA, Vienna (1962).
[1.5] INTERNATIONAL ATOMIC ENERGY AGENCY, Isotope Techniques for Hydrology, Technical Reports Series No. 23, IAEA, Vienna (1964).
[1.6] INTERNATIONAL ATOMIC ENERGY AGENCY, Tritium and Other Environmental Isotopes in the Hydrogeological Cycle, Technical Reports Series No. 73, IAEA, Vienna (1967).
[1.7] INTERNATIONAL ATOMIC ENERGY AGENCY, Guidebook on Nuclear Techniques in Hydrology, Technical Reports Series No. 91, IAEA, Vienna (1968).
[1.8] TUGSAVUL, A., MERTEN, D., SUSCHNY, O., The Reliability of Low-Level Radiochemical Analysis: Results of Intercomparisons Organized by the Agency During the Period 1966-1969, IAEA, Vienna (1970).
[1.9] INTERNATIONAL ATOMIC ENERGY AGENCY, Low-Level Tritium Measurements (Proceedings of a Consultants Group Meeting, Vienna, 24–28 Sept. 1979), IAEA- TECDOC-246, IAEA, Vienna (1981).
[1.10] INTERNATIONAL ATOMIC ENERGY AGENCY, Methods of Low-Level Counting and Spectrometry (Berlin West, 6–10 April 1981), Proceedings Series, IAEA, Vienna (1981).
[1.11] INTERNATIONAL ATOMIC ENERGY AGENCY, Tritium in Some Typical Ecosystems, Technical Reports Series No. 207, IAEA, Vienna (1981).
[1.12] INTERNATIONAL ATOMIC ENERGY AGENCY, Isotopes in Lake Studies, Panel Proceedings Series, IAEA, Vienna (1979).
[1.13] INTERNATIONAL ATOMIC ENERGY AGENCY, Behaviour of Tritium in the Environment (San Francisco, 16–20 Oct. 1978), Proceedings Series, IAEA, Vienna (1979).
[1.14] INTERNATIONAL ATOMIC ENERGY AGENCY, Isotope Hydrology Symposia, Isotope Hydrology Publications, IAEA Isotope Hydrology Network, https://nucleus.iaea.org/sites/ihn/Pages/IHPublications.aspx
[1.15] INTERNATIONAL ATOMIC ENERGY AGENCY, Global Network of Isotopes in Precipitation (GNIP), https://www.iaea.org/services/networks/gnip
[1.16] INTERNATIONAL ATOMIC ENERGY AGENCY, Global Network of Isotopes in Rivers (GNIR), https://nucleus.iaea.org/sites/ihn/Pages/GNIR.aspx
[1.17] INTERNATIONAL ATOMIC ENERGY AGENCY, Interpretation of Environmental Isotope Data in Hydrology (Vienna, 24–28 June 1968), IAEA-TECDOC-116, IAEA, Vienna (1970).
[1.18] INTERNATIONAL ATOMIC ENERGY AGENCY, Environmental Isotope Data No. 10, World Survey of Isotope Concentration in Precipitation (1988–1991), Technical Reports Series No. 371, IAEA, Vienna (1994).

[1.19] INTERNATIONAL ATOMIC ENERGY AGENCY, Statistical Treatment of Data on Environmental Isotopes in Precipitation, Technical Reports Series No. 331, IAEA, Vienna (1992).

[1.20] INTERNATIONAL ATOMIC ENERGY AGENCY, Isotopic Composition of Precipitation in the Mediterranean Basin in Relation to Air Circulation Patterns and Climate, IAEA-TECDOC-1453, IAEA, Vienna (2005).

[1.21] INTERNATIONAL ATOMIC ENERGY AGENCY, Monitoring Isotopes in Rivers: Creation of the Global Network of Isotopes in Rivers (GNIR), IAEA-TECDOC-1673, IAEA, Vienna (2012).

[1.22] ITER, A Pragmatic and Creative Approach (2018), https://www.iter.org/newsline/-/2933

[1.23] INTERNATIONAL ATOMIC ENERGY AGENCY, The Fukushima Daiichi Accident, IAEA, Vienna (2015).

[1.24] INTERNATIONAL ATOMIC ENERGY AGENCY, IAEA Comprehensive Report on the Safety Review of the ALPS Treated Water at the Fukushima Daiichi Nuclear Power Station, IAEA, Vienna (2023), https://www.iaea.org/topics/response/fukushima-daiichi-alps-treated-water-discharge-comprehensive-reports

[1.25] INTERNATIONAL ATOMIC ENERGY AGENCY, Radiation Protection and Safety of Radiation Sources: International Basic Safety Standards, IAEA Safety Standards Series No. GSR Part 3, IAEA, Vienna (2014), https://doi.org/10.61092/iaea.u2pu-60vm

[1.26] UNITED NATIONS, Biological Effects of Selected Internal Emitters—Tritium, Annex C (Report to the General Assembly), Scientific Committee on the Effects of Atomic Radiation (UNSCEAR), UN, New York (2017).

[1.27] INTERNATIONAL ATOMIC ENERGY AGENCY, IAEA Review of Safety Related Aspects of Handling ALPS Treated Water at TEPCO's Fukushima Daiichi Nuclear Power Station, First Interlaboratory Comparison on the Determination of Radionuclides in ALPS Treated Water, IAEA, Vienna (2022), https://www.iaea.org/sites/default/files/first_interlaboratory_comparison_on_the_determination_of_radionuclides_in_alps_treated_water.pdf

[1.28] INTERNATIONAL ATOMIC ENERGY AGENCY, IAEA Report on Preparedness and Response for a Nuclear or Radiological Emergency in the Light of the Accident at the Fukushima Daiichi Nuclear Power Plant, Action Plan on Nuclear Safety Series, IAEA, Vienna (2013).

[1.29] INTERNATIONAL ATOMIC ENERGY AGENCY, Preparedness and Response for a Nuclear or Radiological Emergency, IAEA Safety Standards Series No. GSR Part 7, IAEA, Vienna (2015), https://doi.org/10.61092/iaea.3dbe-055p

[1.30] INTERNATIONAL ATOMIC ENERGY AGENCY, Preparedness and Response for a Nuclear or Radiological Emergency Combined with Other Incidents or Emergencies, EPR-Combined Emergencies, IAEA, Vienna (2020).

[1.31] INTERNATIONAL ATOMIC ENERGY AGENCY, Transfer of Tritium in the Environment After Accidental Releases from Nuclear Facilities, IAEA-TECDOC-1738, IAEA, Vienna (2014).

[1.32] INTERNATIONAL ATOMIC ENERGY AGENCY, The Environmental Behaviour of Radium: Revised Edition, Technical Reports Series No. 476, IAEA, Vienna (2014).

[1.33] INTERNATIONAL ATOMIC ENERGY AGENCY, The Environmental Behaviour of Polonium, Technical Reports Series No. 484, IAEA, Vienna (2017).

[1.34] INTERNATIONAL ATOMIC ENERGY AGENCY, The Environmental Behaviour of Uranium, Technical Reports Series No. 488, IAEA, Vienna (2023).

Chapter 2

PROPERTIES OF TRITIUM

V. VOSTROTIN, F. EYROLLE

2.1. PHYSICAL AND CHEMICAL PROPERTIES OF TRITIUM

Tritium (^3H, commonly abbreviated as T) is the heaviest isotope of hydrogen found in the environment. The tritium atom consists of a nucleus with one proton, two neutrons and one electron. The size of a tritium atom is well represented by its Bohr radius, $a_0 \approx 53$ pm. The nucleus of tritium, with an atomic mass of 3.016 u, has a binding energy of 2.827 MeV per nucleon [2.1]. The tritium nucleus is unstable and radioactive.

The energy required to ionize a tritium atom, $T \rightarrow T^+ + e^-$, is 13.55 eV. The dissociation energy of a molecule formed by two atoms of hydrogen varies slightly for different isotopic modifications of molecular hydrogen and is equal to 4.63 eV for the T–T bond, 4.61 eV for the D–T bond (where D denotes deuterium, ^2H) and 4.57 eV for the H–T bond. These compare with 4.52 eV for the H–H and 4.60 eV for D–D bonds [2.2].

The physical properties of tritium are similar to those of common stable protium (^1H) and stable deuterium. Under normal conditions, tritium is a colourless, highly flammable diatomic gas with the molecular formula T_2. It is possible to make liquid tritium at atmospheric pressure by cooling it below 25 K (−248°C). Liquid hydrogen can be stored in insulated containers under pressure.

Elemental tritium has a high coefficient of diffusion. It readily diffuses through porous substances such as rubber and can also diffuse through metal. Being an isotope of hydrogen, tritium is also highly reactive and, depending on the prevailing physical and chemical conditions, can be involved in many chemical reactions. The environmentally dominant form of tritium is tritiated water in the form of HTO. Other forms, DTO and TTO, are also released by heavy water reactors (HWRs), but these are less common and are considered to be of secondary environmental importance.

Tritium is a promising fuel component for energy production using nuclear fusion. This is because, at high temperatures of the order of 10^8 K, a tritium nucleus (in a confined plasma) can fuse with a deuterium nucleus, producing stable ^4He, a neutron and significant energy:

$$D + T \rightarrow {}^4He + n + 17.6\ \text{MeV} \tag{2.1}$$

This fusion reaction creates high energy output, which is distributed between the neutron, with average energy 14.1 MeV, and the nucleus of ^4He, with average recoil energy of 3.5 MeV [2.3].

The nucleus of the tritium atom is unstable and undergoes beta decay with half-life $T_{1/2} = 12.32 \pm 0.02$ a [2.4] (see also Table 2.1 [2.5, 2.6]) to stable ^3He, producing a beta particle and an antineutrino:

$$^3H \rightarrow {}^3He^+ + e^- + \bar{\nu}_e \tag{2.2}$$

The specific activity A_T of elemental tritium is:

$$A_T = \frac{\ln 2\ N_A}{T_{1/2}\ \mu_T} \approx 3.560 \times 10^{14}\ \text{Bq/g} = 356\ \text{TBq/g} \tag{2.3}$$

where N_A is the Avogadro constant (mol^{-1}) [2.7] and μ_T is the molar mass of elemental tritium (g/mol) [2.4].

Beta particles emitted by the spontaneous decay of tritium are a weakly penetrating radiation owing to their low energy, with a mean value of 5.682 keV and a maximum energy of 18.592 keV [2.5]. In water, the average track length of the beta particle is 0.95 μm and the maximum track length is 7.7 μm [2.6], which is comparable to a typical biological cell nucleus diameter of 6–15 μm and a cell diameter of 10–100 μm. Owing to the low energy of tritium beta particles and their short ranges, they are effectively absorbed by liquid and solid materials.

The radioactive decay of the tritium nucleus results in the transfer of some recoil energy to its progeny, the positively charged ion $^3He^+$. This energy depends on the directions in which the electron and the antineutrino are emitted during the beta decay; correspondingly, the recoil energy of ^3He varies and is characterized by a mean value of 1.0 eV and a maximum of 3.3 eV. This energy is insufficient for either self-ionization of the progeny atom (required energy of the order 10 keV) or for tissue ionization (requiring about 30 eV). Besides the recoil energy, the progeny $^3He^+$ ion also carries an excitation energy of about 11 eV, which can influence the fate of any organic molecule to which the tritium atom is bound, thus leading to its chemical transmutation and the modification of its chemical properties. The main physical properties of tritium are presented in Table 2.1.

TABLE 2.1. MAIN PHYSICAL PROPERTIES OF TRITIUM

Physical parameter	Value	Reference
Binding energy per nucleon (MeV)	2.827	[2.1]
Half-life (s)	388 781 328 ± 631 139	[2.5]
Half-life (year)[a]	12.32 ± 0.02	[2.4]
Progeny	^{3}He	[2.5]
Type of radioactive decay	β decay	[2.5]
Specific activity (Bq/g)	3.560×10^{14}	Eq. (2.3)
Beta-particle average energy (keV)	5.6817 ± 0.0012	[2.5]
Beta-particle maximum energy (keV)	18.5906 ± 0.0032	[2.5]
Maximum CSDA[b] range of beta particles in air (mm)	7.2	[2.6]
Maximum CSDA range of beta particles in water (μm)	7.7	[2.6]

[a] Tropical year, equal to 31 556 926 s [2.4].
[b] CSDA: continuous slowing down approximation.

2.2. ISOTOPIC FRACTIONATION

Tritium is heavier than the main stable isotope of hydrogen, protium, thus increasing the angle between hydrogen bonds of the tritiated water molecule from 104.5° (in $^{1}H_2O$ molecules) to 120° [2.8]. This difference leads to isotopic fractionation between tritium and protium by altering the kinetic and thermodynamic properties of the tritiated water molecules. These properties control each phase change (evaporation, condensation, solidification, fusion) [2.9]. These are exploited for the electrolytic enrichment of tritiated water, which is used to increase the concentration of tritium in water samples. This enrichment leads to improved analytical efficiency and reduced counting time with liquid scintillation measuring instruments.

Other applications using the fractionation of tritium and stable isotopes of hydrogen (protium and deuterium) is in the study of plant growth [2.10] and in discriminating between the responses of C3 plants (e.g. grassland grass, wheat) and C4 plants (e.g. maize, sugarcane) [2.11, 2.12] to climate change [2.13]. It is now widely accepted that the incorporation of tritium within organic matter occurs with low isotopic fractionation [2.8, 2.14–2.16].

2.3. TRITIUM COMPOUNDS IN THE ENVIRONMENT

In the environment, tritium — like protium and deuterium — is present in all compartments of the geosphere: atmosphere, hydrosphere, lithosphere, biosphere and anthroposphere [2.8]. Numerous physicochemical forms of tritium (tritium compounds) are distributed among these compartments and exchange among themselves. The most important of these tritium compounds are illustrated in Fig. 2.1.

The chemical properties and behaviour of tritium are almost identical to those of the main isotope of hydrogen, protium. Whether of natural origin or human made, tritium is generally mobile in the environment and in biological systems, except when isolated for various times into resilient or refractory

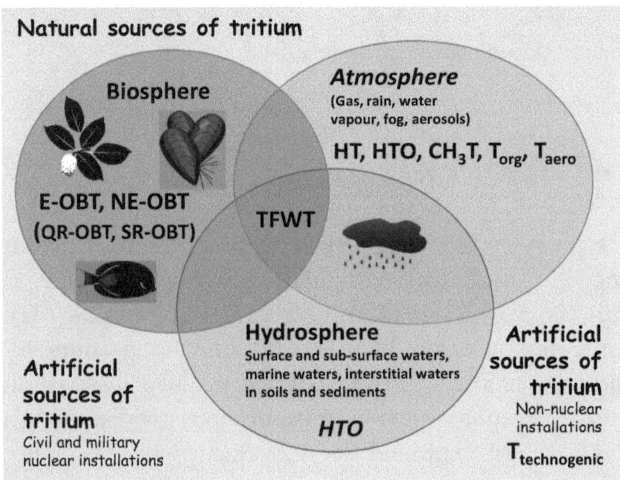

FIG. 2.1. Physicochemical forms of tritium in the environment: HT (tritiated hydrogen), HTO (tritiated water), CH_3T (tritiated methane), T_{org} (gaseous organic tritium), T_{aero} (tritiated aerosols), TFWT (tissue-free water tritium), $T_{technogenic}$ (technogenic tritium), E-OBT (exchangeable organically bound tritium), NE-OBT (non-exchangeable organically bound tritium), QR-OBT (quickly recycled organically bound tritium), SR-OBT (slowly recycled organically bound tritium). Reproduced from Ref. [2.8] with permission courtesy of Elsevier.

materials (e.g. technogenic particulates, woody materials; see Chapter 6). Most environmental tritium exists in one of three chemical forms [2.17]:

(a) Tritiated water. Tritiated water, also known as tritium oxide or 'superheavy water' (11% heavier than H_2O), is the prevalent form of tritium in the natural environment and in living organisms. Tritium is either introduced directly into the environment in this form, or HTO originates from HT or T_2 oxidation facilitated by light (photosynthesis) or bacterial action. Tritium in HTO rapidly exchanges with protium in water, thus promoting homogenization of HTO concentrations in living organisms. The difference in atomic mass of molecules in which the light hydrogen atom is replaced by tritium influences isotopic ratios in physical processes such as evaporation, condensation or solidification. This leads to low level enrichment of tritium relative to hydrogen atoms in the condensed phase. Tritium enrichment, along with ^{17}O enrichment, also occurs in surface waters subjected to evaporation, such as in salt pans.

(b) Gaseous tritium (HT or T_2). This chemical form of tritium is typically associated with airborne releases of this gas from irradiated nuclear fuels at nuclear facilities (e.g. via reprocessing operations) and could become more significant in the future owing to the development of nuclear fusion reactors for electricity production that use tritium as a fuel. The processes of oxidation of HT or T_2 to HTO, and the subsequent behaviour of HTO in the environment, are governed by complex mechanisms. Following their release into the environment, HT and T_2 are dispersed in the local atmosphere and reach the soil surface. Only a small fraction, estimated from less than 0.4% [2.18] to 15–25% [2.19], of HT and T_2 can be converted to HTO in atmospheric air via oxidation by free hydroxyl radicals (OH). In the soil, HT and T_2 are more efficiently converted to HTO via enzymatic reactions and microbial activity. The rate of HT and T_2 oxidation in soil depends on the soil's properties, including water content, porosity and microbiology. The oxidation of hydrogen in the soil column, including HT and T_2, accounts for approximately 75–85% of the global sink of tropospheric hydrogen [2.19]. As a result of the oxidation process, tritiated hydrogen is transformed into tritiated water and enters the water cycle. Given the large amounts of tritium and associated radiological concerns, future designs of power fusion reactors are expected to implement measures to prevent tritium gas releases by capturing and recycling as much as technically feasible.

(c) OBT. In this form, tritium is chemically bonded to elements such as carbon, oxygen, nitrogen and sulphur that are present in organic matter. This results in incorporation of tritium into organic compounds by isotopic exchange and by the synthetic and/or metabolic processes of living matter. The large number of such organic compounds and their different distributions within organisms may explain the heterogeneous distribution of tritium among biological tissues. Persistence of tritium within such compounds depends on the nature of the bond between the tritium and the organic molecule and on the association of organic molecules with biological tissues. Correspondingly, a distinction is made between the following:

- — Exchangeable OBT (E-OBT), where bond strengths for hydrogen atoms bound to oxygen, sulphur or nitrogen atoms allow an exchange between molecular hydrogen and tritium. Molecules containing tritium may be readily accessible for new exchanges. This exchangeable fraction of tritium that is bound in a labile form to biomolecules may easily reach equilibrium with the concentration of tritiated water in the inter- and intracellular environments.
- — Non-exchangeable OBT (NE-OBT), which occurs when tritium is covalently bound to a carbon atom within the molecule. This is a strong and stable bond, as long as the biomolecule itself is neither transformed nor destroyed. In this case, the time when tritium remains incorporated depends on the rate of biomolecular turnover; namely, fast in the case of molecules involved in the energy cycle and slow in the case of structural molecules. Similarly, macromolecules such as DNA and energy reserve molecules may be subject to slow turnover and long tritium retention times.

The partitioning of tritium present in organic matter among tritiated water, exchangeable and non-exchangeable tritium varies according to the respective intake of HTO and OBT, the nature of the organic bonds generating OBT, and the metabolism of individual organism species. Tritium is heavier than ordinary hydrogen and, owing to the kinetic isotope effect, enzymatic reactions involving the heavier isotope are slower than those involving the lighter isotopes [2.17, 2.20].

Besides biointegration of tritium into molecules of 'natural' organic compounds, various technological processes can result in the production of other, technogenic, forms of tritium. Such compounds bearing highly tritiated molecules or particles can be produced by, for example, radiopharmaceutical industries, such as the former industrial sites in the Bristol Channel and Cardiff Bay in the UK (see Chapters 3 and 5). They may also originate from the residues of luminous paints produced by the watchmaking industry, for example, those found in sediments of the upper reaches of the Rhône river [2.21]. Similarly, tritiated residues have been observed near factories manufacturing self-luminous safety lighting and labels in Canada [2.22]. The last two examples indicate that weakly biodegradable (i.e. refractory or recalcitrant) tritiated submicrometre particles were dispersed in the environment, leading to long term environmental fixation of tritium or its compounds. These sources, probably numerous and widespread in countries with nuclear or radioisotope processing industries, are not yet fully identified, so their impact on the environment is poorly known.

Monitoring of diverse forms of tritium in water requires special procedures for sample treatment. Conventionally, the water samples are distilled prior to assay of tritium activity using liquid scintillation spectrometry. However, contribution of organic forms of tritium released from radiopharmaceutical facilities and refractory microparticles may also need to be analysed in unprocessed liquid samples. Solid matrices, such as soils and sediments, need to be combusted prior to analysis [2.8].

2.4. CHARACTERIZATION OF TRITIUM CONTENT

2.4.1. Activity concentration units

As tritium is a radioactive isotope, its content in a sample can be expressed in terms of activity concentration. Depending on the aggregate state and properties of the sample, as well as on the sample preparation procedures, the activity concentration can be reported per volume (Bq/m^3 or Bq/L) for gases and liquids, or per wet or dry mass (Bq/kg) for liquids and solids.

Expressing tritium content in terms of activity concentration in various media facilitates the assessment of tritium intake and concomitant radiation doses and is thus practical and commonly used in radiation protection. However, activity concentration units provide no information on the abundance of tritium relative to hydrogen, which may be required to characterize the tritium contamination of a carrying medium. For example, the tritium activity concentration in the atmosphere, expressed in Bq/m^3 of air, depends on the amount of water vapour contained in the air; consequently, the relationship between activity concentrations in air and water is not straightforward. Similarly, the tritium content in a dry solid material, expressed in Bq/kg, depends on the amount of hydrogen in the dry residue of the sample, while the total amount of tritium in a fresh sample includes also tritium content in the sample water (Bq/L). In such situations, comparison of the tritium content in different media can become more informative when using atomic ratios of numbers of tritium to hydrogen atoms (see Section 2.4.2). Correspondingly, the water vapour content in air samples and hydrogen content (usually the weight fraction of H) in solid compounds have also to be measured.

Analysis of the total hydrogen concentration in samples with small amounts of organic matter (e.g. most soils, riverine or marine sediments) faces additional uncertainties if the OBT content is expressed as an activity concentration. In some cases, the presence of OBT can contribute most of the uncertainty owing to the heterogeneous distribution of organic matter within samples with low (close to the detection limits) hydrogen content. Consideration should also be given to the analysis of soil and sediment samples that may contain significant amounts of either residual inorganic water molecules or hydrogen bound within mineral structures. These can lead to an underestimation of the OBT content due to the dilution of tritium by these unlabelled water components. These are collected with tritium during the combustive extraction process of solid samples [2.23]. Similarly, a single drying of a sample, followed by combustion, can result in an overestimation of OBT content, if residual E-OBT in the dried sample is interpreted as NE-OBT (see Chapter 9).

2.4.2. Tritium unit

Owing to the chemical equivalence of tritium to ubiquitous hydrogen (protium), the tritium unit (TU), representing a ratio of tritium to hydrogen atoms, was introduced and is in common scientific use to conveniently characterize the amount of tritium in a material. The TU arose from the estimate of the natural abundance of tritium in Norwegian surface water, which was taken to approximate the secular equilibrium produced between cosmogenic production and radioactive decay [2.24]. Analytical TUs are mainly used in research involving very small tritium activity concentrations, such as in oceanography and hydrology.

The TU is dimensionless and is defined as the number of tritium atoms per 10^{18} atoms of hydrogen. In liquid water at room temperature, 1 TU corresponds to an activity concentration of approximately 0.119 Bq/L. Other practically relevant equivalence relationships between the analytical TU and the tritium activity concentration in various compounds, computed using data from Refs [2.1, 2.2, 2.5], are shown in Table 2.2.

The relationships between units based on atom ratios and activity concentrations are not straightforward and depend on molar masses (volumes), on the number of hydrogen atoms in a molecule as well as on physical densities and aggregated states of the compound. A systematic and comprehensive analysis of the relationship between TU and activity based units used in hydrological applications for various combinations in water of the three hydrogen isotopes and three environmentally prevalent isotopes of oxygen can be found elsewhere [2.24].

TABLE 2.2. EQUIVALENCE RELATIONSHIPS BETWEEN TRITIUM ACTIVITY CONCENTRATION AND ATOM RATIO OF TRITIUM AND PROTIUM FOR WATER AND SOME LABORATORY GASES AND SOLVENTS

Substance	State	Formula	Temperature (°C)	Activity concentration equivalent of tritium atom ratio of 1 TU		Tritium atom ratio (TU) equivalent of activity concentration	
				(Bq/kg)	(Bq/L)	1 Bq/kg	1 Bq/L
Water	Liquid	H_2O	0.1	1.192×10^{-1}	1.192×10^{-1}	8.390	8.390
			3.96[a]		1.192×10^{-1}		8.390
			20		1.190×10^{-1}		8.405
			25		1.188×10^{-1}		8.414
Hydrogen	Gas[b]	H_2	25	1.065	8.778×10^{-5}	0.938 7	11 392
Methane	Gas[b]	CH_4	25	2.677×10^{-1}	1.755×10^{-4}	3.735	5 698
Ethane	Gas[b]	C_2H_6	25	2.142×10^{-1}	2.633×10^{-4}	4.668	3 798
Methanol	Liquid	CH_4O	20	1.340×10^{-1}	1.060×10^{-1}	7.461	9.433
Ethanol	Liquid	C_2H_6O	20	$1.39\ 8 \times 10^{-1}$	1.104×10^{-1}	7.151	9.060
Acetone	Liquid	C_3H_6O	20	$1.10\ 9 \times 10^{-1}$	8.765×10^{-2}	9.016	11.14
Benzene	Liquid	C_6H_6	20	8.247×10^{-2}	7.248×10^{-2}	12.13	13.80
Toluene	Liquid	C_7H_8	20	9.322×10^{-2}	8.081×10^{-2}	10.727	12.38

[a] Temperature at which the maximal density of liquid water at pressure 101.325 kPa (1 atm) is observed.
[b] At normal atmospheric pressure (101.325 kPa).

REFERENCES TO CHAPTER 2

[2.1] HUANG, W.J., WANG, M., KONDEV, F.G., AUDI, G., NAIMI, S., The AME 2020 atomic mass evaluation (I) — Evaluation of input data, and adjustment procedures, Chinese Phys. C **45** 3 (2021) 030002, https://doi.org/10.1088/1674-1137/abddb0

[2.2] LUO, Y.-R., Comprehensive Handbook of Chemical Bond Energies, CRC Press, Boca Raton, FL (2007), https://doi.org/10.1201/9781420007282

[2.3] UNITED NATIONS, Biological Effects of Selected Internal Emitters—Tritium, Annex C (Report to the General Assembly), Scientific Committee on the Effects of Atomic Radiation (UNSCEAR), UN, New York (2017).

[2.4] KONDEV, F.G., WANG, M., HUANG, W.J., NAIMI, S., AUDI, G., The NUBASE2020 evaluation of nuclear physics properties, Chinese Phys. C **45** 3 (2021) 030001, https://doi.org/10.1088/1674-1137/abddae

[2.5] PURCELL, J.E., SHEU, C.G., Nuclear Data Sheets for A = 3, Nucl. Data Sheets **130** (2015) 1, https://doi.org/10.1016/j.nds.2015.11.001

[2.6] SELTZER, S., et al., Key Data for Ionizing-Radiation Dosimetry: Measurement Standards and Applications, ICRU Rep. 90 (2014), https://doi.org/10.1093/jicru_ndw043

[2.7] TIESINGA, E., MOHR, P., NEWELL, D., TAYLOR, B., CODATA recommended values of the fundamental physical constants: 2018, Rev. Mod. Phys. **93** (2021) 025010, https://doi.org/10.1103/RevModPhys.93.025010

[2.8] EYROLLE, F., et al., An updated review on tritium in the environment, J. Environ. Radioact. **181** (2018) 128–137, https://doi.org/10.1016/j.jenvrad.2017.11.001

[2.9] BAUMGÄRTNER, F., KIM, M.-A., Isotope effects in the equilibrium and non-equilibrium vaporization of tritiated water and ice, Int. J. Rad. Appl. Instr. A **41** 4 (1990) 395–399, https://doi.org/10.1016/0883-2889(90)90149-B

[2.10] EPSTEIN, S., YAPP, C.J., HALL, J.H., The determination of the D/H ratio of non-exchangeable hydrogen in cellulose extracted from aquatic and land plants, Earth Planet. Sci. Lett. **30** 2 (1976) 241–251, https://doi.org/10.1016/0012-821X(76)90251-X

[2.11] WHITE, J.W.C., “Stable hydrogen isotope ratios in plants: A review of current theory and some potential applications”, in Stable Isotopes in Ecological Research, Ecological Studies, Vol. 68 (RUNDEL, P.W., EHLERINGER, J.R., NAGY, K.A., Eds), Springer, New York (1989) 142–162 https://doi.org/10.1007/978-1-4612-3498-2_10

[2.12] FLANAGAN, L.B., BAIN, J.F., EHLERINGER, J.R., Stable oxygen and hydrogen isotope composition of leaf water in C_3 and C_4 plant species under field conditions, Oecologia **88** (1991) 394–400, https://doi.org/10.1007/BF00317584

[2.13] SCHIEGL, W.E., Deuterium content of peat as a paleoclimatic recorder, Science **175** 4021 (1972) 512–513, https://doi.org/10.1126/science.175.4021.512

[2.14] McFARLANE, J.C., Tritium fractionation in plants, Environ. Exp. Bot. **16** 1 (1976) 9–14, https://doi.org/10.1016/0098-8472(76)90028-9

[2.15] KIM, M.A., BAUMGÄRTNER, F., Tritium fractionation in anomalous water bound to environmental samples, J. Environ. Radioact. **36** 2 (1997) 111–127, https://doi.org/10.1016/S0265-931X(96)00083-5

[2.16] BAUMGÄRTNER, F., KIM, M.-A., Hydrogen isotope partitioning during plant growth, Appl. Radiat. Isot. **48** 6 (1997) 721–725, https://doi.org/10.1016/S0969-8043(96)00301-6

[2.17] INSTITUT DE RADIOPROTECTION ET DE SÛRETÉ NUCLÉAIRE, Tritium and the Environment, IRSN, Fontenay-aux-Roses (2010).

[2.18] KIM, S.B., STUART, M., BREDLAW, M., FESTARINI, A., BEATON, D., HT to HTO conversion and field experiments near Darlington Nuclear Power Generating Station (DNPGS) site, J. Environ. Radioact. **132** (2014) 73–80, https://doi.org/10.1016/j.jenvrad.2014.02.001

[2.19] YVER, C.E., et al., A new estimation of the recent tropospheric molecular hydrogen budget using atmospheric observations and variational inversion, Atmos. Chem. Phys. **11** 7 (2011) 3375–3392, https://doi.org/10.5194/acp-11-3375-2011

[2.20] BELOT, Y., ROY, M., MÉTIVIER, H., Le Tritium de l'Environnement à l'Homme, Les Editions de Physique, Les Ulis (1996).

[2.21] GONTIER, G., SICLET, F., Le tritium organique dans les écosystèmes d'eau douce: évolution à long terme dans l'environnement des centres nucléaires de production d'électricité français, Radioprotection **46** 4 (2011) 457–491, https://doi.org/10.1051/radiopro/2011139

[2.22] MIHOK, S., et al., Tritium dynamics in soils and plants grown under three irrigation regimes at a tritium processing facility in Canada, J. Environ. Radioact. **153** (2016) 176–187, https://doi.org/10.1016/j.jenvrad.2015.12.025

[2.23] KIM, S.B., BAGLAN, N., DAVIS, P.A., Current understanding of organically bound tritium (OBT) in the environment, J. Environ. Radioact. **126** (2013) 83–91, https://doi.org/10.1016/j.jenvrad.2013.07.011

[2.24] GROSSE, A.V., JOHNSTON, W.M., WOLFGANG, R.L., LIBBY, W.F., Tritium in nature, Science **113** 2923 (1951) 1–2, https://doi.org/10.1126/science.113.2923.1

Chapter 3

SOURCES OF TRITIUM

V. VOSTROTIN

3.1. NATURAL SOURCES OF TRITIUM

There are three main sources of natural tritium on the Earth: (a) production in the atmosphere by galactic cosmic rays (GCRs), (b) production in the atmosphere by energetic particles originating from solar coronal mass ejections, and (c) direct accretion from the Sun. Tritium produced by natural processes on the Earth is rapidly converted into HTO, which then enters the global hydrological cycle.

The natural production of tritium in the atmosphere is mainly due to nuclear reactions induced by incident radiation, including protons and other charged GCR particles, and secondary neutrons produced by these reactions with nuclei of atmospheric oxygen and nitrogen. Of these, the most important for tritium production is a nuclear reaction between nitrogen and neutrons with energy exceeding the reaction threshold of 4.3 MeV [3.1]:

$$^{14}\mathrm{N}+\mathrm{n}\rightarrow {}^{12}\mathrm{C}+{}^{3}\mathrm{H} \tag{3.1}$$

The contribution of solar particles to the formation of cosmogenic tritium is small, equalling 1.6% on average [3.2].

The rate of tritium generation in the atmosphere at a given energy spectrum of GCR particles is mainly influenced by three parameters:

(a) The geomagnetic cut-off rigidity, which quantitatively characterizes the protective properties of the Earth's atmosphere. This parameter enables the prediction of the penetration of the GCR particles through the Earth's magnetosphere, depending on the energy and direction of a particle. The geomagnetic cut-off rigidity depends on the shape and strength of the Earth's magnetic field and is mainly determined by latitude. Owing to the weaker influence of the magnetic field on GCR particles in the polar regions, the rate of tritium formation in these regions is up to seven times higher than in the equatorial region [3.3].

(b) The density of the atmosphere, which exponentially decreases with increasing altitude above the Earth's surface.
(c) The solar activity, which determines the shape and strength of the geomagnetic field of the Sun, acting at distances of up to 100 astronomical units from the Sun and interacting with the Earth's magnetosphere. With a decrease in solar activity, the intensity of the flow of GCR particles reaching the Earth and its atmosphere increases and vice versa.

Using recent data [3.4], the contemporary model of cosmogenic tritium generation in the Earth's atmosphere [3.5] results in estimates of global annual tritium production varying from 220 to 330 g owing to the solar activity cycle. The estimated annual average value of 280 g corresponds to the activity of produced tritium of approximately 100 PBq/a. Earlier estimates of the natural tritium production rate [3.6] resulted in lower values of the tritium production rate in the atmosphere and ranged from 150 to 200 g/a. For example, estimates of the global inventory of natural tritium in the hydrological cycle based on earlier studies resulted in values ranging from 3.2 kg to 3.6 kg [3.7–3.9]. With an average rate of formation of tritium atoms in the atmosphere equal to 0.32 atoms $\cdot$ $cm^{-2} \cdot s^{-1}$ [3.4], the production rate and the average global inventory are estimated as 258 g/a and 4.5 kg, respectively [3.10]. A similar approach using the average production rate [3.11] results in an estimate of 4.9 kg for the global inventory.

Estimates of the various rates of natural production of tritium atoms in the Earth's atmosphere found in the literature [3.3–3.5, 3.11–3.15] are presented in Table 3.1.

3.2. ANTHROPOGENIC SOURCES OF TRITIUM

In the period from 1952 to 1980, the series of atmospheric thermonuclear test explosions introduced approximately 1.8×10^{20}–2.3×10^{20} Bq (equivalent to 520–664 kg) of tritium into the environment [3.2, 3.9, 3.16]. Currently, in addition to residual activities of these test explosions, the most important anthropogenic sources of environmental tritium are the following:

— NPPs (especially with pressurized HWRs (PHWRs));
— Facilities for reprocessing of spent nuclear fuel;
— Facilities producing tritium for military and civilian purposes;
— Industrial and manufacturing applications of tritiated compounds for peaceful purposes in medicine and industry, such as tritium labelled pharmaceuticals and luminescent signs and paints;
— Nuclear waste storage.

TABLE 3.1. ESTIMATED RATES OF NATURAL PRODUCTION OF TRITIUM ATOMS IN THE EARTH'S ATMOSPHERE (ATOMS · $cm^{-2} \cdot s^{-1}$)

Year of publication	Entire atmosphere			Polar regions			Reference
	Low solar activity	Moderate solar activity	High solar activity	Low solar activity	Moderate solar activity	High solar activity	
1954	—[a]	0.12 ± 0.04	—	—	—	—	[3.11]
1961	—	0.5 ± 0.3[b] 0.28 ± 0.05[c]	—	—	—	—	[3.12]
1966	0.20 ± 0.09	0.19 ± 0.09	0.16 ± 0.09	—	—	—	[3.13]
1967	—	0.20 ± 0.05[d]	—	—	—	—	[3.14]
1979	—	0.26	—	—	—	—	[3.15]
1995	—	0.26	—	—	—	—	[3.3]
2009	—	0.32	—	—	—	—	[3.5]
2020	0.41	0.35	0.27	0.92	0.72	0.51	[3.4]

[a] —: data not available.
[b] Based on the geochemical inventory.
[c] Based on cosmic ray data.
[d] Based on direct measurements.

In the near future, another source of anthropogenic tritium is likely to be created by emerging fusion power generating technologies, including research and prototype devices and facilities under construction. An example is the ITER facility in Cadarache (France), which is currently under construction.[1]

3.2.1. Nuclear weapon tests

Anthropogenic tritium was globally dispersed during above ground nuclear tests, mainly of thermonuclear weapons, conducted by the USA, the USSR and, at a smaller scale, France, the UK and China. According to the estimates of Cauquoin et al. [3.8], fission devices produced tritium at 8 g/Mt, whereas tests of thermonuclear fusion devices produced tritium at 2.07 kg/Mt. According to UNSCEAR [3.9], the 504 atmospheric tests conducted in the period from 1945 to 1980 resulted in a total yield of tritium of about 520 kg (1.86×10^8 TBq).[2] Approximately 95% of all anthropogenic tritium releases occurred from 1952 to 1962. These releases were unevenly dispersed between the Northern and Southern Hemisphere and were distributed by the complex global atmospheric circulations operating in the environment, as well as the location of testing. Consequently, estimation of the residual amount of tritium from the nuclear tests is not straightforward and requires sophisticated modelling (see Section 4.2.2).

The timing and total annual releases of anthropogenic tritium following atmospheric nuclear weapon testing shows a first onset in 1952, with a rapid rise to a maximum in 1963, followed by a sharp decline (Fig. 3.1). The USSR and the USA contributed most to the overall tritium releases owing to their nuclear weapon testing programmes at 65% and 29%, respectively (see Fig. 3.2). The total contribution to tritium releases from nuclear tests undertaken by other countries (China, France and the UK) amounted to approximately 6% [3.9] (cf. Fig. 3.2).

3.2.2. Nuclear power plants

The total number of nuclear power reactors[3] in operation worldwide in 2021 was 442 [3.17], including 94 units in the USA, 56 in France, 50 in China, 38 in the Russian Federation and 33 in Japan. A comprehensive worldwide list of power reactors in operation and being built, as well as their capacities and electric power supplied, is provided in Table 3.2.

---

[1] https://www.iter.org/construction/construction

[2] See Chapter 4 for alternative estimates.

[3] In this chapter, statistics for the major nuclear power generating facilities is provided. New emerging types of small device, such as small modular reactors or microreactors, are not considered, as their contribution to global release of tritium is currently considered insignificant.

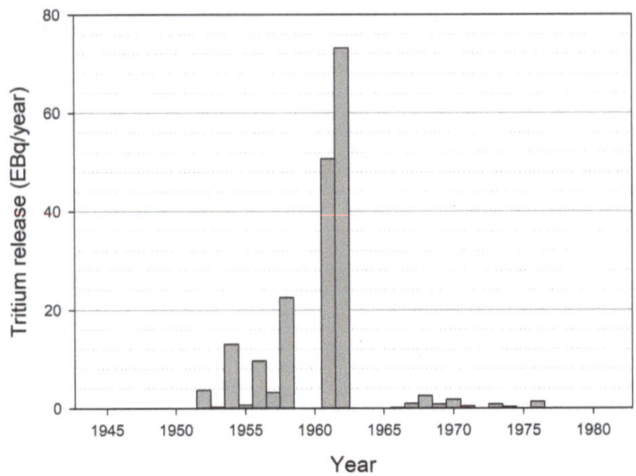

FIG. 3.1. Tritium releases into the atmosphere due to above ground tests of nuclear weapons during the period 1945–1980. The estimated total released activity of tritium was ~1.86×10^8 TBq (~520 kg) [3.9].

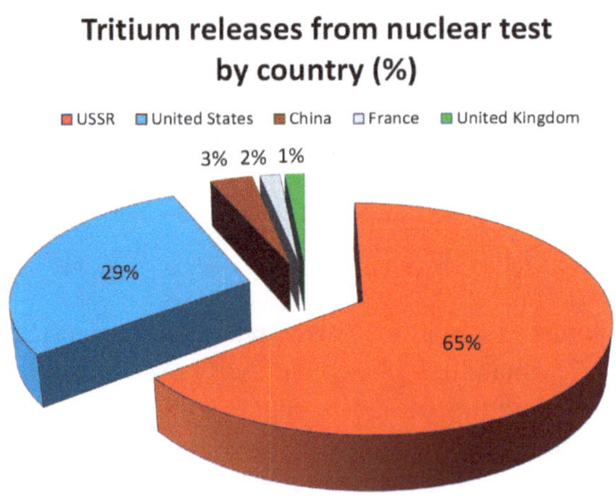

FIG. 3.2. Fractions of the total tritium activity (estimated at 1.86×10^8 TBq) released into the air due to atmospheric nuclear testing from 1945 to 1980 by country [3.9].

TABLE 3.2. GLOBAL OVERVIEW OF POWER REACTORS AND NUCLEAR SHARE ON 31 DECEMBER 2020 [3.17]

Country	Reactors in operation		Reactors under construction		Nuclear electricity supplied	
	Number of units	Net capacity (MW(e))	Number of units	Net capacity (MW(e))	(TWh)	% of total
Argentina	3	1 641	1	25	10.0	7.5
Armenia	1	415	n.a.[a]	n.a.	2.6	34.5
Bangladesh	n.a.	n.a.	2	2 160	n.a.	n.a.
Belarus	1	1 110	1	1 110	0.3	1.0
Belgium	7	5 942	n.a.	n.a.	32.8	39.1
Brazil	2	1 884	1	1 340	13.2	2.1
Bulgaria	2	2 006	n.a.	n.a.	15.9	40.8
Canada	19	13 624	n.a.	n.a.	92.2	14.6
China	50	47 528	13	12 565	344.7	4.9
Czech Republic	6	3 934	n.a.	n.a.	28.4	37.3
Finland	4	2 794	1	1 600	22.4	33.9
France	56	61 370	1	1 630	338.7	70.6
Germany	6	8 113	n.a.	n.a.	60.9	11.3
Hungary	4	1 902	n.a.	n.a.	15.2	48.0
India	22	6 255	7	4 824	40.4	3.3
Iran, Islamic Republic of	1	915	1	974	5.8	1.7
Japan	33	31 679	2	2 653	43.1	5.1
Korea, Republic of	24	23 150	4	5 360	152.6	29.6
Mexico	2	1 552	n.a.	n.a.	10.9	4.9
Netherlands	1	482	n.a.	n.a.	3.9	3.3

TABLE 3.2. GLOBAL OVERVIEW OF POWER REACTORS AND NUCLEAR SHARE ON 31 DECEMBER 2020 [3.17] (cont.)

Country	Reactors in operation		Reactors under construction		Nuclear electricity supplied	
	Number of units	Net capacity (MW(e))	Number of units	Net capacity (MW(e))	(TWh)	% of total
Pakistan	5	1 318	2	2 028	9.6	7.1
Romania	2	1 300	n.a.	n.a.	10.6	19.9
Russian Federation	38	28 578	3	3 459	201.8	20.6
Slovakia	4	1 837	2	880	14.4	53.1
Slovenia	1	688	n.a.	n.a.	6.0	37.8
South Africa	2	1 860	n.a.	n.a.	11.6	5.9
Spain	7	7 121	n.a.	n.a.	55.8	22.2
Sweden	6	6 882	n.a.	n.a.	47.4	29.8
Switzerland	4	2 960	n.a.	n.a.	23.0	32.9
Türkiye	n.a.	n.a.	2	2 228	n.a.	n.a.
United Arab Emirates	1	1 345	3	4 035	1.6	1.1
UK	15	8 923	2	3 260	45.7	14.5
Ukraine	15	13 107	2	2 070	71.5	51.2
USA	94	96 553	2	2 234	789.9	19.7
Total	442	392 612	52	54 435	2 553.2	n.a.

[a] n.a.: not applicable.

The distribution of worldwide operational reactor units by type and total net electric capacity is shown in Table 3.3 [3.17]. The typical tritium release rates per unit of generated electric power for different types of nuclear reactor [3.16] are shown in Table 3.4.

TABLE 3.3. DISTRIBUTION OF OPERATIONAL REACTORS BY TYPE AND ELECTRIC CAPACITY PRODUCED IN 2019 [3.17]

Reactor type	Number of reactors	Total net electrical capacity (MW(e))
Boiling light water cooled and moderated reactor	63	64 122
Fast breeder reactor	3	1 400
Gas cooled, graphite moderated reactor	14	7 725
Light water cooled, graphite moderated reactor	12	8 358
Pressurized heavy water moderated and cooled reactor	48	23 923
Pressurized light water moderated and cooled reactor	302	287 084
	442	392 612

TABLE 3.4. SPECIFIC TRITIUM DISCHARGE RATES FROM DIFFERENT TYPES OF NUCLEAR REACTOR [3.16]

Reactor type	Tritium release rates ($TBq \cdot GW(e)^{-1} \cdot a^{-1}$)	
	In the atmosphere	In water
Pressurized light water moderated and cooled reactor (zircaloy cladding)	3.7	25.9
Boiling light water cooled and moderated reactor	1.85	3.7
Pressurized heavy water moderated and cooled reactor	740	185
Gas cooled, graphite moderated reactor	7.4	11.1

TABLE 3.5. NUMBER OF OPERATIONAL PHWRs AND GENERATED GROSS ELECTRIC CAPACITY IN THE COUNTRIES WHERE SUCH REACTORS ARE OPERATED [3.17]

Country	Number of PHWRs	Gross electric capacity (MW(e))
Argentina	3	1 772
Canada	19	14 512
China	2	1 456
India	18	4 460
Korea, Republic of	3	1 832
Pakistan	1	100
Romania	2	1 411
Total	48	25 543

As seen from the data in Tables 3.3 and 3.4, by the end of 2020, there were 48 PHWRs globally, which produced the greatest environmental releases of tritium to the air from NPPs. Electric energy generation produced by HWR facilities amounted to 6.2% in 2019. Table 3.5 presents the distribution of countries by number of HWRs and electric capacity available [3.17].

The following sections provide more detailed information on tritium releases in the atmosphere and water across various countries and regions.

3.2.2.1. Canada

The electric capacity of HWRs operating in the Canadian NPPs Bruce A and Bruce B in 2020 accounted for 57% of the capacity from all HWRs worldwide. Data on HTO releases from the Canadian NPPs between 2002 and 2016 [3.18–3.32] are shown in Figs 3.3 and 3.4.

Each of the two NPPs consists of four nuclear reactors. The electric capacity of the Bruce A NPP in 2019 amounted to 3320 MW(e) and 3507 MW(e) were generated by Bruce B. From Figs 3.3 and 3.4, it follows that the average specific release rates of tritium to the atmosphere were 182 GBq · $MW(e)^{-1}$ · a^{-1} for the Bruce A NPP and 139 GBq · $MW(e)^{-1}$ · a^{-1} for the Bruce B NPP, with an average specific rate of tritium releases as liquid effluents equal to 54 GBq · $MW(e)^{-1}$ · a^{-1} and 159 GBq · $MW(e)^{-1}$ · a^{-1}, respectively. Thus, the specific rate of releases of

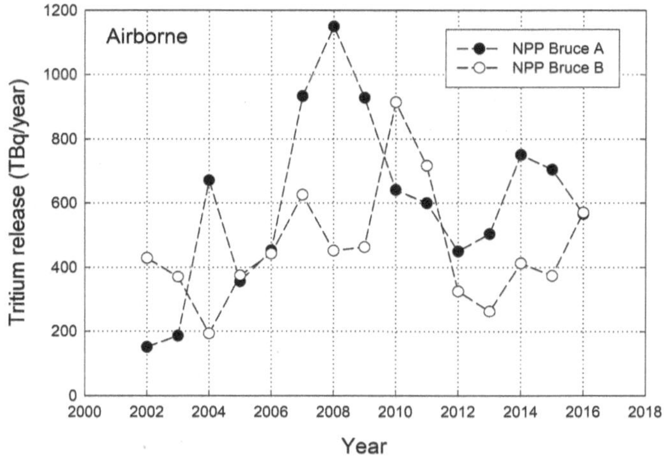

FIG. 3.3. Atmospheric tritium releases from the Bruce A and Bruce B NPPs (Canada) from 2002 to 2016 [3.18–3.32].

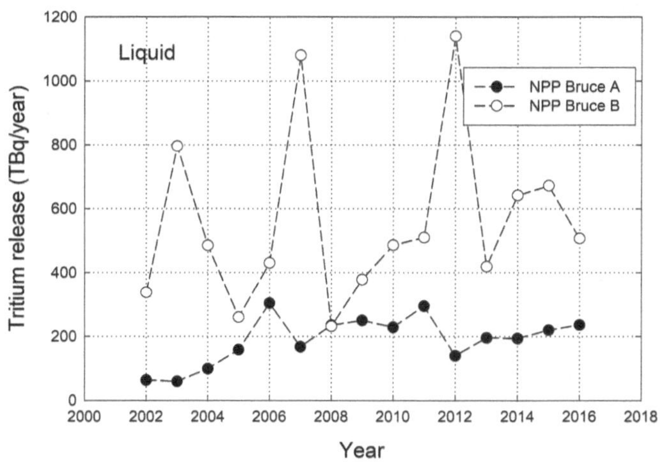

FIG. 3.4. Tritium releases to water from the Bruce A and Bruce B NPPs (Canada) from 2002 to 2016 [3.18–3.32].

tritium to the atmosphere from the Bruce A NPP was on average 1.3 times higher, while tritium releases into water were 2.9 times lower than those from Bruce B NPP. On the basis of the data on specific capacities of the Bruce A and Bruce B NPPs, the annual release of tritium from the Canadian HWRs is estimated at 2320 TBq/a for releases to the atmosphere and 1550 TBq/a into water. Data show that up to 75% of the tritium released from the Canadian HWRs was in the form of DTO.

3.2.2.2. India

In India, the power units TAPS-3 and TAPS-4 of the Tarapur NPP are equipped with heavy water nuclear reactors with an electric capacity of 540 MW(e) each. According to the Indian Atomic Energy Regulatory Board [3.33], operation of these units in 2016 and 2017 resulted in a total release of 84 TBq of tritium to the atmosphere and of 45.7 TBq to water by each unit. Correspondingly, specific release rates for these two HWRs amounted to 78 GBq · $MW(e)^{-1} \cdot a^{-1}$ to airborne and 42 GBq · $MW(e)^{-1} \cdot a^{-1}$ for liquid releases. Compared with the similar PHWR NPP, Bruce A, the specific release rates of tritium from the two HWR units of the Tarapur NPP are lower by 2.3 times for atmospheric releases and by 1.3 times less for release of tritiated water (cf. Section 3.2.2.1).

3.2.2.3. Republic of Korea

The tritium release rates from four NPPs in the Republic of Korea (Kori, Hanbit, Wolsong, Hanul) during the period 2011–2015 have been reported by Kong et al. [3.34]. The Wolsong NPP consists of four PHWR units and has a gross electric capacity of 2570 MW. This NPP produces annually from 135 to 188 TBq of airborne tritium releases, which amounts to an average specific airborne release rate of 59 GBq · $MW(e)^{-1} \cdot a^{-1}$. The three other NPPs, having pressurized light water moderated and cooled reactors (PWRs), release annually from 10 to 18 TBq into the atmosphere, practically independent of their gross electric capacities of 3380 MW(e) (NPP Kori), 6210 MW(e) (NPP Hanbit) and 6220 MW(e) (NPP Hanul).

The reported tritium liquid release rates, originating from the Wolsong NPP, declined from 92 TBq/a in 2011 to 26 TBq/a in 2015. The liquid discharge rates for the three other NPPs with PWRs were in the range 34–78 TBq/a. These also could not be correlated to their gross electric generating capacity.

3.2.2.4. United States of America

Reports on airborne and liquid releases of radionuclides from 68 NPPs in the USA have been published by the US Nuclear Regulatory Commission [3.35]. The total annual release rates of tritium from all US NPPs between 2008 and 2017 are presented in Fig. 3.5. Over this period, the annual release rates varied insignificantly, being on average equal to 269 ± 18 TBq/a to the atmosphere and 1598 ± 106 TBq/a to water (see Fig. 3.5). The overall tritium releases to water from US NPPs (2008–2017) systematically exceeded airborne releases by approximately six times.

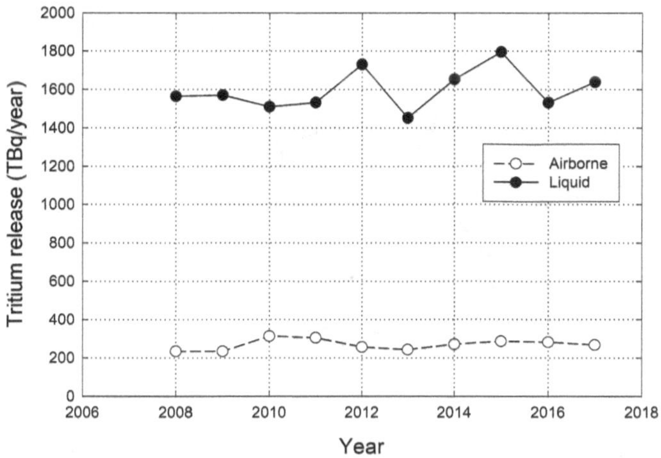

FIG. 3.5. Tritium releases to the atmosphere and to water from NPPs in the USA from 2008 to 2017 [3.35].

Nuclear energy reactors in the USA and Canada are of distinctively different types — light water reactors (LWRs) in the USA and HWRs in Canada. Despite the technical differences, a comparison of tritium releases from US and Canadian NPPs show that the total annual releases of tritium in water were approximately equal, and the total annual tritium releases to air from the US NPPs were approximately an order of magnitude lower than similar releases from the Canadian NPPs. At the same time, electric power generation by the US NPPs in 2018 exceeded the power generation by Canadian NPPs by 8.5 times (807 078 GWh versus 95 037 GWh) [3.36]. These differences can be attributed to the different prevailing types of nuclear reactor; HWR designs are currently used in Canada, unlike the USA, which uses only LWRs to generate electricity [3.17].

3.2.2.5. Europe — countries of the European Union and the United Kingdom

The countries of the European Union (EU) and the UK published information on radionuclide releases into the environment from NPPs and nuclear fuel cycle facilities over 1995–2017 [3.37]. During this period, these countries released 17 200 TBq (48.2 g) of tritium to the atmosphere and 81 500 TBq (228 g) to water (Figs 3.6 and 3.7).

As seen from Fig. 3.6, the largest contributions to airborne tritium releases were from NPPs that operated in the UK (71%), Romania (21%) and France (4%). Atmospheric tritium releases in Romania in 2007–2017 were due to the operation of two HWRs at the Cernavoda NPP. Spain, Belgium and Germany contributed a total of 3% of the total amount of tritium, while operation of the

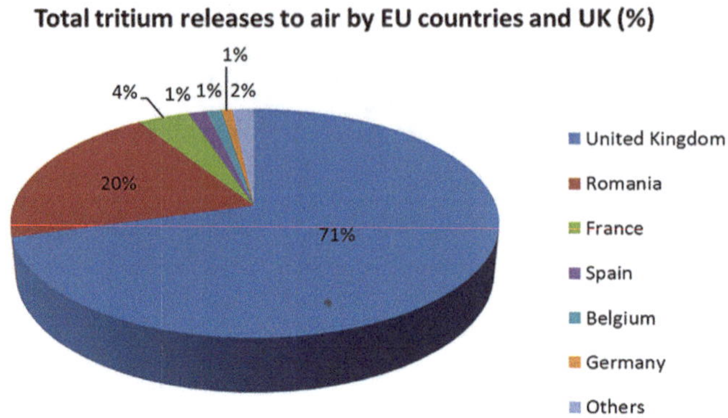

FIG. 3.6. Airborne tritium releases from NPPs of EU countries and the UK between 1995 and 2017. The total tritium release amounted to 17 200 TBq (48.2 g) [3.37].

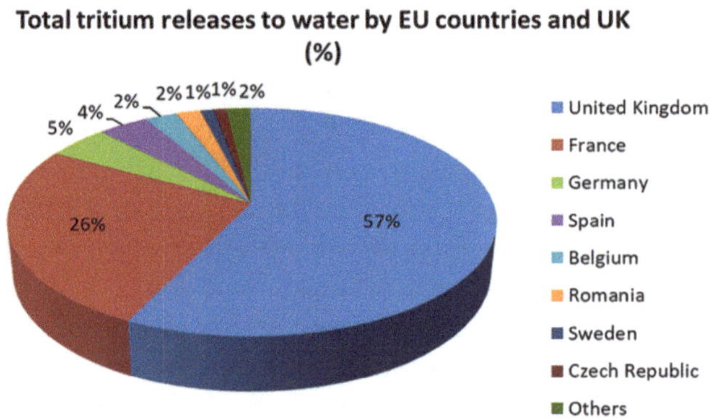

FIG. 3.7. Tritium releases to water from NPPs of the EU countries and the UK (1995–2017). The total tritium release amounted to 81 500 TBq (228 g) [3.37].

NPPs in ten other EU countries accounted for approximately 2% of the total atmospheric release.

As seen from Fig. 3.7, among European countries, the largest contributors of tritium released to water from 1995 to 2017 were the UK (57%), France (26%) and Germany (5%).

The total annual tritium releases to the atmosphere and water from the EU and UK NPPs are presented in Figs 3.8 (airborne) and 3.9 (aquatic). The figures show that in the period 1995–2005, the largest amounts of tritium releases to the atmosphere were produced by the NPPs in the UK, followed by a substantial decline after 2005, when these releases became comparable to the releases from

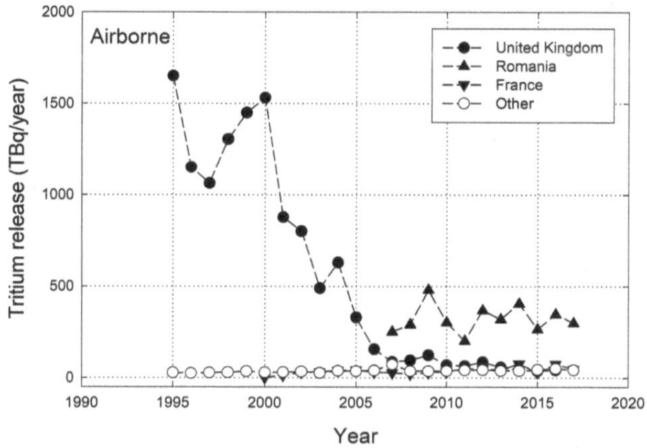

FIG. 3.8. Airborne tritium releases from European NPPs for 1995–2017 [3.37].

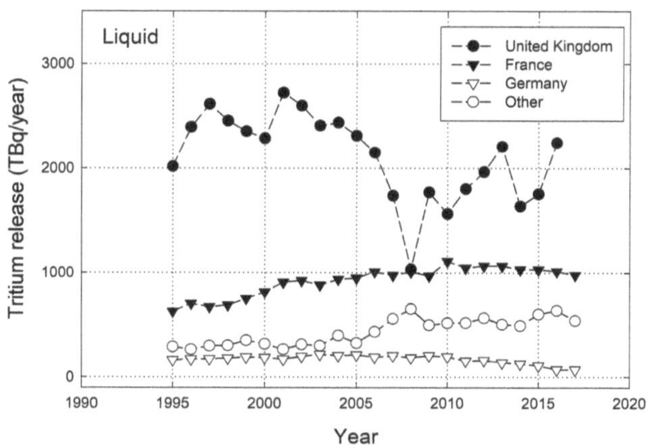

FIG. 3.9. Tritium releases into water from European NPPs for 1995–2017 [3.37].

other European NPPs. From 2007, the NPP in Romania, with two HWRs, appears to be the largest source of tritium entering the atmosphere.

The UK and France had the highest release rates to the aquatic (marine and freshwater) environment from 1995 to 2017 (Fig. 3.9). Comparison of the data from Figs 3.5 and 3.9 suggests that tritium release rates into water from all NPPs in the UK and the USA between 2008 and 2016 were approximately at the same level, ranging from 1500 to 2000 TBq/a.

3.2.2.6. Japan

According to the Japan Nuclear Energy Safety Organization Report [3.38], 18 NPPs were registered to discharge tritium into water from 2003 to 2012. The total annual tritium discharges from Japanese NPPs are presented in Fig. 3.10.

It follows from Fig. 3.10 that reported levels of tritium in liquid effluents from Japanese NPPs between 2003 and 2011 amounted to 300–430 TBq/a. Thereafter, in 2012, the levels of tritium released from all Japanese NPPs were reduced to 50 TBq/a, reflecting the Government's decision to shut down many NPPs in 2012–2014 as a result of the accident at the Fukushima Daiichi NPP in 2011.

Following the accident at the Fukushima Daichi NPP in 2011, persistent and systematic mitigation and decontamination actions implemented at the plant resulted in the accumulation of significant volumes of wastewater. This contained radioactive material, including tritium [3.39]. To facilitate disposal and reduction of the accumulated wastewater (more than one million cubic metres), the Advanced Liquid Processing System (ALPS) was installed and put into operation to remove long living radionuclides and to ensure that residual activity concentrations met regulatory requirements for safe disposal. The ALPS was shown to effectively remove many radionuclides but not tritium [3.40, 3.41], so in March 2023 the ALPS processed water stored at the Fukushima Daiichi NPP

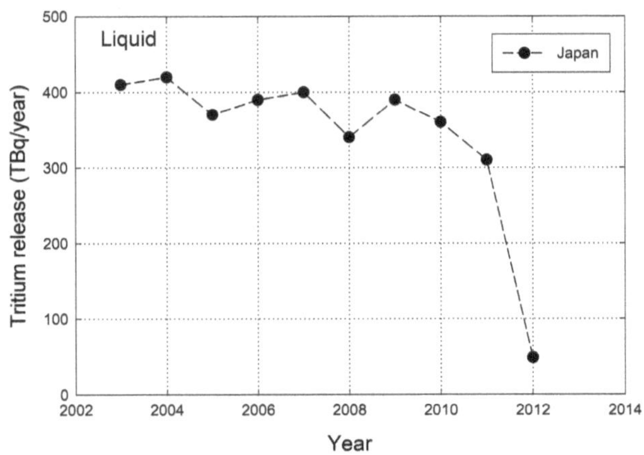

FIG. 3.10. Tritium releases into water from Japanese NPPs for 2003–2012 [3.38].

site still contained an estimated 700 TBq of tritium, which is equivalent to about 13 g of tritium in the form of HTO.[4]

The national regulatory authorities of Japan and the Fukushima Daiichi NPP operator developed and initiated in 2023 a procedure for safe disposal of the ALPS treated water, containing mostly tritium, into the ocean. According to the procedure, which was started in August 2023[5], the ALPS treated water will be diluted and released gradually over the following years to ensure radiological safety for human beings and the environment. The IAEA assists the Japanese Government and the regulator in planning, implementing and monitoring the releases [3.42].

3.2.2.7. Russian Federation

Information on tritium releases from the NPPs in the Russian Federation comes from local reports of NPPs [3.42] and reports of the Research and Production Association 'Typhoon' (Typhoon RPA) of the Russian Federal Service for Hydrometeorology and Environmental Monitoring [3.43]. According to these reports, 230 TBq of tritium was released to the atmosphere and 253 TBq of tritium was released to water from eight Russian NPPs in the period from 2004 to 2017. Table 3.6 presents data on liquid releases of tritium from three NPPs in different periods, based on information from Typhoon RPA [3.43]. Unfortunately,

TABLE 3.6. LIQUID RELEASES OF TRITIUM FROM RUSSIAN NPPs BETWEEN 2004 AND 2017 [3.43]

Nuclear power plant	Time period	Tritium release rate to water (TBq/a)
Kalininskaya	2004–2007	18–24.9
Novovoronezhskaya	2009–2017	9.5–36.3[a]
Kolskaya	2010–2017	10.6–16.4

[a] These figures disagree with reports of the Novovoronezhskaya NPP for the period between 2010 and 2017, which indicated annual liquid releases in the range 0.8–2.9 TBq.

---

[4] https://www.tepco.co.jp/en/hd/decommission/information/newsrelease/reference/pdf/2023/reference_20230928_01-e.pdf

[5] https://www.iaea.org/interactive/timeline/104850

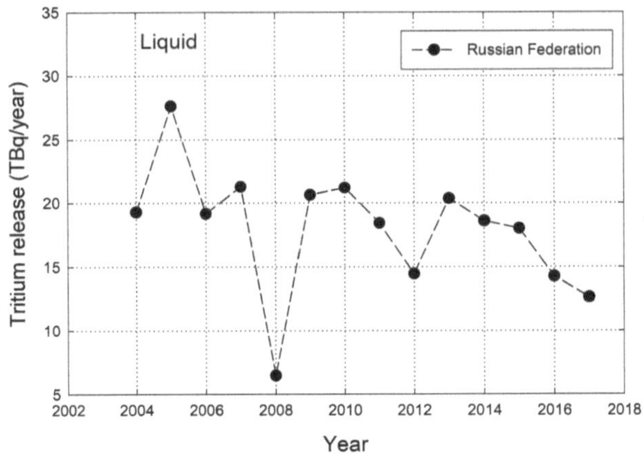

FIG. 3.11. Reported tritium discharges into water from Russian NPPs for 2004–2017. The data shown are for different numbers of NPPs in different periods (see Table 3.6) [3.43].

the NPP and Typhoon RPA reports are incomplete and cover different periods. The summary information on total annual tritium releases to water reported for Russian NPPs between 2004 and 2017 is presented in Fig. 3.11. However, the data shown in the figure are for two NPPs in 2004–2007, five NPPs in 2008–2009, seven NPPs in 2010–2015 and 2017 and eight NPPs in 2016. Because the data are incomplete, they do not reflect the variation of the total tritium release rates in the Russian Federation.

3.2.2.8. China

The average annual release rates of tritium into the air and water from several Chinese NPPs (Qinshan, Daya Bay, Ling Ao) were reported by Yang et al. [3.44, 3.45] and Zhang et al. [3.46]. In Table 3.7, the average annual release rates and the total estimated activity of tritium released from these NPPs in the period from 1993 to 2009 are shown. The data in the table demonstrate that the two HWRs of the Qinshan III NPP appeared as the main source of tritium emissions into the air, making up to 94% of the total release rate of the whole NPP. The annual average rates of tritium releases into water from all NPPs, except for Qinshan I, ranged from 17.1 to 42.9 TBq/a and did not show explicit dependence on the type of reactor.

TABLE 3.7. AVERAGE RATES AND ESTIMATED TOTAL ACTIVITY OF TRITIUM RELEASES FROM CHINESE NUCLEAR POWER PLANTS IN THE PERIOD FROM 1993 TO 2009 [3.44]

NPP	Reactor type	Power capacity (MW(e))	Period	Airborne releases		Liquid releases	
				Average rate (TBq/a)	Total[a] (TBq)	Average rate (TBq/a)	Total[a] (TBq)
Qinshan I	PWR	310	1993–2009	0.654	11.1	3.0	51
Qinshan II	PWR	2 × 650	2002–2009	0.164	1.3	17	137
Qinshan III	PHWR	2 × 700	2003–2009	36.9	258	34	235
Qinshan (total)	n.a.[b]	n.a.	1993–2009	n.a.	271	n.a.	422
Daya Bay	PWR	2 × 984	1994–2009	0.957	15.3	43	654
Ling Ao	PWR	2 × 990	2002–2009	0.635	5.1	41	324

[a] Estimated as a product of the average annual release rate and the reported period length.
[b] n.a.: not applicable.
Note: PHWR: pressurized heavy water reactor; PWR: pressurized water reactor.

More recent data on airborne and liquid releases of tritium from the Qinshan NPPs were provided by Yang et al. [3.45] for the period from 2007 to 2016 and are shown in Figs 3.12 (releases in air) and 3.13 (releases in water). Yang et al. [3.45] compared tritium releases from the Qinshan NPPs to the world-averaged data published by UNSCEAR [3.2, 3.9, 3.47] and noted that the specific release rates of tritium per unit of electrical power capacity from the LWR of the Qinshan I NPP into the air and water were the same as the global averages. The specific tritium release rates per unit electric power capacity from the HWRs of Qinshan III into air and water were an order of magnitude lower than the global averages. The LWRs of the Qinshan II NPP were found in an intermediate position — tritium release rates per unit of electrical power capacity were several times lower than the global averages for tritium released into the air and close to the global averaged values for tritium releases into water.

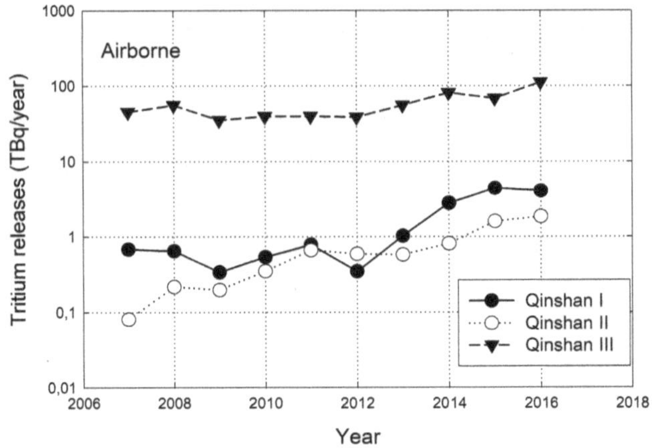

FIG. 3.12. Airborne tritium releases from the Qinshan NPPs (China) in 2007–2016 [3.45].

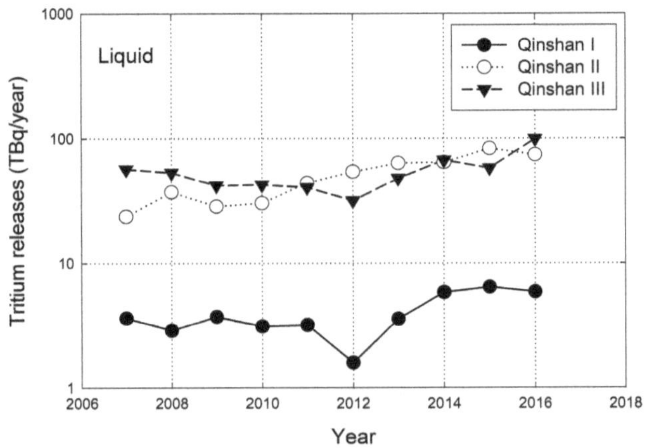

FIG. 3.13. Tritium releases into water from the Qinshan NPPs (China) in 2007–2016 [3.45].

3.2.3. Spent fuel reprocessing plants

According to the IAEA Integrated Nuclear Fuel Cycle Information System (NFCIS) [3.48], there were seven (including the Sellafield facility, where fuel reprocessing operations ceased in 2022[6]) spent fuel reprocessing plants (SFRPs) either in operation or being commissioned worldwide after 2020 (see Table 3.8).

The data in Table 3.8 and in Ref. [3.49] indicate that by 2020 the leading sources of tritium emissions in the world nuclear industry were three SFRPs: La Hague (France), Sellafield (UK) and Mayak Production Association (Mayak PA; Russian Federation).

The end of fuel reprocessing operations in the Sellafield complex and planned start of operations at the Rokkasho reprocessing plant in Japan in 2025[7] may change the relative importance of facilities producing major releases of tritium into the environment.

TABLE 3.8. COMMERCIAL SPENT FUEL REPROCESSING PLANTS IN OPERATION OR PLANNED FOR COMMISSIONING AFTER 2020 [3.48]

Country	Facility name	Facility status	Design capacity (t HM/a)[a]	Start of operation	End of operation
France	La Hague, UP2-800	In operation	1000	1996	
France	La Hague, UP3	In operation	1000	1990	
Japan	Rokkasho Reprocessing Plant	Under construction	800	2025	
Russian Federation	RT-1, Mayak PA	In operation	400	1971	
Russian Federation	Pilot Demonstration Centre for LWR SNF reprocessing, Krasnoyarsk	In operation	231	2016	

[6] https://www.gov.uk/government/news/job-done-sellafield-plant-safely-completes-its-mission

[7] See https://www.jnfl.co.jp/en/release/press/2022/detail/20230210-1.html and https://www.jnfl.co.jp/en/release/press/2022/detail/20230210-1.html

TABLE 3.8. COMMERCIAL SPENT FUEL REPROCESSING PLANTS IN OPERATION OR PLANNED FOR COMMISSIONING AFTER 2020 [3.48] (cont.)

Country	Facility name	Facility status	Design capacity (t HM/a)[a]	Start of operation	End of operation
Russian Federation	RT-2, Krasnoyarsk	Under construction	800	—[b]	
Russian Federation	Test demonstration facility for LWR SNF reprocessing, Krasnoyarsk	Under construction	250	2022	
UK	Sellafield Magnox Reprocessing	In operation[c]	1000	1964	2022

[a] t HM: tonnes of heavy metal.
[b] —: data not available.
[c] The planned end of fuel reprocessing operations was shifted from 2020 to 2022.
Note: LWR: light water reactor; SNF: spent nuclear fuel.

3.2.3.1. *European spent fuel reprocessing plants*

At the time of writing (2021), the spent fuel reprocessing facilities in Europe with the highest capacity were at La Hague and Sellafield. The former has capacity of 2000 t HM/a (tonnes of heavy metal per year) and the latter of 2500 t HM/a.

Tritium in irradiated nuclear fuel is mainly retrieved during reprocessing when the fuel is sheared and dissolved. It is found in the form of tritiated water in liquid discharges, most of which is released into the sea. At the La Hague plant, for example, approximately, 30 g (~10^4 TBq) of tritium is discharged annually per 1600 t of processed nuclear material and at Sellafield approximately 8 g (~2800 TBq) has been released annually [3.6]. To some extent, owing to tritium decay, the amount of released tritium depends on the duration of the holding (cooling) time prior to fuel treatment.

The total annual tritium releases to the atmosphere and to water reported for La Hague and Sellafield between 1995 and 2017 are shown in Figs 3.14 and 3.15, respectively [3.37]. These show that the highest releases to air, but the lowest releases to water, were from the Sellafield complex.

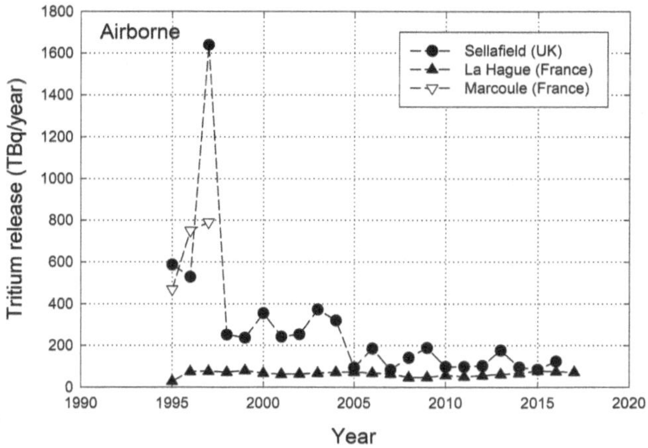

FIG. 3.14. Tritium atmospheric releases from nuclear fuel cycle facilities in the UK and France for 1995–2017 [3.37].

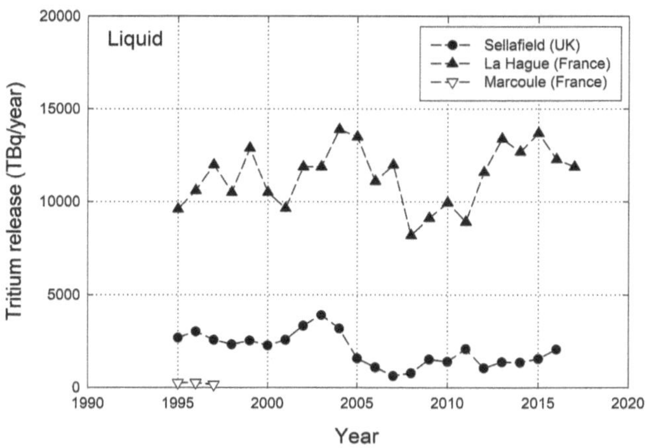

FIG. 3.15. Tritium releases to water from nuclear fuel cycle facilities in the UK and France for 1995–2017 [3.37].

3.2.3.2. Reprocessing plant in the Russian Federation

The reprocessing plant at Mayak PA (Ozersk, Russian Federation) has a capacity of 400 t HM/a. According to the annual reports of Typhoon RPA for the period from 2003 to 2017 [3.43], Mayak PA was the main source of tritium emissions into the environment in the Russian Federation. The total annual release rates of tritium from the site to the atmosphere and water during the specified period are shown in Figs 3.16 and 3.17.

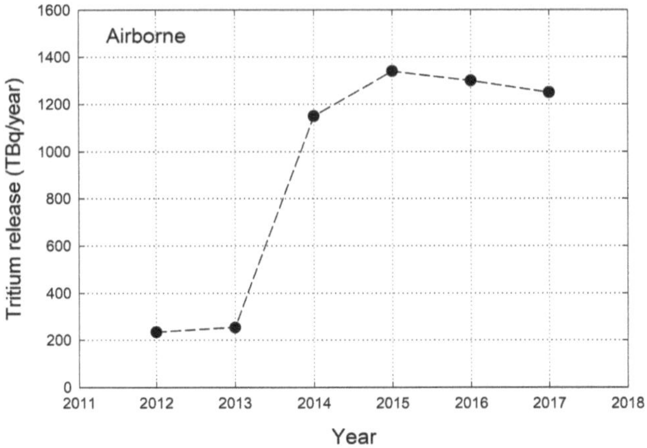

FIG. 3.16. Releases of tritium to the atmosphere from Mayak PA (Russian Federation) [3.43].

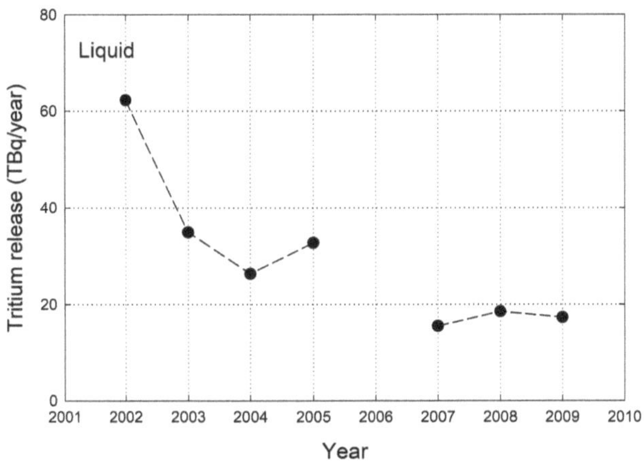

FIG. 3.17. Releases of tritium to water from Mayak PA (Russian Federation) [3.43].

3.2.4. Nuclear research and tritium production facilities

Five countries (USA, USSR/Russian Federation, UK, France, China) that developed and tested thermonuclear weapons possess specialized facilities of two types that release tritium into the environment: (a) tritium production facilities and (b) nuclear weapon and research centres. Information about tritium releases and discharges generated by these facilities and centres is presented below.

3.2.4.1. United States of America

As a multi-faceted nuclear research centre, the Lawrence Livermore National Laboratory (LLNL) in the USA performs operations that result in releases of tritium to the environment. According to LLNL Environmental Reports [3.50], in the period from 1981 to 2017, LLNL released up to 1200 TBq of tritium into the atmosphere. The LLNL reports indicate accidental releases, which occurred in 1984, 1985, 1990 and 1991. They also provide some sparse data on tritium releases from the Sandia National Laboratory, located in New Mexico, indicating releases of 6.9 TBq of airborne tritium in 1993, of which 4.9 TBq was discharged in the form of HTO. The data show (Fig. 3.18) that the highest tritium release rates from LLNL were in the period between 1981 and 1992, with a distinctive maximum in 1984 (270 TBq/a). The amount of atmospheric tritium releases in the period from 1992 to 2017 was on average 3.4 TBq/a.

The level of tritium discharges into water produced by LLNL between 1984 and 2017 was in the range from 6.4×10^{-4} to 0.133 TBq/a — negligible compared with airborne releases.

Operations of the tritium facility at Savannah River National Laboratory (SRNL), located in South Carolina, also resulted in releases of tritium. The total annual tritium releases to the atmosphere and to water reported for SRNL between 2001 and 2017 are presented in Fig. 3.19 [3.51].

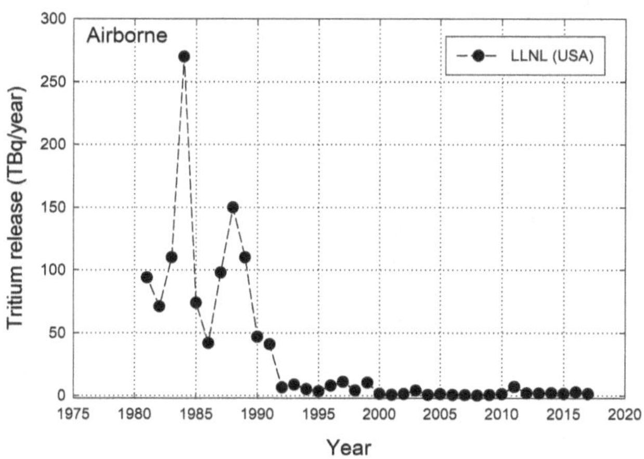

FIG. 3.18. Tritium releases to the atmosphere from LLNL (USA) for 1984–2017 [3.50].

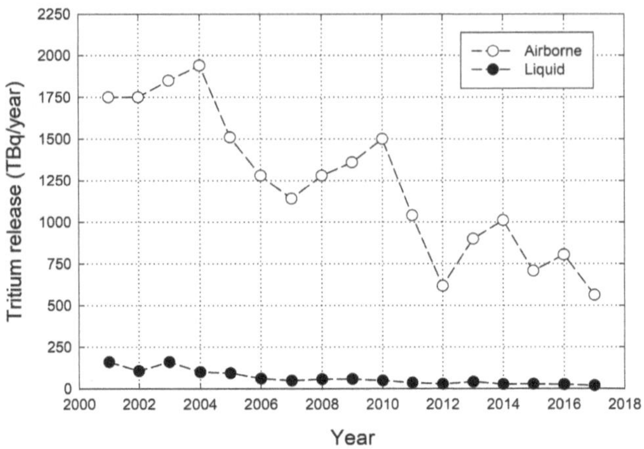

FIG. 3.19. Airborne and liquid releases of tritium from the SRNL facility (USA) for 2001–2017 [3.51].

3.2.4.2. Russian Federation

Mayak PA, located in the Southern Urals, was the first facility in the USSR to produce plutonium for nuclear weapons. A uranium–graphite reactor for producing plutonium and a radiochemical plant for its extraction from irradiated uranium were put into operation in 1948. In total, seven nuclear reactors operated at Mayak PA, five of which were decommissioned. The reprocessing of spent nuclear fuel from nuclear power generation, research and other types of reactor was started in 1977 [3.52].

Mayak PA is a complex of nuclear facilities comprising several divisions. It has produced environmental releases of tritium arising from reactor operations, tritium production facilities and from an SFRP (RT-1). According to Typhoon RPA data (see Section 3.2.3), the total releases and discharges cannot be differentiated by source.

The All-Russian Research Institute of Experimental Physics (VNIIEF) was founded in 1946 in Sarov, historically known as Arzamas-16. VNIIEF was the main centre for research and development of nuclear weapons in the USSR. Currently, VNIIEF is engaged in a broad spectrum of advanced research and development activities, including support of the reserves of the Russian nuclear stockpile and the design of nuclear warheads. The All-Russian Research Institute of Technical Physics (VNIITF), located in Snezhinsk in the Southern Urals at a distance of about 40 km from Mayak PA, was established in 1955 as a back-up facility for VNIIEF.

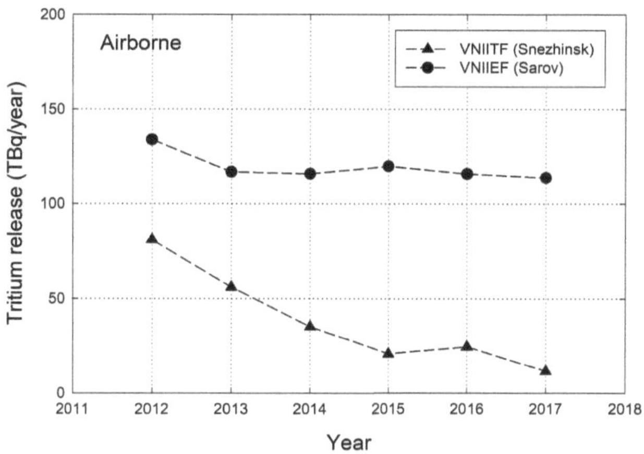

FIG. 3.20. Airborne tritium releases from nuclear weapon facilities in the Russian Federation, VNIITF in Snezhinsk (triangles) and VNIIEF in Sarov (circles), for 2012–2017 [3.43].

According to Typhoon RPA reports [3.43], in the period from 2012 to 2017 the levels of tritium airborne releases from VNIIEF and VNIITF were in the range 110–135 TBq/a and 10–80 TBq/a, correspondingly. There is no information about tritium liquid discharges. The variation in the total annual tritium releases to the atmosphere reported for VNIIEF and VNIITF is shown in Fig. 3.20.

3.2.4.3. United Kingdom

The government owned Atomic Weapons Establishment (AWE), located in Aldermaston, is a centre for the design, production, management, decommissioning and disposal of nuclear warheads. According to reports of UK governmental agencies [3.53], the level of airborne tritium releases from AWE was in the range 0.35–19.4 TBq/a between 1995 and 2017. The total annual airborne releases of tritium reported for the AWE facility indicates that tritium releases into the air have decreased by more than an order of magnitude from 2006 onwards (Fig. 3.21). The level of liquid tritium releases from the site was much lower, in the range 3.72×10^{-4} to 2.84×10^{-2} TBq/a.

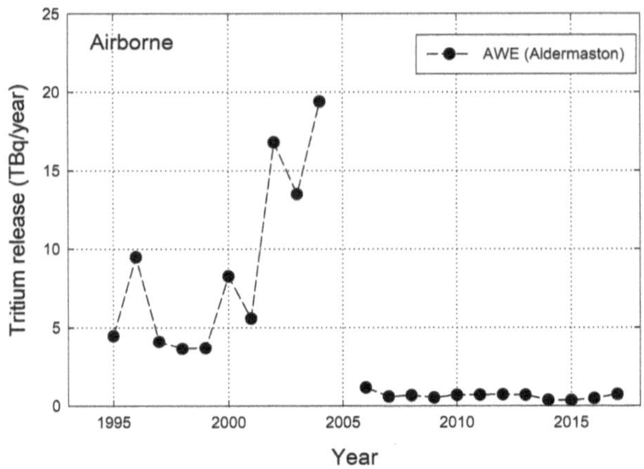

FIG. 3.21. Airborne tritium releases from the AWE facility (Aldermaston, UK) for 1995–2017 [3.53].

3.2.4.4. France

The Valduc Centre for Nuclear Studies of the French Atomic Energy and Alternative Energies Commission (CEA)[8] is involved in military programmes, including nuclear weapon development, production, maintenance and dismantling. According to Ref. [3.54], the tritium releases to the atmosphere from the CEA Valduc site were in the range 24.3–282 TBq/a between 2012 and 2016 (Fig. 3.22), showing that discharges of tritium to the local atmosphere decreased by an order of magnitude over the period from 2013 to 2016.

[8] https://www.cea.fr/Pages/le-cea/les-centres-cea/valduc.aspx

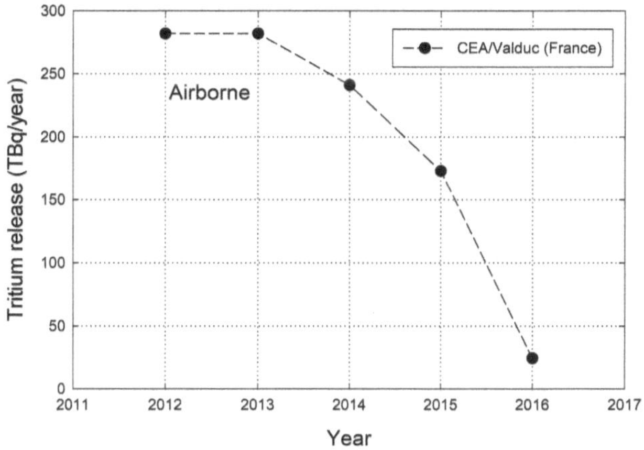

FIG. 3.22. Airborne tritium releases from the CEA Valduc facility in France for 2012–2016 [3.54].

3.2.4.5. Canada

Chalk River Laboratories (CRL) is a multifaceted facility with advanced laboratories, whose work activities range widely from services and development related to the nuclear industry, to other science areas such as physics, metallurgy, chemistry, biology and engineering. CRL is the principal site for major research and development to support and advance nuclear technology, in particular, Canadian deuterium–uranium (CANDU) reactor technology[9] and more recently, the development of small modular reactors. Until the shutdown of its nuclear reactor in 2018, CRL produced a large share of the world's supply of ^{99}Mo, which is used to produce the medical radioisotope ^{99m}Tc and a large share of the world's supply of radiopharmaceuticals.[10]

Airborne tritium releases from CRL are determined mainly by the National Research Universal (NRU) facility. The main contributors to the total HTO levels discharged from CRL are the waste management areas through Perch Creek, followed by the Process Outfall monitoring location and groundwater discharges from sources inside the boundaries of the controlled area. Total airborne releases and liquid discharges of tritium from the CRL site from 2015 to 2020 are shown in Fig. 3.23 [3.55]. The decrease in releases from 2018 to 2020 are largely attributed to the shutdown of the NRU reactor in March 2018 and, consequently, to the decreased contributions of the NRU reactor facility to the Process Outfall.

---

9 https://www.cnsc-ccsn.gc.ca/eng/reactors/research-reactors/chalk-river/

10 https://www.cbc.ca/news/science/chalk-river-makes-1st-isotopes-in-15-months-1.941869

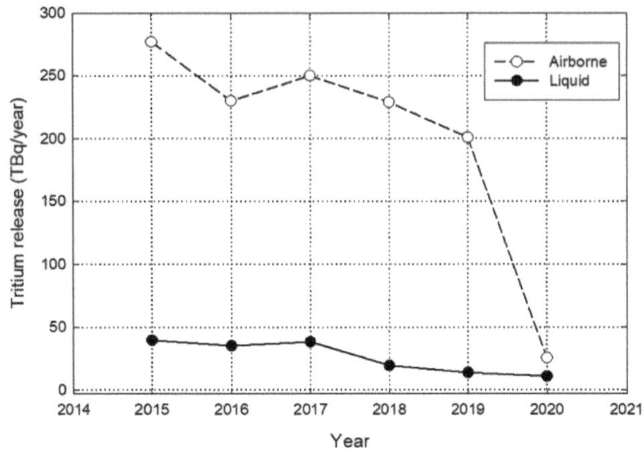

FIG. 3.23. Tritium releases to the atmosphere and water from CRL for 2015–2020 [3.55].

3.2.5. Accidental releases of tritium from nuclear industries

According to UNSCEAR [3.56], significant accidental releases from tritium production facilities have been reported to occur from LLNL in 1970 and from SRNL in 1974–1984 [3.57] (see Table 3.9) The accidentally released activity decreased with time from 11 000–18 000 TBq in the early 1970s to 300 TBq in 1984. The chemical forms of the released tritium were predominantly elemental hydrogen (gas) or HTO or their mixture. Monitoring demonstrated that elemental tritium was gradually converted in the environment to HTO.

The Chernobyl accident in 1986 resulted in the release of an estimated 37 TBq of tritium into the atmosphere. This was an estimate based on the relative fission yields of tritium (0.01%) and ^{137}Cs (6.22%) and a total released ^{137}Cs activity estimated at 9250 TBq [3.58].

The accident at the Fukushima Daiichi NPP in 2011 resulted in substantial releases of radionuclides, including tritium. The impact of the accident was estimated in the UNSCEAR 2016 Report [3.56], which states:

> "After the Fukushima accident, tritium concentrations in precipitation collected 170–700 km southwest from the Fukushima Daiichi Nuclear Power Station and in plant water collected in its vicinity were substantially elevated (up to a factor of a few tens) compared with pre-accident levels… the amount of tritium released and deposited over the north-west Pacific Ocean was in the range of 0.1–0.5 PBq [100–500 TBq]."

TABLE 3.9. ACCIDENTAL RELEASES OF TRITIUM IN THE USA FROM LAWRENCE LIVERMORE NATIONAL LABORATORY (LLNL) AND SAVANNAH RIVER NATIONAL LABORATORY (SRNL) [3.56, 3.57]

Site	Year	Airborne tritium release (TBq)	HTO (%)
LLNL	1970	11 000	<1
SRNL	1974	18 000	<1
SRNL	1975	6 700	0.6
SRNL	1981	1 200	>99
SRNL	1983	2 100	1
SRNL	1984	300	70

In the more recent UNSCEAR 2020/2021 Report, using data published by Kaizer et al. [3.59], it is estimated that the tritium input to the western Pacific was in the range 300–700 TBq [3.60].

3.2.6. Non-nuclear industries

There are several commercial facilities worldwide that use or have used tritium to create tritium labelled products, such as radiopharmaceuticals for nuclear medicine, tritium powered self-luminous watches, dials, trinkets and light signs. Information on tritium releases from such facilities is not readily available, so this section provides only exemplary information for a few large facilities in the UK and Canada, which have reported their releases to the environment.

In the UK, a radiopharmaceutical manufacturing site (most recently owned by GE-Healthcare) on the South Wales coast, east of Cardiff, operated from 1980 until 2010. This facility manufactured radiopharmaceutical products that were used for diagnostic and therapeutic nuclear medicine procedures. It was owned by various companies during its lifetime and ceased manufacturing operations in 2010. By 2019, following a concerted decommissioning programme, the Cardiff site formally ceased to be a nuclear licensed site.

The total annual atmospheric and liquid tritium discharges from the Cardiff site for the period 1995–2017 [3.53] are given in Fig. 3.24 (see also Sections 5.5.5 and 5.6.3). During this period, 6860 TBq and 2800 TBq of tritium were discharged

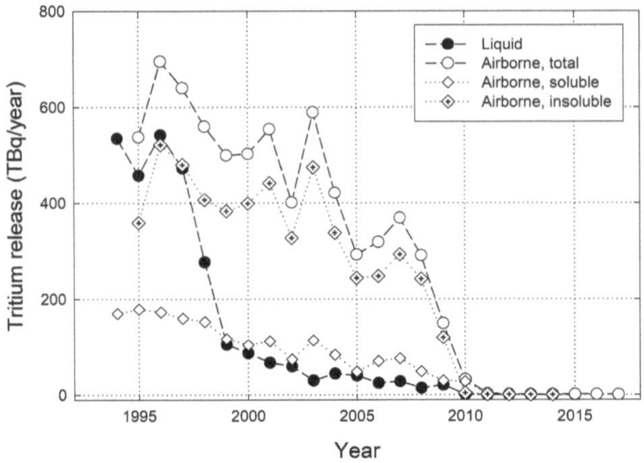

FIG. 3.24. Tritium releases to the atmosphere and to water from the Cardiff facility (UK) for 1994–2017 [3.53].

to the atmosphere and to water from the site, respectively. The integrated fraction of insoluble compounds in the airborne releases was 77% [3.53].

In Canada, SRB Technologies, Inc. (SRBT) is the world's largest manufacturer of self-luminous emergency lighting and exit signs, which are powered by tritium. The company has developed and produced tritium powered light sources since the 1960s. The production of the tritium-containing items is accompanied by releases of tritium to the environment. The total annual atmospheric tritium releases from SRBT are shown in Fig. 3.25. During the period from 2000 to 2010, 56 300 TBq of tritium was released from the site into the local atmosphere in the form of HTO and HT [3.61]. The fraction of HTO in the airborne releases increased from 7% in 2000 to 37% in 2019.

3.2.7. Prospective releases from future fusion reactors

Since the 1950s, fusion energy has been presented as an attractive and potentially highly efficient energy source that will not be accompanied by a substantial production of long lived radioactive waste. On the basis of many decades of research and engineering, the way to industrial fusion energy generation is being sought by using large fusion reactors (e.g. tokamaks, stellarators) to fuse mixed deuterium–tritium fuel.[11] Such fusion power generation will require significant amounts of tritium and will likely be accompanied by releases of tritium into the environment.

---

[11] See https://nucleus.iaea.org/sites/fusionportal

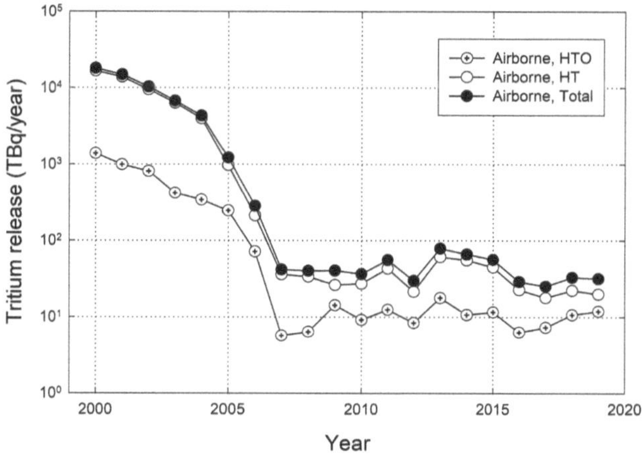

FIG. 3.25. Tritium releases to the atmosphere from SRB Technologies (Canada) for 2000–2019 [3.61].

To understand and assess potential tritium releases from future fusion energy facilities, experiments at prototype devices provide necessary data to optimize tritium recovery and reactor construction. For example, to assess tritium releases from the future ITER device, experiments with tritium have been carried out at the Joint European Torus (JET) facility in the UK since the 1980s [3.62, 3.63].

The first experiment at JET, in which the plasma in a tokamak was fuelled with tritium, was carried out in November 1991 using 0.1 g (36 TBq) of tritium. The design construction and safety justification of the tritium handling plant for the final D–T phase at JET using 90 g (32 100 TBq) of tritium has been essentially completed. Valuable experience in tritium handling, clean-up and waste disposal has been gained, which was directly relevant to the use of tritium in future JET operation and in controlled fusion research in general [3.62, 3.64].

The subsequent experiments, conducted until 2022, have targeted the impact of isotope mass on key plasma physics mechanisms such as plasma–wall interactions and energy transport. The largest potential source of tritium releases in oxide form is the JET baking plant. The experiments estimated that if the partial pressure of tritium in the vessel was assumed to be 10^{-4} hPa during operation and glow discharge cleaning occurred for 3000 h/a, then the permeation through the bellows in the absence of inhibition because of oxide layer formation would be 120 TBq of tritium per year. However, with a more realistic inhibition factor of 100, a bypass clean-up loop could be used to recover the tritium, and any losses to the atmosphere for full D–T operation could be reduced to around 2 TBq. Tritium permeation through this route during the first tritium experiment was too small to be reliably measured [3.65].

When cryodistillation is used for the separation of hydrogen isotopes, tritium and deuterium are recovered for re-injection. Protium (hydrogen) is discharged having trace quantities of tritium. Conservatively, assuming a product purity of 10^{-7} (cf. the design value of 5×10^{-9}) leads to an estimated annual discharge of 7.5 TBq of HT from a daily throughput of 30 g [3.65]. In general, JET experiments suggest that tritium accounting in fusion systems will be difficult and probably of low accuracy. The projection of the results of JET tritium experiments to future nuclear fusion power facilities, such as DEMO, remain largely uncertain. Given the current tritium confinement technology, Larsen and Babineau [3.65] estimated a tritium release rate from ITER of 1 g/a per 500 MW fusion power generation. The authors also stated that ITER may be permitted to release 0.6–1.8 g/a (216–650 TBq/a) during operations, with the higher value expected during years when open vessel maintenance is undertaken.

Until ITER starts power production with tritium, which is planned after 2035, JET tritium and deuterium–tritium experiments have offered the opportunity for fusion scientists to investigate physics relevant to burning plasmas for ITER and for the planned prototype fusion power plant DEMO.

The expected power generating operations of ITER and other fusion facilities worldwide would raise the demand for commercially available tritium, which is currently formed solely by the Canadian heavy water CANDU type nuclear energy reactors. Approximately half of those CANDU units are planned for decommissioning during this decade, thus raising concerns over an oncoming deficit of tritium to fuel ITER and the subsequent power generating fusion facilities [3.66].

3.3. CONCLUSIONS

Atmospheric nuclear weapon tests carried out in the middle to later part of the 20th century had a global environmental impact. The total amount of tritium emitted into the atmosphere (approximately 520 kg or 1.86×10^8 TBq) caused a pronounced spike in the global tritium background in terms of activity concentrations in rainwater from 0.1 Bq/L to hundreds of becquerels per litre. The radioactive decay of the released tritium led to the present global background level, which has now almost returned to levels (~1 Bq/L) existing before the atmospheric nuclear weapon tests of the 1950s and 1960s.

Contemporary sources of global anthropogenic tritium originate from the emissions and discharges of technogenic tritium from 453 nuclear reactors worldwide. According to IAEA data, these nuclear reactors annually discharge approximately 19 500 TBq of tritium into the air and around 12 300 TBq into water. At the same time, the 49 PHWRs contribute 94% to the total air emissions of tritium and 37% to the total water discharges. The largest number of HWRs (24) are in Canada. By contrast, the air and water emissions from the 301 PWRs contributed 5.4% and 60% to air emissions and water discharges, respectively.

Nuclear spent fuel reprocessing plants are the major sources of tritium discharged to water. Every year, the reprocessing facility in La Hague discharges about 12 000 TBq into the English Channel; that is comparable to the annual water discharge of tritium from all nuclear reactors.

Military research facilities produce mostly local air emissions of tritium at levels ranging from 0.35 TBq/a (AWE, UK) to 120 TBq/a (VNIIEF, Russian Federation). The tritium production facility of the SRNL (USA) currently produces local air emissions and water discharges of tritium at levels of 750 TBq/a and 20 TBq/a, respectively. Mayak PA, which comprises reactors, tritium production and an SFRP, currently produces local air emissions and water discharges of tritium at levels of 1200 TBq/a and 17 TBq/a, respectively.

Industrial releases to air and water from the former radiopharmaceutical company in Cardiff were reduced significantly from 1998 to 2010. Starting from 2010, annual water discharges was at the level of 1 TBq; air emissions at that time became negligible. At the SRBT facility in Canada, which produces luminous signs, present day local airborne releases are in the range 30–50 TBq/a.

The accident at the Fukushima Daiichi NPP in 2011 and subsequent mitigation and decontamination activities have resulted in the temporary accumulation of radioactively contaminated wastewater. This wastewater was cleaned from most activation and fission products using the ALPS. After this treatment, an estimated 700 TBq of tritium (as reported in March 2023) (see Section 3.2.2.6) in ALPS treated wastewater resides in on-site storage tanks and requires special safe handling and acceptable disposal. The Fukushima Daichi NPP operator, under the oversight of the Japanese Government, the national regulator, the IAEA and other agencies, undertakes dilution and controlled discharges of the treated water into the Pacific Ocean. This procedure is expected to continue for several decades [3.41].

Assuming future success of the research and development of controlled fusion and construction of fusion energy generating facilities, a nuclear fusion reactor with 800 MW(e) power containing 300 g of tritium in the fuel is estimated to produce annual air emissions of around 75 TBq/a. The amount of tritium that will be released from a fusion power plant is currently a matter of conjecture, although an approximate estimate of 1 g/a is often mentioned. Computational studies using a tritium permeation modelling code have predicted the tritium permeation losses from a 517 MW fusion power plant producing and consuming about 29 kg/a of tritium to range from 0.05 to 6.18 g/a, depending on the assumed operating parameters [3.65].

REFERENCES TO CHAPTER 3

[3.1] HUANG, W.J., WANG, M., KONDEV, F.G., AUDI, G., NAIMI, S., The AME 2020 atomic mass evaluation (I). Evaluation of input data, and adjustment procedures, Chin. Phys. C **45** 3 (2021) 030002, https://doi.org/10.1088/1674-1137/abddb0

[3.2] UNITED NATIONS, Sources and Effects of Ionizing Radiation, Vol. I: Sources, 2008 Report of the United Nations Scientific Committee on the Effects of Atomic Radiation (UNSCEAR) to the General Assembly, UN, New York (2008).

[3.3] MASARIK, J., REEDY, R.C., Terrestrial cosmogenic-nuclide production systematics calculated from numerical simulations, Earth Planet. Sci. Lett. **136** 3 (1995) 381, https://doi.org/10.1016/0012-821X(95)00169-D

[3.4] POLUIANOV, S.V., KOVALTSOV, G.A., USOSKIN, I.G., A new full 3-D model of cosmogenic tritium ^3H production in the atmosphere (CRAC:3H), J. Geophys. Res. Atmos. **125** 18 (2020) e2020JD033147, https://doi.org/10.1029/2020JD033147

[3.5] MASARIK, J., BEER, J., An updated simulation of particle fluxes and cosmogenic nuclide production in the Earth's atmosphere, J. Geophys. Res. Atmos. **114** (2009) D11103, https://doi.org/10.1029/2008JD010557

[3.6] INSTITUT DE RADIOPROTECTION ET DE SÛRETÉ NUCLÉAIRE, Tritium and the Environment, IRSN, Fontenay-aux-Roses (2010).

[3.7] MOMOSHIMA, N., et al., Tritium in the atmospheric environment, J. Nucl. Radiochem. Sci. **8** 2 (2007) 117, https://doi.org/10.14494/jnrs2000.8.117

[3.8] CAUQUOIN, A., et al., The global distribution of natural tritium in precipitation simulated with an Atmospheric General Circulation Model and comparison with observations, Earth Planet. Sci. Lett. **427** (2015) 160, https://doi.org/10.1016/j.epsl.2015.06.043

[3.9] UNITED NATIONS, Sources and Effects of Ionizing Radiation, Vol. I: Sources, 2000 Report of the United Nations Scientific Committee on the Effects of Atomic Radiation (UNSCEAR) to the General Assembly, UN, New York (2000).

[3.10] OMS, P.-E., et al., Inventory and distribution of tritium in the oceans in 2016, Sci. Total Environ. **656** (2019) 1289, https://doi.org/10.1016/j.scitotenv.2018.11.448

[3.11] KAUFMAN, S., LIBBY, W.F., The natural distribution of tritium, Phys. Rev. **93** 6 (1954) 1337, https://doi.org/10.1103/PhysRev.93.1337

[3.12] CRAIG, H., LAL, D., The production rate of natural tritium, Tellus **13** (1961) 85–105, https://doi.org/10.1111/j.2153-3490.1961.tb00068.x

[3.13] NIR, A., KRUGER, S.T., LINGENFELTER, R.E., FLAMM, E.J., Natural tritium, Rev. Geophys. **4** 4 (1966) 441, https://doi.org/10.1029/RG004i004p00441

[3.14] TEEGARDEN, B., Cosmic-ray production of deuterium and tritium in the Earth's atmosphere, J. Geophys. Res. **72** 19 (1967) 4863 https://doi.org/10.1029/JZ072i019p04863

[3.15] O'BRIEN, K., Secular variations in the production of cosmogenic isotopes in the Earth's atmosphere, J. Geophys. Res. Space Phys. **84** A2 (1979) 423, https://doi.org/10.1029/JA084iA02p00423

[3.16] INTERNATIONAL ATOMIC ENERGY AGENCY, Management of Waste Containing Tritium and Carbon-14, Technical Reports Series No. 421, IAEA, Vienna (2004).

[3.17] INTERNATIONAL ATOMIC ENERGY AGENCY, Nuclear Power Reactors in the World, Reference Data Series No. 2, IAEA, Vienna (2021).

[3.18] BRUCE POWER, Annual Summary and Assessment of Environmental Radiological Data for 2002, Bruce Power, Tiverton (2003).

[3.19] BRUCE POWER, Annual Summary and Assessment of Environmental Radiological Data for 2003, Bruce Power Limited Partnership, Tiverton (2004).

[3.20] BRUCE POWER, Annual Summary and Assessment of Environmental Radiological Data for 2004, Bruce Power Limited Partnership, Tiverton (2005).

[3.21] BRUCE POWER, Annual Summary and Assessment of Environmental Radiological Data for 2005, Bruce Power Limited Partnership, Tiverton (2007).

[3.22] BRUCE POWER, Annual Summary and Assessment of Environmental Radiological Data for 2006, Bruce Power Limited Partnership, Tiverton (2007).

[3.23] BRUCE POWER, Annual Summary & Assessment of Environmental Radiological Data for 2007, Bruce Power Limited Partnership, Tiverton (2008).

[3.24] BRUCE POWER, Annual Summary & Assessment of Environmental Radiological Data for 2008, Bruce Power Limited Partnership, Tiverton (2009).

[3.25] BRUCE POWER, Annual Summary and Assessment of Environmental Radiological Data for 2009, Bruce Power Limited Partnership, Tiverton (2010).

[3.26] BRUCE POWER, Annual Summary and Assessment of Environmental Radiological Data for 2010, Bruce Power Limited Partnership, Tiverton (2011).

[3.27] BRUCE POWER, Annual Summary and Assessment of Environmental Radiological Data for 2011, Bruce Power Limited Partnership, Tiverton (2012).

[3.28] BRUCE POWER, 2012 Environmental Monitoring Program Report, Bruce Power Limited Partnership, Tiverton (2013).

[3.29] BRUCE POWER, 2013 Environmental Monitoring Program Report, Bruce Power Limited Partnership, Tiverton (2014).

[3.30] BRUCE POWER, 2014 Environmental Monitoring Program Report, Bruce Power Limited Partnership, Tiverton (2015).

[3.31] BRUCE POWER, 2015 Environmental Monitoring Program Report, Bruce Power Limited Partnership, Tiverton (2016).

[3.32] BRUCE POWER, 2016 Environmental Monitoring Program Report, Bruce Power Limited Partnership, Tiverton (2017).

[3.33] NUCLEAR POWER CORPORATION OF INDIA LIMITED, 37th Annual Report 2023–2024, NPCIL (2024), https://www.npcil.nic.in/WriteReadData/userfiles/file/Annual_Report_2023_2024_26082024_01.pdf

[3.34] KONG, T.Y., KIM, S., LEE, Y., SON, J.K., MAENG, S.J., Radioactive effluents released from Korean nuclear power plants and the resulting radiation doses to members of the public, Nucl. Eng. Technol. **49** 8 (2017) 1772, https://doi.org/10.1016/j.net.2017.07.021

[3.35] NUCLEAR REGULATORY COMMISSION, Radioactive Effluent and Environmental Reports (2024), https://www.nrc.gov/reactors/operating/ops-experience/tritium/plant-info.html

[3.36] INTERNATIONAL ATOMIC ENERGY AGENCY, Power Reactor Information System (PRIS), Vienna, https://pris.iaea.org/pris/

[3.37] EUROPEAN COMMISSION, RAdioactive Discharges Database (RADD), https://europa.eu/radd/index.dox

[3.38] JAPAN NUCLEAR ENERGY SAFETY ORGANIZATION, Operational Status of Nuclear Facilities in Japan, 2013 Edition, JNES, Tokyo (2014).

[3.39] POVINEC, P.P., HIROSE, K., AOYAMA, M., TATEDA, Y., Fukushima Accident: 10 Years After, Elsevier, Amsterdam (2021).

[3.40] INTERNATIONAL ATOMIC ENERGY AGENCY, IAEA Review of Safety Related Aspects of Handling ALPS Treated Water at TEPCO's Fukushima Daiichi Nuclear Power Station. Report on First Interlaboratory Comparison on the Determination of Radionuclides in ALPS Treated Water (2023), https://www.iaea.org/sites/default/files/iaea_analysis_of_k4c_report_sh_pmg.pdf

[3.41] INTERNATIONAL ATOMIC ENERGY AGENCY, IAEA Comprehensive Report on the Safety Review of the ALPS Treated water at the Fukushima Daiichi Nuclear Power Station (2023), https://www.iaea.org/sites/default/files/iaea_comprehensive_alps_report.pdf

[3.42] ROSENERGOATOM, Environmental Reports, Rosenergoatom JSC, Moscow (in Russian), https://www.rosenergoatom.ru/safety_environment/vozdeystvie-na-okruzhayushchuyu-sredu/ekologicheskie-otchety-ao-kontsern-rosenergoatom

[3.43] TYPHOON RPA, Collection of Yearbooks "Radiation situation on the territory of Russia and neighbouring states" (in Russian), https://www.rpatyphoon.ru/products/pollution-media.php

[3.44] YANG, D.J., CHEN, X.Q., LI, B., Tritium release during nuclear power operation in China, J. Radiol. Prot. **32** 2 (2012) 167, https://doi.org/10.1088/0952-4746/32/2/167

[3.45] YANG, J., GU, Z., ZHAO, Y., LIAN, B., WANG, Y., Analysis on the emission of tritium from Qinshan nuclear power base of China in recent ten years, IOP Conf. Ser. Earth Environ. Sci. **227** (2019) 052031, https://doi.org/10.1088/1755-1315/227/5/052031

[3.46] ZHANG, Q., LIU, T., YANG, L., MENG, D., SONG, D., Monitoring and management of tritium from the nuclear power plant effluent, IOP Conf. Ser. Earth Environ. Sci. **108** (2018) 042057,
https://doi.org/10.1088/1755-1315/108/4/042057

[3.47] UNITED NATIONS, Exposures from nuclear power production, Annex B, 1988 Report of the United Nations Scientific Committee on the Effects of Atomic Radiation (UNSCEAR) to the General Assembly, UN, New York (1988).

[3.48] INTERNATIONAL ATOMIC ENERGY AGENCY, Integrated Nuclear Fuel Cycle Information System,
https://infcis.iaea.org

[3.49] INTERNATIONAL ATOMIC ENERGY AGENCY, Status and Trends in Spent Fuel Reprocessing, IAEA TECDOC-1467, IAEA, Vienna (2005).

[3.50] LAWRENCE LIVERMORE NATIONAL LABORATORY, Site Annual Environmental Report, LLNL, Livermore, CA,
https://saer.llnl.gov

[3.51] SAVANNAH RIVER SITE, Annual Environmental Reports,
https://www.srs.gov/general/pubs/ERsum/index.html

[3.52] ANSPAUGH, L., DEGTEVA, M., VASILENKO, E., Mayak Production Association: Introduction, Radiat. Environ. Biophys. **41** 1 (2002) 19–22,
https://doi.org/10.1007/s00411-002-0148-5

[3.53] ENVIRONMENT AGENCY, NATURAL RESOURCES WALES, FOOD STANDARDS AGENCY, Radioactivity in Food and the Environment (RIFE) Reports,
https://www.gov.uk/government/publications/radioactivity-in-food-and-the-environment-rife-reports

[3.54] AUTORITÉ DE SÛRETÉ NUCLÉAIRE, Livre Blanc du Tritium, ASN, Montrouge (2024),
https://www.asn.fr/sites/tritium/

[3.55] CANADIAN NUCLEAR LABORATORIES, Annual Compliance Monitoring Report Effluent Verification Monitoring at Chalk River Laboratories 2020, Rep. CRL-509254-ACMR-2020, Canadian Nuclear Laboratories, Ontario (2021).

[3.56] UNITED NATIONS, Sources, Effects and Risks of Ionizing Radiation, 2016 Report of the United Nations Scientific Committee on the Effects of Atomic Radiation (UNSCEAR) to the General Assembly, with Scientific Annexes, UN, New York (2017).

[3.57] OKADA, S., MOMOSHIMA, N., Overview of tritium: Characteristics, sources, and problems, Health Phys. **65** 6 (1993) 595,
https://doi.org/10.1097/00004032-199312000-00001

[3.58] KOGA, T., MORISHIMA, H., NIWA, T., KAWAI, H., Tritium precipitation in European cities and in Osaka, Japan owing to the Chernobyl nuclear accident, J. Radiat. Res. **32** (1991) 267–276,
https://doi.org/10.1269/jrr.32.267

[3.59] KAIZER, J., AOYAMA, M., KUMAMOTO, Y., MOLNAR, M., PALCSU, L., POVINEC, P.P., Tritium and radiocarbon in the western North Pacific waters: Post-Fukushima situation, J. Env. Radioact. **184–185** (2018) 83–94, https://doi.org/10.1016/j.jenvrad.2018.01.015

[3.60] UNITED NATIONS, Levels and Effects of Radiation Exposure Due to the Accident at the Fukushima Daiichi Nuclear Power Station: Implications of Information Published Since the UNSCEAR 2013 Report, Scientific Annex B, Vol. II (Report to the General Assembly), Scientific Committee on the Effects of Atomic Radiation (UNSCEAR), UN, New York (2022).

[3.61] SRBT, Annual Compliance Reports, http://www.srbt.com/compliance_Reports.php

[3.62] BELL, A.C., "JET tritium experience", Safety in Tritium Handling Technology (MANNONE, F., Ed.), Springer Dordrecht (1993) 217–236, https://doi.org/10.1007/978-94-011-1910-8_11

[3.63] CLERY, D., European fusion reactor sets record for sustained energy, Science **375** (2022) 6581, https://doi.org/10.1126/science.ada1089

[3.64] BELL, A.C., BALLANTYNE, P., CALDWELL-NICHOLS, C., WYKES, M., Routine tritium releases from JET, Fusion Technol. **21** 2P2 (1992) 506–511, https://doi.org/10.13182/FST92-A29797

[3.65] LARSEN, G., BABINEAU, D., An evaluation of the global effects of tritium emissions from nuclear fusion power, Fusion Eng. Des., **158** (2020) 111690, https://doi.org/10.1016/j.fusengdes.2020.111690

[3.66] CLERY, D., Out of gas, Science **376** (2022) 6600, https://doi.org/10.1126/science.add5098

Chapter 4

TRITIUM IN THE ATMOSPHERE

O. CONNAN, F. EYROLLE

4.1. INTRODUCTION

Most tritium of natural origin is produced in the upper atmosphere by nuclear reactions of cosmic radiation with nitrogen and oxygen nuclei. This environmental compartment received a substantial surge of tritium from the atmospheric nuclear weapon tests undertaken particularly in the period 1952–1963, followed by later inputs from tests performed by France and China. The last atmospheric nuclear test was conducted in China on 16 October 1980. Consequently, the global stratosphere and troposphere were both significantly labelled over the decades of testing by air mass circulation and mixing processes, particularly in the Northern Hemisphere, where most of the atmospheric nuclear weapon tests were carried out. Currently, the tritium concentrations in precipitation worldwide have returned to the 'pre-test' levels (<1 Bq/L) owing to radioactive decay, air mass mixing and exchange with the surface ocean water. The presence of an excess of tritium in precipitation relative to the natural background concentration has been monitored at the global scale for more than 60 years. Additionally, from the early 1970s, gaseous and liquid tritium emissions associated with the operation of nuclear facilities, located mostly in the Northern Hemisphere, contributed to atmospheric tritium at both local and regional scales.

Tritiated hydrogen is oxidized to HTO in the atmosphere and in soil and is integrated into the water cycle. HTO is then transferred to the Earth's surface by precipitation and follows water mass transport and recycling within the geosphere, hydrosphere and the biosphere. Dispersion of atmospheric tritium delivered by nuclear tests depended on the circulation of air masses at the global scale. The spatial dispersal of gaseous tritium emissions from point sources located on the ground is mostly controlled by wind strength and direction. Environmental and geographical factors such as season, latitude, altitude or the distance to the ocean influence tritium concentrations in precipitation.

Assessing the radiological impact on human beings and the environment often relies on modelling to simulate the dispersion of tritium (and other radionuclides) in the atmosphere. This is particularly so for industrial emissions, where empirical monitoring data and modelling are used to explain or predict the contamination of the terrestrial environment.

The worldwide GNIP was established in 1960 by the IAEA and WMO. It was designed to monitor atmospheric thermonuclear test fallout through the measurements of hydrogen and oxygen isotopes in precipitation. Peaks in tritium contents in precipitation that exceeded the natural background by a factor of 1000 were registered in the Northern Hemisphere in the early 1960s. These resulted from nuclear weapon test detonations in the atmosphere by the USA, the USSR and, to a much lesser extent, by the UK until the signing of the Partial Test Ban Treaty in 1963.

4.2. TRITIUM SOURCE TERMS

4.2.1. Natural tritium production

Most tritium of natural origin is produced by nuclear reactions between secondary cosmic rays and nitrogen or oxygen nuclei in the upper atmosphere. Production is largest in the upper stratosphere, but some energetic cosmic ray neutrons and protons can also penetrate deeply, producing cosmogenic radionuclides in the lower atmosphere. As for most cosmogenic radionuclides, production not only is altitude and latitude dependent, but also varies with the 11 year solar cycle that modulates cosmic ray penetration through the Earth's geomagnetic field. Tritium produced in the upper atmosphere is oxidized to tritiated water (HTO) and migrates to the troposphere, where it is incorporated into the hydrological cycle.

The calculated global average production rates of natural tritium per unit surface area of the Earth and its total annual production are estimated to be 0.32 atoms $\cdot$ cm^{-2} $\cdot$ s^{-1}, which results in production rates equal to 258 g/a or 92 PBq/a [4.1, 4.2]. With this production, Oms et al. [4.2] estimated the surface area deposition rate on the Earth's surface for tritium of natural origin as 180 Bq $\cdot$ m^{-2} $\cdot$ s^{-1}. Assuming that production balances radioactive decay, the inventory of natural tritium on the Earth is estimated at about 4.5 kg or 1.6 EBq [4.2]. These estimates are always subject to uncertainties due to the calculations and the different models used. For example, Nie et al. [4.3] reported a steady state quantity of 3.5 kg based on estimates by Michel [4.4]. According to UNSCEAR [4.5], on the basis of older data [4.6, 4.7], the natural production of tritium would be 0.12–0.20 atoms $\cdot$ cm^{-2} $\cdot$ s^{-1} at the Earth's surface (96–161 g/a or 35–57 PBq/a), with more likely values of 0.20–1.0 atoms $\cdot$ cm^{-2} $\cdot$ s^{-1} (161–806 g/a or 58–289 PBq/a). A recent publication [4.8] indicates a production rate for moderate solar activity of 0.35 atoms $\cdot$ cm^{-2} $\cdot$ s^{-1}, which corresponds to an annual production rate of 282 g/a or 100 PBq/a.

Besides cosmogenic generation, a minute fraction of natural tritium has a telluric (geogenic) origin, resulting from the reaction between traces of lithium (^{6}Li) contained in rocks and neutrons produced by spontaneous fission of uranium and thorium [4.9]. Telluric and/or geogenic production in a 1 km thick rock is estimated to have a rate of 10 atoms $\cdot$ m^{-2} $\cdot$ s^{-1} (0.56 Bq $\cdot$ m^{-2} $\cdot$ a^{-1}) [4.9, 4.10].

4.2.2. Anthropogenic tritium

Human activity has contributed significantly to the presence of tritium in the environment through military and civil nuclear activities. Worldwide, industrial releases of tritium between 1998 and 2002 were estimated by UNSCEAR [4.5] to be approximately 11.7 PBq/a for gaseous releases and 16.0 PBq/a for liquid releases, resulting in a total average annual release rate of 27.7 PBq/a (~78 g/a). Historically, the main sources of anthropogenic tritium production are, in descending order of magnitude: atmospheric nuclear testing > nuclear fuel reprocessing plants > NPPs.

4.2.2.1. Atmospheric nuclear weapon tests

Nuclear weapon tests have been conducted since 1945 and represent the most significant anthropogenic source of radiation exposure in the environment for the world population. Atmospheric nuclear tests were carried out until 1980, with most occurring before 1963. Test sites were spread across the globe, but most of the atmospheric detonations occurred in the Northern Hemisphere, made by the USA (197 tests), the USSR (219 tests), France (45 tests), the UK (21 tests) and China (22 tests) [4.11] (see Fig. 4.1). Approximately two-thirds of the total estimated yield of 440 Mt produced by 504 atmospheric nuclear tests and combat explosions in 1945–1980 were released in only 25 explosions. Each of these had a yield equal to or greater than 4 Mt and was conducted by the USA and the USSR in 1954–1962 [4.11].

Nuclear weapon tests were progressively stopped with the adoption of the Partial Test Ban Treaty of 1963, the 1974 Threshold Test Ban Treaty and the 1996 Comprehensive Nuclear Test Ban Treaty. Atmospheric tests by the USA, the USSR and the UK stopped in 1963, with France stopping in 1974 and China in 1980. The last atmospheric nuclear weapon test was carried out on 18 October 1980 in China.

In 1962, 118 atmospheric tests were carried out by the USSR and the USA — 78 and 40 tests, respectively. From 1962, following the Partial Test Ban Treaty, the USSR, the USA and the UK agreed to conduct only underground

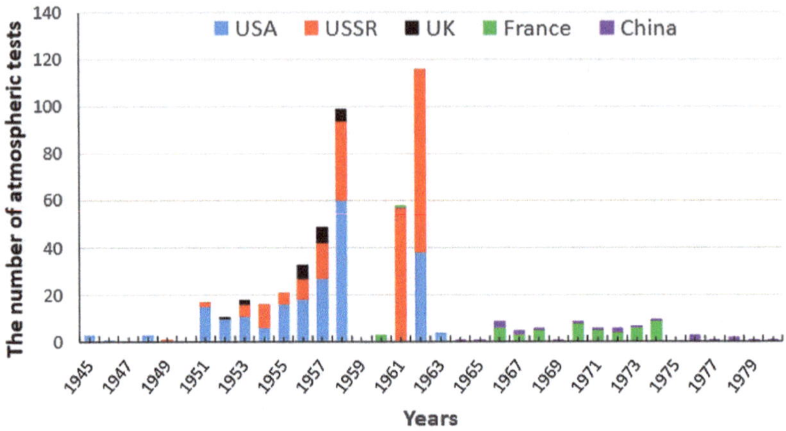

FIG. 4.1. Number of atmospheric nuclear weapon tests performed by the USA, USSR, UK, France and China between 1945 and 1980 (data from table 8 of Ref. [4.12]).

tests. In total, 504 atmospheric tests and combat explosions, as well as 39 safety tests without nuclear yield, were carried out by all nuclear weapon developing countries [4.11]. In total, together with the underground tests, which have continued (1870 in total), more than 2400 nuclear tests have been carried out worldwide (see Fig. 4.2) [4.13]. In addition to the countries already mentioned above, some other countries have carried out underground tests, including India (6 tests between 1974 and 1998), Pakistan (2 tests in 1998) and more recently North Korea (6 tests from 2006 to 2017).

Large quantities of tritium in the forms of HT and HTO were released into the environment from these tests, with the total amount over the period 1945–1980 estimated to be 186 000 PBq, with a maximum around 1963 (cf. Fig. 3.1). The total amount of tritium released from nuclear weapon testing has been previously reported and estimated to be in the range 520–550 kg [4.3, 4.5]. Following the radioactive decay of tritium, the remaining global inventory was estimated to be 22–24 kg in 2016, being distributed between oceans (90%), continental waters (9%) and the atmosphere (1%) [4.2, 4.14]. As a result of atmospheric testing, tritium levels in Northern Hemisphere precipitation increased approximately a thousand times over background concentrations, with a maximum in 1963 [4.4]. This led to a labelling of all compartments of the biosphere and geosphere. Since the end of atmospheric testing, environmental tritium levels have systematically decreased and now approach their natural (pre-testing) levels. Nowadays, the tritium activity concentrations in precipitation and surface water have almost returned to levels below 1 Bq/L [4.15–4.17]. The natural tritium levels in precipitation at medium latitudes in the Northern Hemisphere were estimated

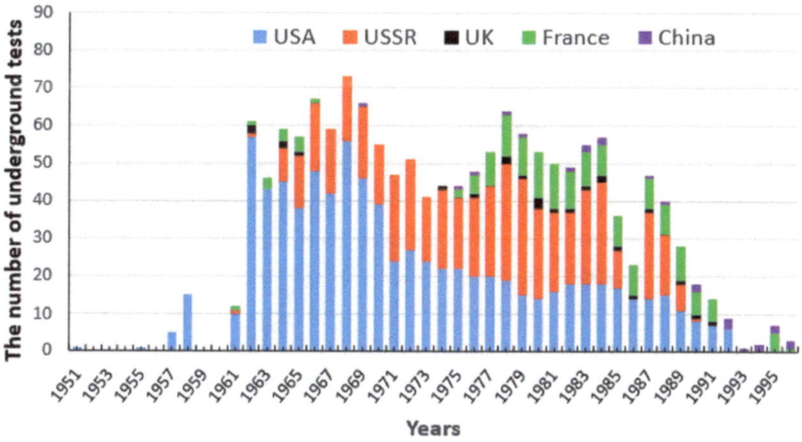

FIG. 4.2. Number of underground nuclear weapon tests performed by the USA, USSR, UK, France and China between 1950 and 1998 (data from table 8 of Ref. [4.12]).

from analyses of tritium content in wine vintages covering the period 1928–1947. The natural background found in these studies was about 0.6 Bq/L [4.9].

4.2.2.2. Atmospheric releases from nuclear facilities

Tritium is released into the atmosphere in the form of tritiated water vapour (HTO or DTO) from the operation of nuclear reactors, especially HWRs, and as tritium gas (HT) and HTO from SFRPs and others nuclear facilities.[1] Tritium is the dominant radionuclide in the releases from nuclear industries. The evolution of NPPs and fuel management, as well as the appearance of new facilities producing and/or consuming tritium (e.g. ITER nuclear fusion project), indicate a trend of increasing tritium releases into the environment.

Estimates of atmospheric releases of tritium arising from the world's nuclear power production vary widely and depend on reactor type, design, construction and the age of the facility. Estimates of tritium release rates can be made using data on contemporary world nuclear energy generation in 2021 (see Table 3.2) and reported normalized tritium release rates in 1993–2015 [4.3]. The estimated global atmospheric release rates of tritium range from 0.9–5.8 PBq/a (2.5–16 g/a), of which about 80–85% originated from PHWRs, which themselves account for only 6.2% of the total net electrical capacity of the world's operational reactors. In France, where 70% of the energy production is derived from nuclear sources

---

[1] See also Section 3.2.2.

(as in 2018), its 56 NPPs release on average 60 TBq/a (~0.17 g) of tritium to air and 1250 TBq/a (~3.5 g) as liquid releases [4.3].

The main SFRPs in Europe (La Hague in France and Sellafield in the UK) released approximately 200–300 TBq/a of tritium in gaseous form between 1995 and 2007 [4.18], with Sellafield accounting for more than 75% of the atmospheric discharges from the three plants. These plants contributed most of the tritium to the atmosphere in the world when compared to all other NPPs. Since 2010, the average annual release of gaseous tritium from La Hague is around 64 TBq/a. It can also be noted that for these reprocessing plants, gaseous releases of tritium represent only a few per cent of the quantities released in liquid form. Other reprocessing plants exist elsewhere in the world, for example, in the Russian Federation and in Japan (see Chapter 3).

Many facilities other than NPPs and SFRPs also release tritium in a gaseous form, such as facilities for military weapons manufacturing and plants for production of tritium powered luminescent items (mainly in Canada). In 2006, gaseous releases originating from items produced by the industry of self-luminous items were estimated to be 300 TBq/a but subsequently decreased to values below 100 TBq/a [4.19]. The production of labelled molecules for medical applications is also associated with tritium releases. Up to 2000, the radiopharmaceutical plant (previously GE-Healthcare) in Cardiff (UK) was releasing tritium compounds at rates of about 500 TBq/a and about 100 TBq/a in 2006 [4.14]. Facilities designing and manufacturing weapons can also be a source of atmospheric releases of gaseous tritium. For example, in France, releases from the facilities at Valduc and Marcoule sites amounted to about 200 TBq/a [4.20]. Examples from other Member States are presented in Section 3.2.4.

After its construction and commissioning, the ITER project, with its 500 MWh fusion reactor, is planned to use and generate large quantities of tritium. This is anticipated to produce additional emissions of tritium into the environment in 2025–2035. The annual consumption of tritium at the facility could be 1.2 kg/a [4.19]. Estimates of gaseous releases are 0.6 g/a (215 TBq/a) in normal operation and 2.5 g/a (897 TBq/a) during periods of maintenance [4.21]. The China Fusion Engineering Test Reactor (CFETR) project, planned for 2020–2030, is similar to the ITER project. According to Nie et al. [4.22], the permitted releases would range from 5 to 33 g/a, equivalent to activities from 1795 to 11 847 TBq/a. Preliminary assessments indicate that the entire CFETR site would hold about 2.5 kg of tritium in steady operation, and the minimum initial startup tritium inventory would be 848 g for each long pulse. The throughput of tritium is expected to be 21.24 g per pulse of which 1.062 g would be consumed/burned [4.23].

In the UK, JET has served as a prototype for ITER. Based at the Culham Centre for Fusion Energy near Oxford, JET has a design power of approximately

16 MW and has used 20 g of tritium for the tritium–deuterium experiment DTE1. Since its construction in 1983, plasma operations took place from May to November 1997. The authorized gaseous tritium releases after 2000 were 100 TBq/a [4.24]. The site planned to have an inventory of 90 g of tritium. In the recent DTE2 experiments, 60 g of tritium, most of which can be recovered, were used [4.25]. The final experimental run (DTE3) with deuterium–tritium fuel was undertaken before closure in 2023. A plan[2] has already been developed to decommission and repurpose JET[3,4].

From 1998 to 2002, it is estimated that the annual average gaseous and liquid releases of tritium were 11 700 and 16 000 TBq (equivalent to 32 and 44 g), respectively, from all nuclear facilities around the world [4.5].

4.2.2.3. *Tritium inventory in the environment (with data referenced to 2016)*

The tritium environmental inventory is dynamic and results from a transitional balance between processes of production and radioactive decay. The overall amount of tritium released by different sources can be estimated from the annual production fluxes (natural and technogenic), assuming steady state conditions when annual production balances annual losses via radioactive decay. Tritium input from nuclear tests is still present in the environment and contributes to the global tritium inventory while gradually reducing due to radioactive decay. Correspondingly, the shares of the different tritium sources in the global inventory are changing, and their values estimated for 2016 are schematically shown in Fig. 4.3. For example, the overall remaining inventory of tritium produced during atmospheric nuclear tests with cumulative input of 520–550 kg following radioactive decay was estimated to be between 22 and 24 kg in 2016 [4.2, 4.4, 4.5, 4.11]. By comparison, for the same year the inventory of cosmogenic tritium was estimated to be 4.5 kg, using an average tritium production rate in the atmosphere of approximately 0.32 atoms $\cdot$ $cm^{-2} \cdot s^{-1}$ [4.1]. Industrial releases of tritium were estimated to be 78 g/a [4.5], representing a steady state stock in 2016 of 1.3 kg. It is also notable that leakages associated with large tritium stocks (e.g. nuclear warheads, clocks, tritium powered luminous items, lamps) are obviously possible during their lifetime but are not quantified.

---

[2] https://euro-fusion.org/eurofusion-news/dte3_intro/

[3] https://www.gov.uk/government/publications/jet-decommissioning-and-repurposing

[4] See also Section 3.2.8 on tritium releases from fusion energy facilities.

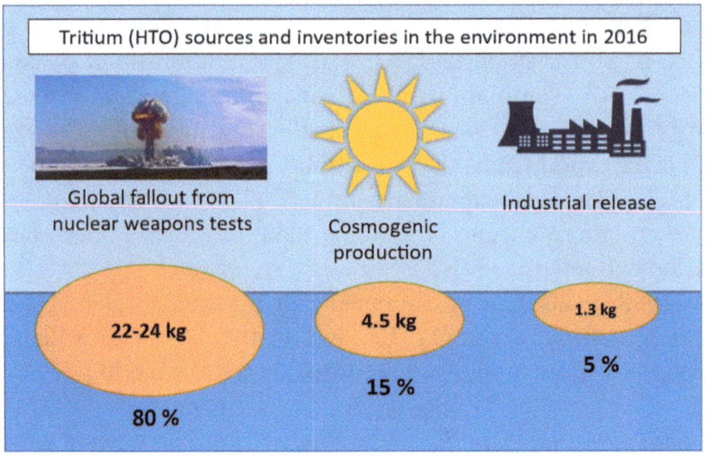

FIG. 4.3. Sources and inventories of tritium in the environment referenced to 2016 for comparison purposes.

4.3. TRANSPORT OF TRITIUM IN THE ATMOSPHERE

4.3.1. Global circulation of air and influence of nuclear weapon tests on atmospheric transport of tritium

The engine of global atmospheric circulation on the Earth is solar insolation, and the differences in temperature between the Equator and the poles drive global air circulation. The mechanisms that set these air masses in motion are complex because they depend on transfers of energy between the oceans and the atmosphere, between the continents and the atmosphere, and between the different layers of the atmosphere. In addition, there is also an influence of the Earth's rotational movement (Coriolis force), which determines the path followed by the air masses. The moist air masses transport water vapour and latent heat of evaporation from the tropics towards the poles, thus influencing the longitudinal distribution of surface air temperature and precipitation.

The global atmospheric circulation can be divided into three circulatory zones in each hemisphere, extending from the Equator to the poles (Fig. 4.4):

(1) The first zone, termed the Hadley cell, is located between the Equator and latitudes of 30° N or 30° S. There, regular winds (trade winds) blow near the surface, from the northeast in the Northern Hemisphere and the southeast in the Southern Hemisphere.

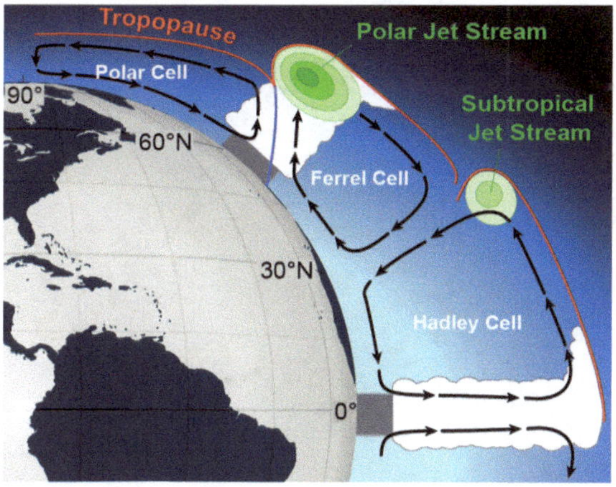

FIG. 4.4. Global atmospheric circulation, showing the Hadley, Ferrel and polar cells and the polar and subtropical jet streams (courtesy of the National Oceanic and Atmospheric Administration, US National Weather Service; https://www.weather.gov/jetstream/).

(2) The second zone, termed the Ferrel cell, is situated at moderate latitudes (30° N–60° N and 30° S–60° S) and is characterized by transient low pressure systems under a generally westerly circulation.
(3) The third zone, the polar cells, are situated north and south of latitudes 60° N and 60° S, respectively, with a surface atmospheric circulation directed generally eastwards.

Between each of these three zones, at altitudes varying between 6 and 15 km, are localized regions of high speed atmospheric flows called jet streams. Located at latitudes of 60° N and 60° S, the polar jet streams are fast moving air flows with volatile and unstable behaviour, while the subtropical jet streams, located at 30° N and 30° S, are slower and more stable (Fig. 4.5).

The successions of weather systems move in a general direction from west to east and are called zonal flows. The separation zone between troposphere and stratosphere (tropopause) varies with latitude. The thickness of the troposphere averages 9 km in the polar region and 17 km in the equatorial region. It is assumed that the lower stratosphere extends to 17 km in polar and 24 km in equatorial regions and in the upper stratosphere to about 50 km in both regions.

Global atmospheric circulation has played a major role in distributing tritium during the period of atmospheric testing of nuclear weapons. It was estimated that the head of the mushroom cloud following a nuclear explosion entered the lower stratosphere when the fission yields exceeded 20 kt and essentially became stratospheric when they exceeded 150 kt. For fission yields

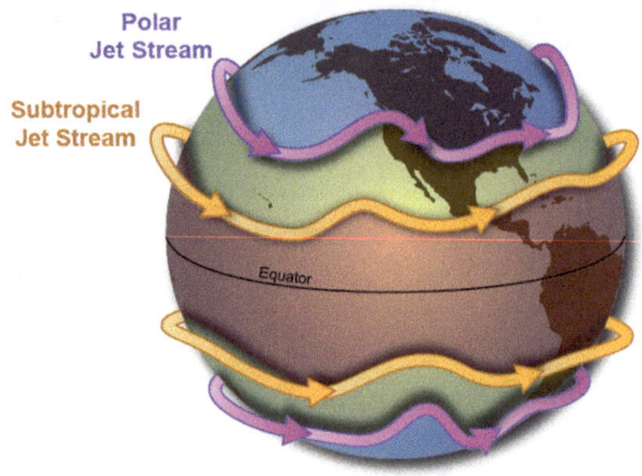

FIG. 4.5. Circulation of polar and subtropical jet streams (courtesy of the National Oceanic and Atmospheric Administration, US National Weather Service; https://www.weather.gov/jetstream/).

above 1 Mt, the cloud reached 25 km. The mean residence time of an aerosol in the lower stratosphere ranges from 3 to 12 months in the polar regions and 8 to 24 months in the equatorial regions.

Owing to the location of nuclear weapon test sites and the impact of atmospheric circulation, the deposition of nuclear fallout was greater in the Northern Hemisphere than in the Southern Hemisphere. According to calculations from ^{90}Sr deposition, the Northern Hemisphere received approximately 76% of the total radioactive fallout [4.11]. The preferential exchange of air between the stratosphere and troposphere at medium latitudes of both hemispheres and the air circulation patterns in the troposphere lead to enhanced deposition of radioactive fallout in the temperate regions and reduced deposition (by a factor of 2–3) in the equatorial and polar regions [4.11]. Large scale atmospheric circulation concentrated the deposition in temperate regions, in the latitude bands between 30° N and 60° N and between 30° S and 50° S [4.11]. The band between 40° N and 50° N is the most affected, with 17% of the fallout (see Fig. 4.6).

To illustrate the movement of contaminated air masses following the detonation of atmospheric nuclear weapons, the example of the last Chinese nuclear test at Lop Nor (40° N) on 16 October 1980 is presented in Fig. 4.7 [4.13]. The plume of radioactive particles moved eastwards, in accordance with the prevailing winds, and remained essentially between 30° N and 50° N. It crossed the Pacific Ocean from 17 to 19 October 1980, when it entered the western coast of Canada, then crossed North America and the Atlantic Ocean between 19 and 22 October, reaching Western Europe on 23 October.

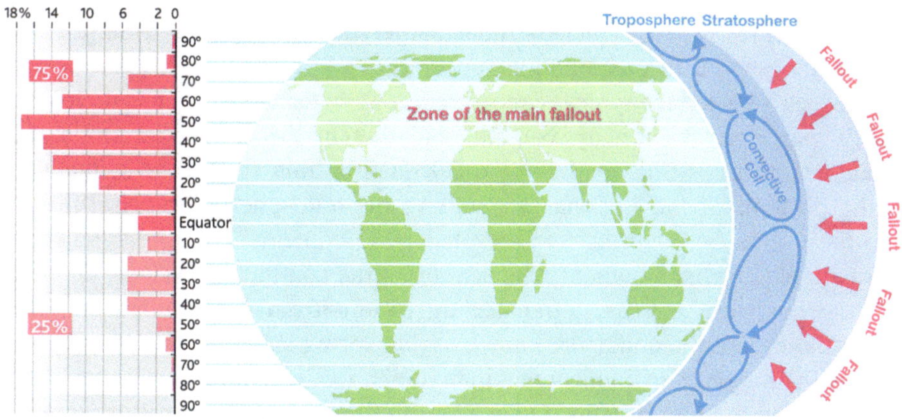

FIG. 4.6. Global distribution of ^{90}Sr deposition as a function of latitude (adapted from Ref. [4.13] with permission courtesy of IRSN).

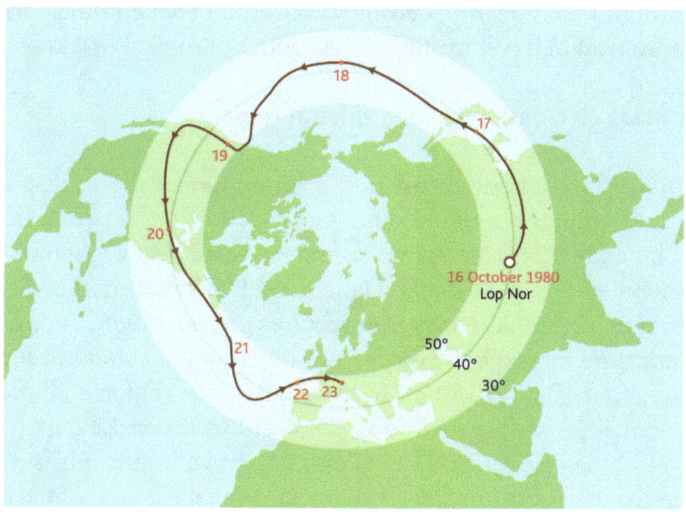

FIG. 4.7. Transport of the tropospheric radioactive cloud from the Chinese Lop Nor nuclear weapon detonation of 16 October 1980 (adapted from Ref. [4.13] with permission courtesy of IRSN).

In addition to the impact of global atmospheric circulation, several other specific mechanisms induce seasonal variations of tritium content in precipitation. First, tritium concentrations in the stratosphere are higher by several orders of magnitude than the natural tritium concentrations in precipitation. According to various studies [4.1, 4.26, 4.27], the content of stratospheric HTO vapour was estimated to be in the range 5×10^5–9×10^5 TU, equivalent to about 3.6–4.5 kg of cosmogenic tritium, which is much larger than the typical tritium content currently observed in precipitation (up to 10 TU). This is mainly due to large differences between the typical concentration of water vapour in the stratosphere (a few parts per million by volume) and that in the lowermost troposphere (from 0.01% at −42°C to 4.24% at 30°C). For this reason, any fluctuation in the position of the tropopause resulting in stratospheric air mass intrusion into the troposphere can lead to elevated HTO content in precipitation. In the Northern Hemisphere, owing to the heating of the continental surface, the height of the tropopause increases, transferring tritium from the stratosphere to the troposphere. This phenomenon, known as the 'spring leak' [4.28], determines the seasonal variation of tritium content in precipitation, reaching a maximum during the spring to summer months. Other identified mechanisms of tritium transport from the stratosphere to the troposphere include the eddy diffusion across the tropopause, which is roughly uniform over the whole Earth surface, the tritium injections by jet streams and the effect of the Hadley cells [4.29]. The Hadley cell circulation would then govern tropospheric tritium concentrations in January in the northern tropical belt and in July in the southern tropical belt.

4.3.2. Air mass circulation at the regional scale

The global atmospheric circulation is an important driver of transport and dispersion of airborne tritium (and other radionuclides) originating from nuclear weapon testing and large scale nuclear accidents such as the Chernobyl accident or the accident at the Fukushima Daiichi NPP. However, the dispersion of tritium at local and regional scales also requires consideration. Today, except for accidents, industrial atmospheric discharges of tritium generally occur at heights ranging from about 10 m to 100 m. In these cases, the regional climatology and local influences for specific sites (e.g. coastal) are the most important factors influencing airborne dispersion. Dependence on prevailing winds is a significant factor and, in the case of tritium, even the local wind at the time of the release will have to be considered. Nuclear sites emitting gaseous tritium use atmospheric modelling to link emissions to the prevailing meteorological conditions and thus the subsequent dispersion of the radionuclide.

The spatial footprint after 60 years of atmospheric releases of gaseous tritium to surface soils near the CRL site has been recently investigated [4.30]. Measurements of OBT show that the tritium retained in the soil organic matter was not related to the distance from the source but was strongly related to the prevailing wind direction (see Section 7.5.5). Similarly, in France, observations conducted over long periods around nuclear sites releasing gaseous tritium showed that elevated tritium contents are recorded in precipitation and, to a lesser extent, in soil organic matter downstream of prevailing winds [4.20].

4.3.3. Tritium levels (HT and HTO) in the atmosphere

The tritium activity in air can be measured in different forms. Atmospheric water vapour containing HTO is generally collected using cold-trap condensers. The gaseous forms of tritium (HT and CH_3T) are also measured by collecting a large volume of air and then separating the chemical forms of tritium using systems of bubblers and activated carbon traps [4.31–4.36]. Measurements of tritium content in precipitation are the most common (see Section 4.4) and are typically made using liquid scintillation counting. Very low level measurements (background samples) are generally made after electrolytic enrichment of tritium in the sample, followed by liquid scintillation counting or gas flow proportional counting (see also Chapter 9).

On a local scale and under the influence of different nuclear facilities (reprocessing plants, PWRs or PHWRs), the proportions and levels of tritium activities in the air, associated with the presence of HT and HTO (in becquerels per cubic metre of air) are likely to vary. These variations may induce either higher HT activities near reprocessing plants or higher HTO activities near CANDU type reactors. The HT and HTO activities will vary widely at the local (0–10 km) and regional (a few hundred kilometres) levels. Their activities depend on the magnitude of atmospheric discharges from the nuclear facilities and on the liquid discharges into the sea or rivers. Exchanges at the water–air interface will contribute to enrichment locally (a few meters or kilometres) and the atmospheric contents. The sampling time will also greatly affect measurement results. For example, a single 1 h measurement downwind of a tritium source will be difficult to compare with a (more typical) monitoring type measurement that integrates sampling over a week or a month. The latter allows lower activity measurements to be made. It is therefore very important in such studies to account for the sampling duration and frequency, especially when comparison is made for data gathered by different groups.

The current radiological background due to HT and HTO in air has been measured at sites not influenced by any nuclear facilities, and activities of HTO, HT and CH_3T in the atmosphere are low. The HTO activities measured in atmospheric water vapour [4.33, 4.36–4.39] between 2002 and 2015 in France and Japan are lower than 0.7 Bq/L of water or below 9 mBq/m^3 of air. The background tritium activities measured in France [4.36] yield an HTO activity concentration of 0.18 ± 0.01 Bq/L of water, corresponding to 1.7 ± 0.2 mBq/m^3 of air. For HT and CH_3T, the activities are lower than 10 mBq/m^3 and 3 mBq/m^3, respectively.

The influence of industrial sources on the tritium activity concentration in air can be demonstrated using published data for some nuclear facilities and power plants in Asia, Europe and North America. For example, tritium measurements made near the reprocessing plant at La Hague [4.36] show that close to the site, HT dominates (85% HT, 15% HTO in the air). In about 20 measurement campaigns performed in 2014–2016, atmospheric air within the plume was continuously sampled for 1 h at distances of several kilometres from the release point. The activity concentrations of HTO ranged from 198 to 1707 mBq/m^3 of air (i.e. up to several hundred becquerels per litre in water vapour), while activity concentrations of HT varied from 918 to 40 030 mBq/m^3 at the same time. Measurements carried out downwind of the installation at distances of about 150 km indicated hardly any detectable activities and were close to the background level. The reported activity concentrations are pertinent to plumes and, therefore, they likely represent maximum values. At the same time, the levels of tritium activity concentrations of HTO and HT measured during 2015–2017 at weekly intervals by the monitoring networks on the same nuclear site varied from 40 to 300 mBq/m^3, with weekly averages of 70 and 140 mBq/m^3, respectively [4.20]. For selected reprocessing plants in the USA, the fraction of HT in the total tritium activity release was 25% and 75%, as reported for the Hanford site and at the SRNL site [4.40]. These data demonstrate that the fraction of hydrogen gas in airborne releases varies for different plants, thus affecting activity concentrations in the atmosphere.

Weekly recordings of tritium in air around the French nuclear research centre at the CEA Valduc site, which released 241 TBq/a in 2014, were found to have HTO activities of 100–3200 mBq/m^3, with an average of 500 mBq/m^3 [4.20]. For HT, the weekly recorded activities in air varied between 120 and 700 mBq/m^3 with an average of 200 mBq/m^3. In the immediate vicinity of the CEA Marcoule nuclear research centre site, which released 130 TBq of tritium in 2013, the average weekly HTO and HT activities in air were 550 mBq/m^3 (maximum of 3200 Bq/m^3) and 270 mBq/m^3 (maximum of 1300 mBq/m^3), respectively. By comparison, in Canada, around the SRBT plant (Pembroke, Ontario), which manufactures tritium powered self-luminous signs, Mihok et al. [4.41] reported tritium activity levels in air of 60–100 000 mBq/m^3 for HTO and 60–70 000 mBq/m^3 for HT over 8–16 h sampling periods, depending on the release.

NPPs are an important source of airborne releases of tritium. According to monitoring data acquired for the French metropolitan authorities, activity concentrations of total tritium (HTO and HT) measured weekly around the NPPs are lower than 250 mBq/m^3, with an average of 70 mBq/m^3 [4.20]. A study around the Cruas NPP in France [4.42] indicated a contribution of 87% in the form of HTO and 13% in other forms (HT, CH_3T), with total tritium activity concentrations lower than 120 mBq/m^3. These values, although low, are an order of magnitude larger than the background level of environmental tritium in the region where the plant is located.

Reported values from other countries are summarized in Table 4.1 and provide exemplary values of the tritium activity concentration in air observed

TABLE 4.1. EXAMPLES OF ENVIRONMENTAL TRITIUM ACTIVITY CONCENTRATIONS IN PROXIMITY TO VARIOUS NUCLEAR FACILITIES

Country, period	Tritium origin	Sample acquisition time	Activity concentration				Reference
			HTO water vapour (Bq/L)	HTO in air (mBq/m^3)	HT in air (mBq/m^3)	CH_3T in air (mBq/m^3)	
Japan, 2003–2005	Background	—[a]	0.69 ± 0.35	6.7 ± 5.4	12.5 ± 6.9	9.0 ± 8.2	[4.38]
Japan, 2002–2004	Background	—	—	6.5	10.7	2.8	[4.33]
Japan, 2003–2006	Background	—	—	9.0 ± 6.5	9.0 ± 1.6	2.0 ± 1.2	[4.37]
Japan, 2003–2006	Background	—	0.25–0.8	3.6 ± 2.3	8.5 ± 2.6	1.9 ± 0.8	[4.39]
Canada, 2013	SRBT plant	8–16 h	190	60–100 000 (mean 1900 ± 5800)	60–65 000 (mean 1400 ± 3900)	—	[4.41]
France, 2014–2015	Background	—	0.18 ± 0.01	1.7 ± 0.2	9.8 ± 3.1	2.5 ± 0.9	[4.36]
France, 2014–2016	Reprocessing plant	1 h	34–167	198–1707	918–40 030	—	[4.20]
France, 2014–2016	Reprocessing plant	1 week	—	70 (mean)	140 (mean)	—	

TABLE 4.1. EXAMPLES OF ENVIRONMENTAL TRITIUM ACTIVITY CONCENTRATIONS IN PROXIMITY TO VARIOUS NUCLEAR FACILITIES (cont.)

Country, period	Tritium origin	Sample acquisition time	Activity concentration				Reference
			HTO water vapour (Bq/L)	HTO in air (mBq/m^3)	HT in air (mBq/m^3)	CH_3T in air (mBq/m^3)	
France, 2014–2016	Research facility CEA Valduc	1 week	—	100–3200 (mean 500)	120–700 (mean 200)	—	
France, 2014–2016	Research facility CEA Marcoule	1 week	—	550	270	—	
France, 2014–2017	All NPPs	1 week	—	<250 (mean 70)	—	—	[4.20]
France, 2016	NPP Cruas	1 week	—	<120	—	—	[4.42]
Republic of Korea, 2004–2008	NPP Wolsung	1 month	—	180–40 900	—	—	[4.43]
Serbia, 1992–1995	Research reactor, VINS	1 week	16–2468	270–60 800	40–1050	—	[4.44]
Canada, 1990	CRL	—	—	7700–347 000	—	—	[4.45]

[a] —: not available.

close to various nuclear facilities around the globe. In the Republic of Korea, airborne tritium measurements are recorded around the Wolsung NPP (HWR) [4.43, 4.46], with monthly environmental HTO contents around the plant varying from 180 to 40 900 mBq/m^3. In Serbia, HTO and HT values were reported by Miljevic et al. [4.44] for the Belgrade region near the Vinca Institute of Nuclear Sciences site over the period 1988–1997 as weekly average values. The measured activities fluctuated between 40 and 10 500 mBq/m^3. Wood and Workman [4.45] measured HTO activities in the range 7700–347 000 mBq/m^3 (however, mostly below 20 000 mBq/m^3) at a distance of 500 m from the NRU reactor of CRL in Canada.

In summary, the background levels for HTO and HT, not affected by industrial activities, are currently below 10 mBq/m^3. For HTO water vapour in air, this corresponds to less than 0.2 Bq/L of water. Where there is an influence from nuclear facilities, environmental activity concentrations of tritium may increase within a distance of a few kilometres from a release point, but large variations between the sites exist, depending on the type of industrial facility. The sampling period (from 1 h to a few weeks or months) will then strongly influence the level of activity that can be measured. On the basis of the results of monitoring networks, activities of a few hundred millibecquerels per cubic metre of air for HTO/DTO can be measured, rising to a few thousands millibecquerels per cubic metre around heavy water nuclear reactors. Similarly, tritium levels of a few thousand millibecquerels per cubic metre are recorded around specific facilities (e.g. sites producing self-luminous items and spent nuclear fuel reprocessing and waste sites) from samples in the plume and with short sampling periods [4.36, 4.41].

4.4. TRITIUM CONTENT IN PRECIPITATION

4.4.1. Monitoring of tritium in precipitation at the global scale

Tritium activities in monthly precipitation are routinely monitored at stations of the GNIP, which is the worldwide isotope monitoring network of hydrogen and oxygen isotopes in precipitation. The network was launched in 1960 by the IAEA and WMO and it operates in cooperation with many organizations worldwide. An online version of the GNIP database is hosted by the IAEA.[5] At its inception, the objective was to establish a systematic collection of basic data on the isotopic composition of precipitation at the global scale to determine temporal and spatial variations of oxygen and hydrogen isotopic composition in

[5] https://nucleus.iaea.org/wiser

precipitation. It also enabled the monitoring of atmospheric thermonuclear test fallout through the measurements of tritium content in precipitation. Since the onset of GNIP operations, monthly precipitation samples were collected from over 1000 stations in more than 125 countries [4.15, 4.47].

Observation periods vary among stations, and some cover the period from the late 1950s to the late 2010s, thus providing unique data on the evolution of tritium content in the atmosphere. In this chapter, the data collected at the stations shown in Table 4.2 are used.

TABLE 4.2. LOCATIONS OF GNIP NETWORK STATIONS [4.47]

Station	Country	WMO code	Latitude (°)	Longitude (°)	Altitude (m)
		Southern Hemisphere			
Halley Bay	Antarctica	8902200	−75.6	−20.6	30
Vernadsky	Antarctica	8906300	−65.1	−64.0	20
Invercargill	New Zealand	9384400	−46.4	168.3	2
Kaitoke	New Zealand	9341701	−41.1	175.2	200
Gough Island	UK (South Atlantic)	6890600	−40.4	−9.9	54
Melbourne	Australia	9486800	−37.8	145.0	28
Cape Town	South Africa	6881600	−34.0	18.6	44
Easter Island	Chile (Pacific Ocean)	8546900	−27.2	−109.4	41
Ascension Island	UK (Atlantic Ocean)	6190000	−7.9	−14.4	15
		Northern Hemisphere, south of 68°			
Ocala	USA	7220501	29.2	−82.1	26
Albuquerque	USA	7236500	35.1	−106.6	1619

TABLE 4.2. LOCATIONS OF GNIP NETWORK STATIONS [4.47] (cont.)

Station	Country	WMO code	Latitude (°)	Longitude (°)	Altitude (m)
Hatteras	USA	7230400	35.3	−75.6	3
Punta Delgada	Portugal	851200	37.8	−25.7	175
Madrid	Spain	822201	40.4	−3.7	667
Ottawa	Canada	7162800	45.3	−75.7	114
Thonon-les-Bains	France	748501	46.4	6.5	385
Stuttgart	Germany	1073900	48.8	9.2	314
Brest	France	711001	48.4	−4.6	80
Valentia	Ireland	395300	51.9	−10.3	9
Northern Hemisphere, north of 68°					
Danmarkshavn	Greenland	432000	76.8	−18.7	12
Resolute Bay	Canada	7192400	74.7	−95.0	67
Murmansk	Russian Federation	2211300	69.0	33.1	46
Isfjord	Norway	100500	78.1	13.6	6
Tanagard	Norway	107501	70.4	28.2	130
Nord	Greenland	431000	81.6	−16.7	35

4.4.2. Past and present tritium activity levels in precipitation

The temporal evolution of tritium activity concentration in rainwater at some GNIP stations is shown in Figs 4.8–4.11. For the Northern Hemisphere, the following stations were selected: Hatteras (East coast of USA, marine influence), Ottawa (Canada), Stuttgart (Germany), Thonon-les-Bains (France) and Valentia

(Ireland, marine influence). For the Southern Hemisphere, the following stations were selected: Kaitoke (New Zealand), Cape Town (South Africa), Melbourne (Australia) and Easter Island (Chile, South Pacific Ocean). The earliest available data are from 1953 (Ottawa) and 1957 (Valentia). From 1951 to 1963, atmospheric nuclear weapon tests had been conducted by the USA, the USSR and the UK, with most tests taking place in 1958, 1961 and 1962. These three countries then stopped atmospheric tests and switched to underground testing.

In the Northern Hemisphere, there was a rapid increase in tritium activity in the late 1950s due to nuclear tests. This was followed by the major peak over the period 1963–1964, with tritium levels above several hundred becquerels per litre in rainwater, and then a continuous decline to the present period. The same trend was observed for Ottawa, Hatteras, Valentia, Thonon-Les-Bains and Stuttgart. The detailed evolution for Ottawa, Stuttgart and Valentia shows peaks for 1954, 1958 and 1959, and a succession of peaks each year between 1962 and 1965 (Fig. 4.8(c)). In the period 1958–1962, the largest number of atmospheric tests with high yield were carried out; consequently, the highest activities in precipitation were measured in 1962–1963. There are significant differences between the stations of Hatteras (USA) and Valentia (Ireland), which are both influenced by the diluting influence of marine air masses. These two stations exhibit tritium activities consistently lower than those measured on the mainland (see Figs 4.8(a) and 4.8(d)).

A global decrease in activity has been continuous since about 1965, due to the cessation of atmospheric tests (the last tests were carried out by China in 1980) as well as radioactive decay and isotope exchange with the ocean. In the Northern Hemisphere, activities have stabilized between about 0.5 and 2 Bq/L after the 2000s, with slight variations depending on the marine or continental location of the station collecting monthly precipitation, the seasonal cycle and the latitude. Notable also are the marked seasonal cycles at the stations in Stuttgart, Ottawa, Valentia, for example (Fig. 4.8(d)).

In the Southern Hemisphere (see Fig. 4.9), temporal changes are consistent with those in the Northern Hemisphere, but with less marked variations. The maximum levels of tritium have been and remain almost two orders of magnitude lower than in the Northern Hemisphere, with maximum values that were of the order of 10 Bq/L in 1965–1966. In the Southern Hemisphere, by the mid-1970s, activities were generally lower than 2 Bq/L. The seasonal cycle, although evident, is less marked than in the Northern Hemisphere. It can be noted that for the Easter Island station (South Pacific), the peaks of 1964 and 1966 are clear, which is less obvious for the stations located in New Zealand (Kaitoke, Invercargill) or in South Africa (Cape Town), as seen in Fig. 4.9(c). From the 1990s onwards, the content of tritium in Southern Hemisphere precipitation stabilized below 0.2 Bq/L (Fig. 4.9(c)).

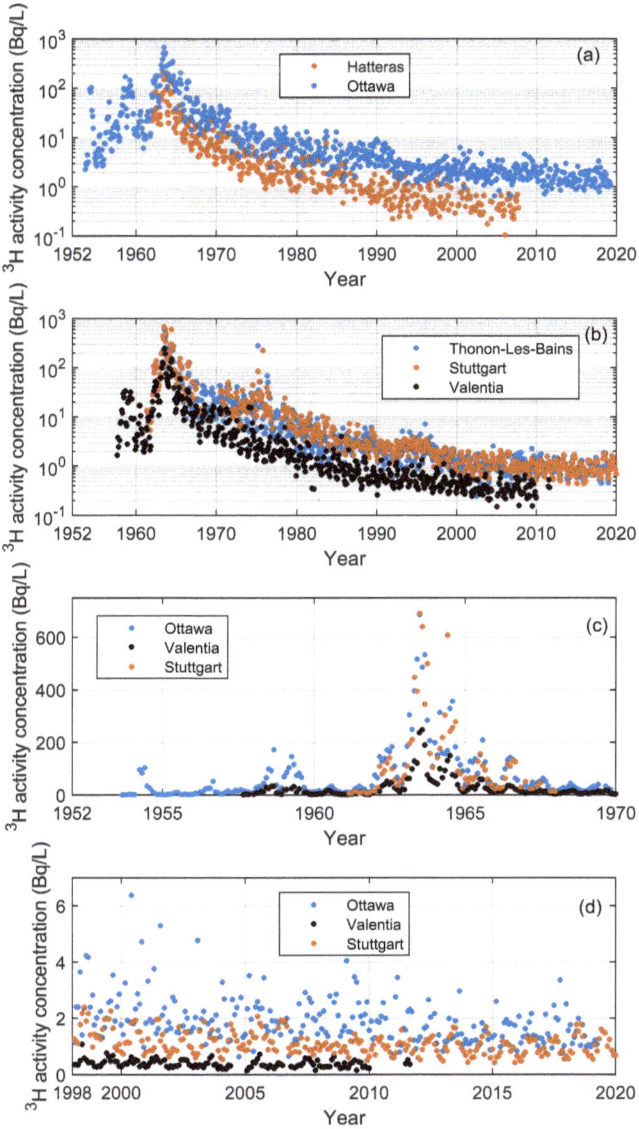

FIG. 4.8. Measured tritium concentrations in precipitation at the following GNIP stations in the Northern Hemisphere [4.15]: (a) Hatteras and Ottawa for 1952–2019; (b) Thonon-les-Bains, Stuttgart and Valentia for 1952–2019; (c) Ottawa, Valentia and Stuttgart 1952–1970; (d) Ottawa, Valentia and Stuttgart for 1998–2019.

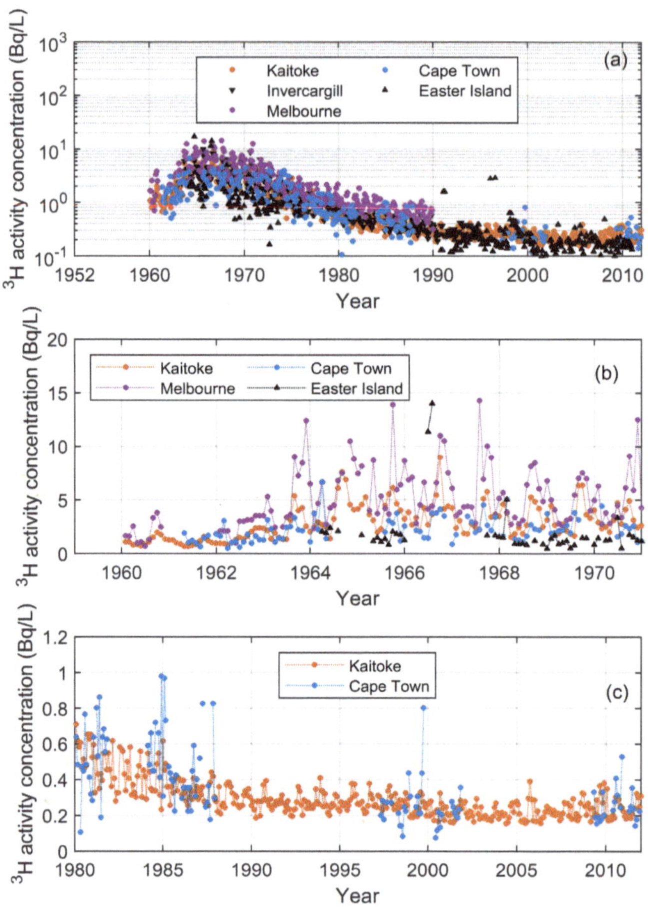

FIG. 4.9. Measured tritium concentrations in precipitation at the following GNIP stations in the Southern Hemisphere [4.15]: (a) Kaitoke, Melbourne, Invercargill, Cape Town and Easter Island for 1952–2011; (b) Kaitoke, Melbourne, Cape Town and Easter Island for 1959–1971; (c) Kaitoke and Cape Town for 1980–2013.

Average tritium values in rainwater were calculated on the basis of the most recent available data for some stations (Table 4.3). For the Southern Hemisphere, the activities are from 0.22 to 0.34 Bq/L, while in the Northern Hemisphere the values vary from 0.34 to 1.56 Bq/L (and are generally lower than 1 Bq/L). For the Northern Hemisphere, the lowest value was registered for Valentia, a station strongly influenced by the Atlantic Ocean, while the highest value was observed at Ottawa, a station most probably affected by technogenic tritium emissions from Chalk River Laboratories and the SRBT facility.

TABLE 4.3. AVERAGE ^{3}H ACTIVITY CONCENTRATIONS IN PRECIPITATION AT THE GNIP STATIONS IN THE SOUTHERN AND NORTHERN HEMISPHERE AND IN ANTARCTICA [4.15]

Station	Country	Hemisphere	Observation period		Sample size	^{3}H activity concentration (Bq/L)
			from	to		
Kaitoke	New Zealand	Southern	2009-01-15	2011-12-15	36	0.24 ± 0.06
Ascension Island	UK	Southern	2009-03-31	2012-07-31	17	0.32 ± 0.74
Cape Town	South Africa	Southern	2009-04-15	2012-10-15	29	0.25 ± 0.09
Gough Island	UK	Southern	2010-01-15	2013-08-15	44	0.22 ± 0.07
Madrid	Spain	Northern	2010-01-15	2015-12-15	56	0.57 ± 0.27
Thonon-Les-Bains	France	Northern	2010-01-15	2018-12-15	108	0.95 ± 0.38
Ottawa	Canada	Northern	2010-01-15	2019-03-15	111	1.56 ± 0.55
Stuttgart	Germany	Northern	2010-01-15	2019-12-15	119	0.99 ± 0.33
Valentia	Ireland	Northern	2009-01-15	2015-12-15	15	0.34 ± 0.11
Halley Bay	Antarctica	Southern	2009-01-01	2010-12-31	24	1.42 ± 0.76
Vernadsky	Antarctica	Southern	2007-01-01	2010-07-31	25	0.44 ± 0.11

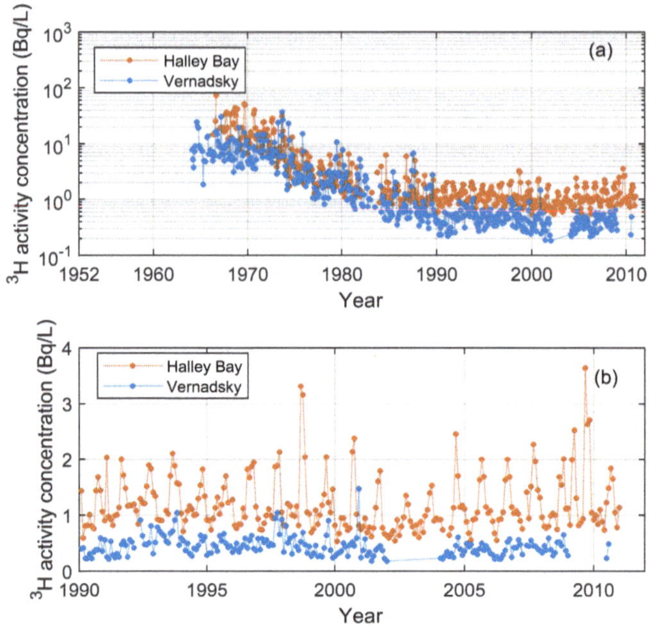

FIG. 4.10. Measured tritium concentrations in precipitation at the GNIP stations Halley Bay and Vernadsky in Antarctica [4.15] for: (a) 1952–2012 and (b) 1990–2012.

Finally, Fig. 4.10 presents the data recorded at two stations in Antarctica, Halley Bay and Vernadsky. The overall evolution is very similar to what has been described previously for the Southern Hemisphere. For these two stations, high values were recorded between 1966 and 1969 (maximum 73 Bq/L for Halley Bay and 30 Bq/L for Vernadsky), followed by a decrease, as observed elsewhere in the world. From the latest measurements (Table 4.3 and Fig. 4.10), the tritium activities already appear stabilized. However, it is notable that the Halley Bay data show systematically higher values than those measured at Vernadsky, and are the highest among the stations in the Southern Hemisphere. According to Fourré et al. [4.26], the Antarctic zone is very strongly affected by stratospheric–tropospheric exchanges due to particularly dry air during the southern winter. Another effect is the limited exchange with marine air masses related to the polar vortex for most continental stations, which ultimately results in higher tritium activities in precipitation. Although relatively close to Halley Bay, the Vernadsky station is located on a peninsula close to the sea, while Halley Bay is located on the mainland, about 60 km away, which is sufficient to cause significant differences observed between those two stations.

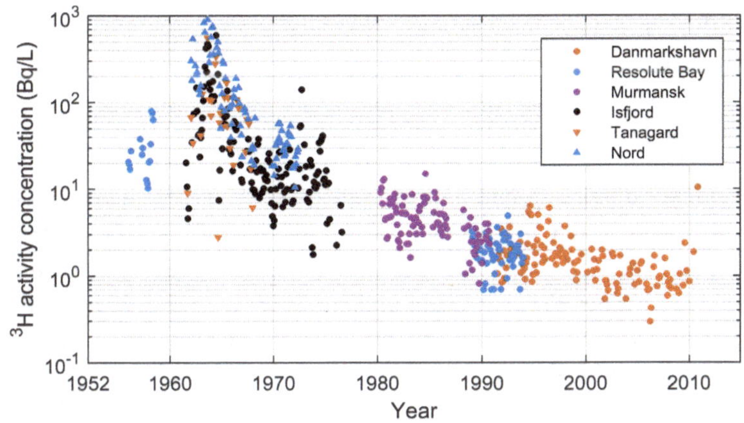

FIG. 4.11. Measured tritium concentrations in precipitation at GNIP stations north of 68° N (Danmarshavn, Resolute Bay, Murmansk, Isfjord, Tanagard, Nord) for 1952–2012 [4.15].

Data obtained for the Arctic zone (latitude between 68° N and 82° N) are presented in Fig. 4.11. In 2009, the latest measurements indicated activities of approximately 1 Bq/L, which are lower than those in Antarctica (cf. Fig. 4.10).

As an interim conclusion, since the end of atmospheric nuclear weapon testing, environmental tritium levels have slowly returned to the values characterizing the period before nuclear tests in the atmosphere; namely, values below 1 Bq/L [4.9] for the Northern Hemisphere and around 0.2 Bq/L for the Southern Hemisphere [4.48, 4.49]. Variations are observed between stations in the same hemisphere in relation to the various influences discussed further below. Recently, Terzer-Wassmuth et al. [4.50] developed a high spatial resolution predictive model for tritium activity in global precipitation based on contemporary measurements (2008–2018); the model predictions depended on latitude, longitude, altitude, distance to shore and air temperature. The model and the resulting maps reflect the differences in expected tritium activity in rainwater from different parts of the world.

4.4.3. Sources of variations in tritium activity levels in precipitation

Locally, tritium activities in precipitation can exhibit large variations due to the direct influence of nuclear tests or nuclear facilities, but also to the mixing of air masses between the stratosphere and the troposphere and the origin of the air mass (marine or continental) at the time of precipitation. The tritium activities recorded in precipitation depend on the latitude, altitude and the distance from the sea. In marine and oceanic environments, water vapour exchange takes place between the sea and the atmosphere, and this reduces tritium activity in the air owing to dilution. This leads to precipitation having lower tritium activities [4.27, 4.28, 4.49, 4.51–4.53].

4.4.3.1. Seasonal trends

Tritium activities in precipitation can vary with season mainly owing to the seasonal exchange of air masses between the stratosphere and the troposphere [4.28, 4.49, 4.52, 4.54, 4.55]. Tritium levels in Northern Hemisphere precipitation show maximum values during spring and summer and minima in winter, which vary by a factor of 2–4, depending on the site. Other parameters, such as latitude, marine or continental influence, can slightly influence amplitudes and time shifts of such seasonal variations.

The seasonal cycles can be explained by exchange of air masses between the stratosphere and troposphere. When the continental surface heats up in the spring, the height of the tropopause increases, causing a larger than normal tritium influx from the stratosphere to the troposphere. This increases its overall tritium concentration and leads to an increase in the tritium activity concentrations measured in rainwater in spring and summer, which is known as the spring leak phenomenon. This is more explicit at medium latitudes, in the baroclinic zones and tropopause discontinuities [4.27, 4.54]. The seasonal variations of tritium contents in precipitation recorded at Stuttgart, Thonon-Les-Bains and Valentia after 2000 are shown in Fig. 4.12. The average ratio of activity concentrations in June (maximum) and January (minimum) is 1.9 for the two continental stations (Stuttgart and Thonon-Les-Bains) and 1.5 for Valentia, which has an influence from marine air masses. In the Southern Hemisphere, the annual seasonal variation cycle is shifted by approximately half a year, showing maxima between September and December, with seasonal variations being lower than in the Northern Hemisphere. Mean seasonal cycles of tritium content for the Melbourne and Kaitoke stations for data available after 1980 are shown in Fig. 4.12. The ratio between the average tritium levels for May (minimum) and November–December (maximum) is 1.3–1.4 for these two stations.

4.4.3.2. Marine and continental effect on tritium activities in precipitation

The distance of sampling stations to the ocean has an impact on the tritium concentration in precipitation, which is known as the continental effect [4.27, 4.49, 4.52, 4.56]. At coastal sites, the lower atmosphere is directly influenced by the ocean. Moisture in maritime air masses has reduced levels of tritium mostly owing to its origin and isotope exchange at the sea–air interface. Consequently, evaporation from the ocean surface leads to reduced tritium concentrations in comparison with surface water on continents. Precipitation formed from such moisture will preserve

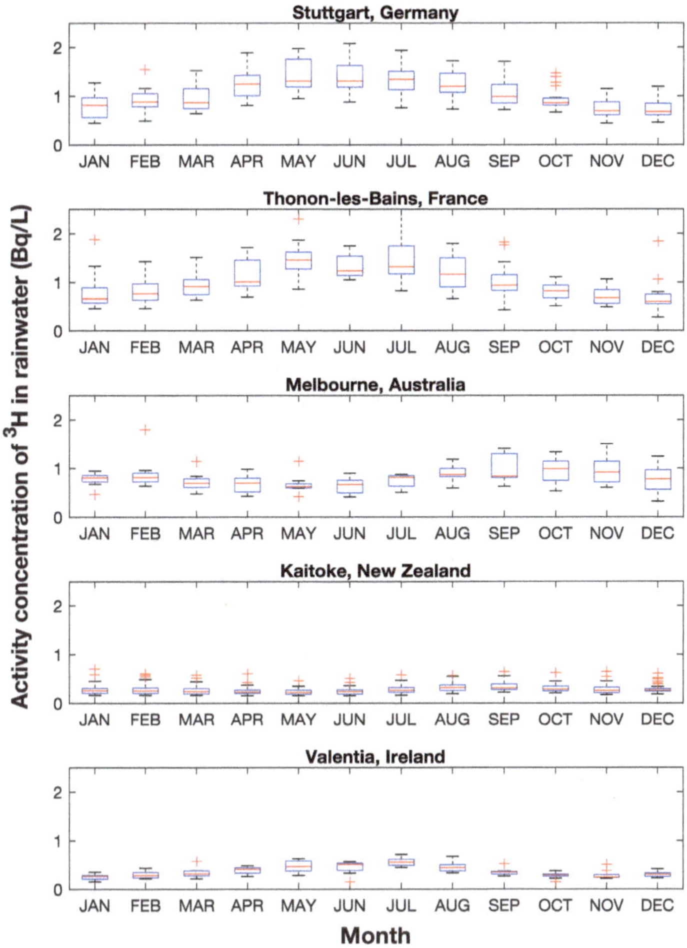

FIG. 4.12. Monthly variations of activity concentrations of ^3H in rainwater, as recorded in Europe in Stuttgart (Germany), Thonon-Les-Bains (France) and Valentia (Ireland) since 2000 and in the Southern Hemisphere in Kaitoke (New Zealand) and Melbourne (Australia) since 1980 [4.15]. The data are shown as boxplots, where boxes span between the 25th and 75th percentiles; medians are shown as red lines inside the boxes and outliers are indicated as red crosses.

these low tritium levels [4.49, 4.57]. As stated in Section 4.4.2, the influence of maritime air masses decreases with increasing distance to the coast, resulting in an increase in tritium activity in precipitation when moving away from the sea [4.27, 4.28, 4.49, 4.51–4.53]. Tritium enrichment at continental sites can also result from molecular exchange with re-evaporated continental water, which has a higher tritium concentration than the water vapour of the air masses above. For inland regions, tritium activity concentration in air and its variability can result from the relative contributions of water vapour from coastal, oceanic and continental sources. Marine water vapour is depleted in tritium, whereas continental water vapours are more tritium enriched and also have a continuous supply of tritium from the upper troposphere over the continents.

Tritium activities from marine and continental stations at similar latitudes in the Northern Hemisphere consistently show notable differences, as indicated earlier (Fig. 4.13). For Western Europe, Brest (France), Punta Delgada (Portugal, Azores Islands) and Valentia (Ireland) represent marine stations, while Stuttgart (Germany) and Thonon-Les-Bains (France) represent continental sites. The difference is distinct and significant. Similar locations with contrasting activities in North America are Ocala and Hatteras (east coast marine stations) and Albuquerque and Ottawa (continental site) (Fig. 4.13(b)).

The Southern Hemisphere has relatively limited long term tritium records in precipitation, but some useful findings provide insights. Tadros et al. [4.49] showed that tritium activity concentrations were higher in central Australia (Alice Springs) compared with coastal sampling sites in Australia, most probably

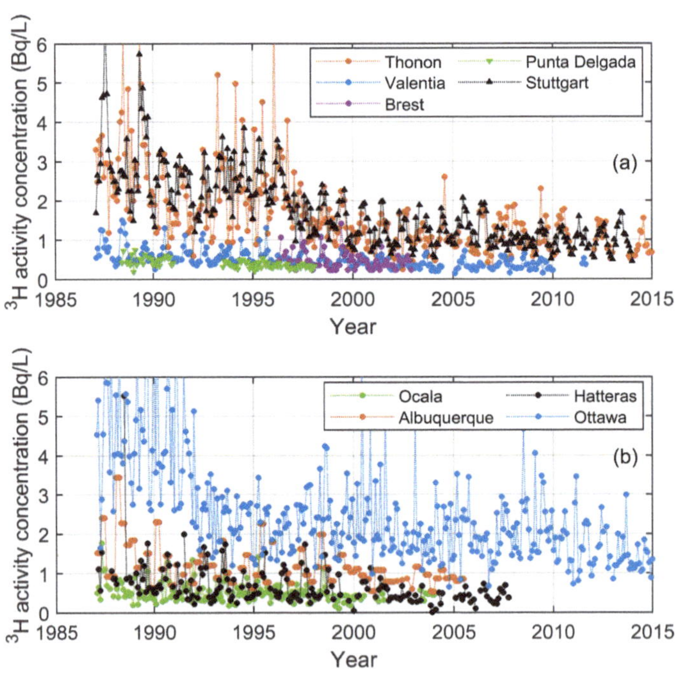

FIG. 4.13. Measured tritium concentrations in precipitation at GNIP stations [4.15] in Western Europe (a) and in North America (b) for 1987–2014.

owing to the continental effect. Between 1972 and 1980, monthly mean values ranged from 0.9 to 2.4 Bq/L at the Alice Springs site in central Australia, while at the coastal site in Brisbane activity concentrations were in the range 0.4–1 Bq/L.

4.4.3.3. Latitudinal trend

Tritium activities in precipitation are known to strongly depend on latitude. At latitudes north of 30° N, systematically higher tritium contents were recorded compared with tropical areas or with locations in the Southern Hemisphere. This dependence can be attributed to the following well established causes:

- The impact of nuclear weapon testing was largest in the Northern Hemisphere, where most tests occurred.
- The cosmogenic tritium production and the stratosphere–troposphere flux depend on latitude and location, occurring principally in the region of the subtropical jet streams (cf. Fig. 4.5).
- At low and medium latitudes, the ocean (with very low tritium activity) plays a major role in diluting tritium in air masses. The high evaporation flux exchange at the sea–air interface causes a reduction in tritium activity in the atmosphere [4.27].

The main atmospheric nuclear weapon tests and subsequent dispersion occurred mostly in the Northern Hemisphere, following the USSR's tests in the Arctic at the Novaya Zemlya archipelago, with a combined yield of 240 Mt, and the USA's tests in the northern Pacific (2–16° N), with a total yield of 153 Mt. Together, these two countries accounted for 90% of the total yield of atmospheric nuclear tests for all countries [4.11]. Several studies [4.58–4.60] have demonstrated the relationship between latitude and tritium activity concentrations in rainwater. Tritium activity concentration in precipitation increases at an exponential rate from the Equator to the poles, doubling every 13° of latitude (see Fig. 4.14 and Table 4.4). At low latitudes and at tropical stations, activities are generally lower by approximately a factor of five [4.28]. For the Southern Hemisphere, Tadros et al. [4.49] showed that mean annual tritium activities in precipitation increase with latitude along the transect from north to south-east Australia (this increase was shown on measurements made during the nuclear testing period (1973–1974), and during the recent period 2005–2011). The latitudinal gradient of tritium over Australia is a consequence of the seasonal stratosphere–troposphere HTO exchange, which occurs in the region of the subtropical jet stream, at about 35–40° S. In Antarctica, relatively higher activities have been recorded, because these areas are more sensitive to intrusion of tritium enriched stratospheric air masses [4.26].

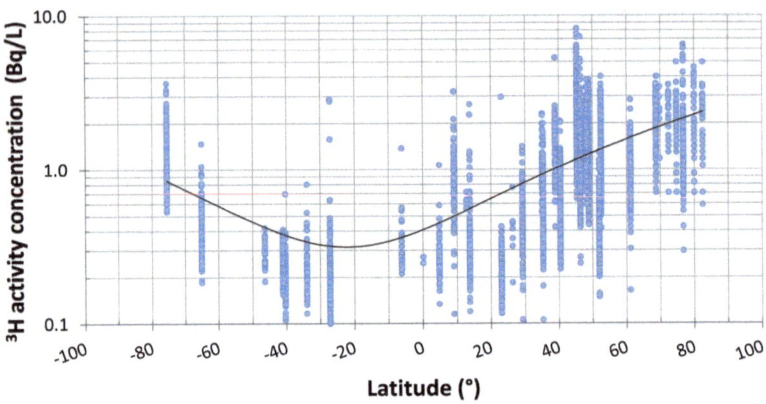

FIG. 4.14. Tritium concentrations in precipitation at the GNIP stations [4.15] for 1990–2019 as a function of latitude (see Table 4.4 for coordinates).

TABLE 4.4. GNIP STATIONS PROVIDING THE DATA SHOWN IN FIG. 4.14, SORTED BY INCREASING LATITUDE [4.47]

GNIP station name	Country or territory	WMO code	Latitude (°)	Longitude (°)	Altitude (m)
Halley Bay	Antarctica	8902200	−75.6	−20.6	30
Vernadsky (Argentine Is.)	Antarctica	8906300	−65.1	−64.0	20
Invercargill	New Zealand	9384400	−46.4	168.3	2
Kaitoke	New Zealand	9341701	−41.1	175.2	200
Gough Island (South Atlantic O.)	UK	6890600	−40.4	−9.9	54
Cape Town (Airport)	South Africa	6881600	−34.0	18.6	44
Easter Island (Pacific O.)	Chile	8546900	−27.2	−109.4	41
Darwin	Australia	9412000	−12.4	130.9	26
Jakarta	Indonesia	9674500	−6.2	106.8	8
Entebbe (Airport)	Uganda	6370500	0.1	32.5	1155

TABLE 4.4. GNIP STATIONS PROVIDING THE DATA SHOWN IN FIG. 4.14, SORTED BY INCREASING LATITUDE [4.47] (cont.)

GNIP station name	Country or territory	WMO code	Latitude (°)	Longitude (°)	Altitude (m)
Bogota	Colombia	8022200	4.7	−74.1	2547
Addis Ababa	Ethiopia	6345000	9.0	38.7	2360
Bamako	Mali	6129100	12.7	−8.0	381
Bangkok	Thailand	4845500	13.7	100.5	2
Havana (CPHR)	Cuba	7837401	23.1	−82.2	137
Bahrain	Bahrain	4115001	26.3	50.6	2
Ocala (Florida)	USA	7220501	29.2	−82.1	26
Albuquerque (New Mexico)	USA	7236500	35.05	−106.62	1619
Hatteras (North Carolina)	USA	7230400	35.3	−75.6	3
Washington (D.C.)	USA	7240500	38.9	−77.0	4
Madrid-Retiro	Spain	822201	40.4	−3.7	667
Ottawa (Ontario)	Canada	7162800	45.3	−75.7	114
Thonon-les-Bains	France	748501	46.4	6.5	385
Vienna (Hohe Warte)	Austria	1103500	48.2	16.4	198
Stuttgart	Germany	1073900	48.8	9.2	314
Valentia (Observatory)	Ireland	395300	51.9	−10.3	9
Berlin	Germany	1038400	52.5	13.4	48
Anchorage (Alaska)	USA	7027300	61.2	−150.0	35
Naimakka	Sweden	206000	68.7	21.5	403
Hall Beach (NWT[a])	Canada	7108100	68.8	−81.3	8
Pond Inlet (NWT)	Canada	7108101	72.4	−78.0	54

TABLE 4.4. GNIP STATIONS PROVIDING THE DATA SHOWN IN FIG. 4.14, SORTED BY INCREASING LATITUDE [4.47] (cont.)

GNIP station name	Country or territory	WMO code	Latitude (°)	Longitude (°)	Altitude (m)
Resolute Bay (Canadian Arctic)	Canada	7192400	74.7	−95.0	67
Danmarkshavn (Greenland)	Denmark	432000	76.8	−18.7	12
Eureka (NWT)	Canada	7191700	80.0	−85.6	10
Alert (NWT)	Canada	7108200	82.5	−62.3	62

[a] NWT: Northwest Territories.

4.4.4. Impact of nuclear industry releases on tritium contents in precipitation

4.4.4.1. Nuclear power plants

Gaseous emissions of tritium vary greatly depending on the type of technology associated with NPPs (Table 4.5). Most gaseous discharges are produced by PHWRs. These reactors account for about 90% of tritium emissions to the atmosphere, while representing only 10% of all reactor units worldwide. PWRs and boiling light water reactors (BWRs) contribute less than 10% of the emissions to the atmosphere [4.18]. HWRs operate in Canada, India, Pakistan, China, Argentina, Romania and the Republic of Korea. The impact of nuclear industries on tritium contents in precipitation has been explored by using the datasets from rainwater monitoring in France and Canada, as examples.

In France, for all operating nuclear reactors (18 NPPs and 56 reactors), the emissions of tritium to air in 2017 were estimated to be between 21 and 46 TBq/a. In Canada, which in 2017 had five NPPs with 19 HWRs, the releases of tritium to atmosphere in 2017 were estimated to be 2516 TBq.[6]

The impact of atmospheric tritium discharges from NPPs can be highly variable, as measurements of locally collected rainwater show. In France, over the period 2011–2014, the monitoring of rainwater around NPPs generally

[6] https://nuclearsafety.gc.ca/eng/resources/publications/reports/annual-reports/index.cfm

TABLE 4.5. TRITIUM ACTIVITIES IN PRECIPITATION AROUND NUCLEAR POWER PLANTS IN FRANCE, CANADA AND REPUBLIC OF KOREA, AND COMPARISON OF TRITIUM GASEOUS RELEASE FOR PWRs AND HWRs

Period	Country, facility	Reactor type	Rainwater (Bq/L)	Remarks
2011–2014	France, NPPs (58 reactor units)[a]	PWR	1.5 (range 4–11)	Monthly average
2009	Canada, NPPs (19 reactor units)[b]	HWR		Annual average
	Bruce NPP		258	
	Pickering NPP		240	
	Darlington NPP		28	
	Gentilly NPP		910	
1998–2015	Republic of Korea, Wolsong NPP	4 HWRs, 2 PWRs		Monthly average
			0.3–1090	0.6–22 km from the NPP
			0.8–10	22 km from the NPP

[a] Gaseous release of tritium for 2017 was in the range 21–46 TBq/a.
[b] Gaseous release of tritium for 2017 was 2516 TBq/a.

indicated activity concentration values close to background level (1.5 Bq/L), but occasionally activity concentrations in the range 4–11 Bq/L were detected [4.61]. In Canada, in 2009, activities of several hundred becquerels per litre were measured in rainfall near NPPs, with annual averages of 258 Bq/L at Bruce Power, 240 Bq/L at Pickering, 28 Bq/L at Darlington and 910 Bq/L at Gentilly. The above data emphasize the large variations in tritium discharges according to the reactor technology used (French PWRs and Canadian HWRs).

Chae et al. [4.43] measured tritium content in rainwater around the Wolsong NPP site in Korea, which has four PHWRs and two PWRs. The activities in rainwater varied from 0.3 to 1090 Bq/L. For the least affected monitoring site, the measured activity concentrations were reportedly in the range 0.8–10 Bq/L, with a mean of 2.3 Bq/L.

4.4.4.2. Spent fuel reprocessing plants

In 2019, there were active tritium emitting reprocessing facilities in different countries, including those at La Hague in France, at Tokai and Rokkasho in Japan, at Sellafield in the UK and at Ozersk in the Russian Federation. In the past, India (Trombay, Tarapur, Kalpakkam), Russian Federation (Seversk, Zheleznogorsk), Pakistan (Nilmore) had facilities either for handling nuclear fuel from civil reactors or for production of weapon-grade plutonium. Other countries such as China, Israel and the Democratic People's Republic of Korea are known to have operated reprocessing facilities, but there is currently no reliable information about their activities. For further information and data on sources of tritium from nuclear facilities around the world, see Chapter 3.

Atmospheric releases of tritium in France from the La Hague reprocessing plant were 60 TBq in 2018, varying between 55 and 78 TBq/a in the 2012–2018 period. In the UK, at the Sellafield site, atmospheric emissions were 90 TBq in 2018 (84–176 TBq/a for the 2012–2018 period). At a distance of 2 km north of the La Hague reprocessing plant, measurements of tritium in rainwater were made every 48 h for 19 months in 2017–2019 (196 measurements). This site is regularly affected by gaseous tritium releases from the plant (with up to 87% in the form of HT), which is produced when irradiated fuel is dissolved [4.36]. These releases can impact the activity in rainwater. The HTO activities in rainwater on this site ranged from <1 Bq/L to 25.5 Bq/L [4.62]. During a monthly sampling campaign in 2017, the average tritium level in precipitation was found to be 7.1 Bq/L [4.61]. By comparison, at the Tokai Mura site in Japan, Matsuura et al. [4.63] reported annual tritium activity concentrations in rainfall ranging from 0.8 to 8.9 Bq/L.

4.4.4.3. Other facilities

Other industrial facilities also produce airborne emissions of tritium. These include: (i) factories producing tritium powered luminescent consumer products (e.g. signs, watches) (ii) radiopharmaceutical facilities producing tritium labelled compounds for nuclear medicine and research in medicine and biology, and (iii) military centres developing and maintaining nuclear weapons. Tritium used in such industries can be partially released into the environment. In the future, there is also a potential for significant releases of tritium during operation of fusion reactors. Further details on such releases are provided below.

In France, tritium emissions from the industries outlined above are estimated at 272 TBq [4.61]. The French CEA, with research centres in Marcoule and Valduc (research and development unit for the French nuclear strategic weapons), is the main emitter of gaseous tritium to the atmosphere. These CEA centres emitted approximately 210 TBq of tritium to air in 2017. The impact of those emissions on rainwater is recorded as 2.5–30.6 Bq/L around Marcoule. Near Valduc, the annual average tritium activity in rainwater is 21.8 Bq/L, with monthly maximum values between 45 and 126 Bq/L, depending on the monitoring station [4.61].

In Canada, SRBT produces luminous consumer products powered by tritium; this is accompanied by the emission of tritium to the atmosphere. Release rates of tritium from this company have decreased significantly since the 2000s, and in 2019, SRBT's airborne emissions of tritium were 31.7 TBq [4.64]. The average tritium concentration measured by eight precipitation monitors located around the SRBT facility in 2019 was 33 Bq/L, with a range of 5–200 Bq/L.

Atomic Energy of Canada Limited, via Canadian Nuclear Laboratories (CNL), operates a tritium facility at its CRL site located approximately 200 km northwest of Ottawa. This facility was involved in the development of PHWR CANDU technology. A variety of nuclear facilities, including the NRU reactor (closed in 2018), were operated on the CRL site from 1945 to 2018. In 2012, the average rate of tritium release to the atmosphere from the facility was around 3.7 TBq per week, and from 2012–2014, the measured tritium activity concentration in precipitation varied between 20 Bq/L and 1000 Bq/L, with average monthly tritium activities of 17.8 ± 12.9 Bq/L in 2011 [4.30].

In the USA, the Savannah River Site (SRS), located in South Carolina, was established in the early 1950s to produce materials used to create nuclear weapons. Over the following decades, five nuclear reactors produced these materials, and several support facilities continue to operate today although the reactors ceased operation in 1988. Current main activities involve treating and processing nuclear waste, environmental cleanup and remediation, tritium processing and protecting nuclear material. Over the 2010–2018 period, the average annual tritium release from SRS was 892 TBq [4.65]. In 2018 and 2019, tritium releases were, respectively, 1480 and 342 TBq. Monitoring surveys of rainwater on the site indicate average values of 69.6 Bq/L, with a range of 16–141 Bq/L. Outside SRS, recorded values ranged between 2.2 and 19.7 Bq/L [4.65].

LLNL (USA) is a research laboratory of the National Nuclear Security Administration within the US Department of Energy (DOE). The laboratory serves as a scientific resource to the US Government and a partner to industry and academia. LLNL operations have the potential to release a variety of constituents into the environment through the atmosphere, surface waters and groundwater pathways. LLNL emits tritium to the atmosphere from multiple sources, with three facilities having measurable emissions. In 2019, a total of 4.8 TBq of tritium was released to the air (2.1 TBq in 2020). The highest measured tritium activity in rainwater in 2019 was 4.6 Bq/L [4.66].

In the Russian Federation, the Mayak PA nuclear complex was originally built between 1945 and 1948 as a military plutonium production facility and later evolved to an SFRP, including tritium and radioisotope production units.[7] Varying levels of tritium in rain and snow around the site have been reported by Vostrotin et al. [4.67]. For daily samples collected in 2007, values of 20–2086 Bq/L (average 329 ± 81 Bq/L) and 7–1160 Bq/L (average 382 ± 42 Bq/L) were reported for two sites near Mayak PA (about 2.5 km from the site). In 2014–2015, values of <8–145 Bq/L were measured (average 19 ± 9 Bq/L). The results of tritium measurement in monthly precipitation in 2014 collected at 6 km distance from Mayak PA were within the range 20–58 Bq/L (average 40 ± 4 Bq/L). In February 2014, tritium in one of the snow samples collected at 7.5 km northwest from Mayak PA was 145 Bq/L. The maximum tritium registered in a snow sample in the end of November 2015 was 3500 Bq/L, that could be evidence of a local air emission of tritium from Mayak PA into the environment. These results in rainwater and snow indicate the presence of variable atmospheric tritium emissions at the Mayak PA site.

4.4.5. Impact of accidental releases of tritium into the atmosphere

4.4.5.1. Three Mile Island

The Three Mile Island NPP, located in Pennsylvania, USA, was equipped with two 900 MW(e) PWRs. On 28 March 1979, an accident occurred, which resulted in partial melting of the core of the Unit 2 reactor, with a release of radioactivity into the containment structure and only low releases to the environment. Tritium release to the atmosphere was estimated at 24.3 TBq or 0.07 g. No data were reported for the environment [4.68].

4.4.5.2. Lawrence Livermore National Laboratory and Savannah River Site

Large releases of tritium from production facilities have been reported at LLNL during its 53 years of operation in 1953–2005. Peterson and Raskob [4.69] calculated that a total of 29 300 TBq was released, with 75% of this amount originating from two accidents in 1965 and 1970. Routine emissions during this period contributed slightly more than 3700 TBq gaseous tritium and about 2800 TBq of tritiated water vapour.

SRS reported several incidents. Between 1959 and 1974, most inadvertent releases were related to reactor operation and no data were reported. For the 1974–1988 period, there have been a series of 13 important accidental releases

---

[7] See Sections 3.2.3.2 and 6.6.6.2.

from the tritium facilities, with a reported total of 38 323 TBq (4605 TBq in HTO form) emitted to atmosphere [4.70]. Annually measured concentrations of tritium in rainwater for monitoring locations around SRS between 1982 and 1986 ranged between 30 and 700 Bq/L.

4.4.5.3. Chalk River Laboratories

During the period of operation of CRL (1944–2017), two major accidents were reported, in 1952 and in 1958. Little information is available on tritium releases during these accidents. Subsequent emissions of 500 TBq/a were reported [4.71] and several tritium experiments have been conducted with tritium emissions to air of 4.2 TBq between 1986 and 1987 [4.72]. In rainwater, values of 74 Bq/L were reported. In 1994, a similar experiment was conducted with a release of 13 TBq [4.73].

4.4.5.4. French and Canadian experiments

In 1986, the French CEA carried out an experiment where 256 TBq of HT was injected to the atmospheric air from a 40 m stack over a 2 min period [4.74]. The objective of the experiment was to investigate the rate of formation of atmospheric HTO from an HT release. Since HTO is approximately 10 000 times more radiotoxic than HT, the formation of small amounts of HTO would dominate the radiological impact of an HT release. Data analysis showed no evidence for the rapid conversion of HT to atmospheric HTO, which was consistent with the results of parallel Canadian experiments carried out in 1986 and 1987. The study established that surface soil oxidation of HT followed by re-emission of the oxidized tritium to the atmosphere would be the dominant process controlling the atmospheric concentration of HTO and hence the impact of an HT release. These experiments were instructive in establishing a contextual background for the smaller atmospheric releases from nuclear facilities and reported accidental releases.

4.4.5.5. Windscale fire

On 10 October 1957, at the Windscale site of the nuclear complex in Sellafield in northwestern England, the graphite moderated, air cooled reactor referred to as Pile No. 1 caught fire and burned for 2–3 days, releasing fission product fallout that spread across the UK and parts of Europe. This fire was preceded by a series of incidents that released fission products to the environment. As for the Chalk River accidents in a similar period, only limited published information exists on releases to the environment. However, it is certain that tritium gas releases would also have occurred, as tritium was produced from the routine irradiation of lithium in magnesium–lithium alloy in aluminium cans. It is estimated, albeit with a large uncertainty, that 1000 TBq of tritium as HTO may have been emitted [4.75]. Later, doses to local populations — mostly from ^{90}Sr, ^{210}Po and ^{131}I, but not from tritium — were published [4.76].

4.4.5.6. Chernobyl accident

As a result of the Chernobyl accident on 26 April 1986, a poorly quantified amount of tritium was released to the environment. The amount was estimated as 37 TBq (0.10 g) for the atmospheric releases [4.3, 4.77] and it was shown that this quantity is very low compared with the release of other radionuclides. Therefore, tritium is rarely mentioned as released in the accident [4.78, 4.79]. However, several studies have been conducted on the impact of this accident on tritium activities in rain. Koga et al. [4.77] studied tritium precipitation in 23 European cities (principally, at the GNIP stations in Germany) and in Osaka (Japan) before and after the accident. For example, in Osaka, elevated ^{137}Cs levels were detected in precipitation, but no increase of tritium activity was detected in Japan. In European cities, no real tritium peak was observed following the accident that exceeded values within the 'normal' fluctuation range. Florkowski et al. [4.10] also reported that there was no increase in the levels of tritium in monthly precipitation or in the near-ground water vapour in Krakow (Poland) in the days and months following the accident. In Krakow, values below 5 Bq/L were recorded in April 1986, which are not significant in relation to seasonal changes.

At the end of April 1986, at some monitoring stations in Austria, the tritium levels in rainwater were about 4 times higher than normal [4.80, 4.81]. At the Bregenz station, levels of 15 Bq/L were measured in the rainwater in April 1986, which was interpreted to indicate a signal related to the Chernobyl accident. The corresponding levels in water vapour of the air were $<$100 Bq/L.

In May 1986, tritium in precipitation in Ukraine and in the west of the Russian Federation had increased by a factor of 2 to 3 compared with 1985 [4.82]. According to GNIP data, only a few stations (Saratov, Gorkij, Minsk) saw this significant increase, in the month of May. It is concluded that an effect on tritium levels following the Chernobyl accident was confirmed only in some stations in the European part of the USSR in the area around the NPP during May 1986, but the increases over normal concentrations were very low.

4.4.5.7. Fukushima Daiichi nuclear power plant accident

The total inventory of tritium in the Fukushima Daiichi NPP in March 2011 was estimated to be 3400 TBq or 9.9 g [4.83]. Part of the inventory was released to the atmosphere, followed by its deposition on the ground and in the ocean with precipitation or by dry deposition. The estimates of the fraction of tritium released in the ocean were in the range 0.28–1.40 g, namely, about 100–500 TBq [4.3, 4.79].

A survey of tritium levels in river water in November 2011 was not able to identify a clear effect of the accident [4.84, 4.85]. Ueda et al. [4.86] studied tritium contents in 18 rivers in the Fukushima Prefecture between July 2011 and November 2014. Tritium activity concentrations above the background levels were mostly found in water samples collected in 2011. These, however, decreased with time after the accident and reached almost background levels in 2013. During the period of measurement (2011–2014), tritium activities in the Hiso River ranged from 0.7 to 2.1 Bq/L in 2011 and from 0.4 to 0.6 Bq/L in 2014. Kakiuchi et al. [4.85] collected plant samples just after the accident outside the 20 km evacuation zone and measured activities of TFWT in the samples. The maximum tritium activity concentration in April 2011 was 167 Bq/L. Activity levels depended on the distance to the Fukushima Daiichi NPP and decreased between March and August 2011. In August 2011, activity levels were below 10 Bq/L at a distance of 20–40 km from the NPP. The maximum value of TFWT measured in plants collected at distances of 30–80 km from the NPP in about ten samples was used to estimate atmospheric HTO, which was found to be 5600 Bq/L [4.85].

Matsumoto et al. [4.87] reported tritium levels in precipitation during March–May 2011 for six sites located 170–700 km southwest of the NPP. These tritium concentrations were elevated compared with the pre-accident levels, which were estimated to be 0.7 Bq/L [4.87]. Among 47 rainwater samples, 21 were above background, with a maximum of 19 Bq/L in the first samples collected in Tsukuba (about 150 km from the accident site) after the accident. The average measured values decreased with increasing distance from the Tsukuba collection point (nearest station, 170 km) to that at Misasa (the most distant station at around 700 km) as: 2.5, 1.5, 1.4, 1.2, 0.7 and 0.8 Bq/L. In the more distant localities, the samples collected immediately after the accident revealed a tritium excess in the range 0.8–1.2 Bq/L above the local background. However, it can be noted that the levels of tritium in rainwater measured in Japan in this study between March and May 2011 were only slightly higher after the accident and had returned to the pre-accident background values two months later. According to these data, tritium activity concentrations in the atmosphere at the accident site over the first weeks after the accident were estimated to be 1500 Bq/m^3 [4.87]. Measurements of tritium have been also carried out in the surface waters of the Pacific Ocean. The measured tritium concentrations ranged from 0.05 to 0.15 Bq/L, exceeding the pre-accident values by a factor of three (maximum).

4.5. TRITIUM TRANSFER AT THE AIR–SOIL INTERFACE

Once released into the atmosphere, HT gas is oxidized to HTO either very slowly in the air or rapidly in the soil by terrestrial biogenic processes. In soil, microorganisms convert the gas to HTO, which can be incorporated into plants [4.88–4.91]. Any HTO that is formed in the soil column becomes part of the soil pore water and can be re-emitted into the atmosphere through evapotranspiration processes as water vapour. Thus, it is expected that enhanced tritium concentrations in continental, near-ground air can be observed at a local or regional scale (e.g. see Ref. [4.92]). Related information can be also found in Chapter 7, where the behaviour of tritium in the terrestrial environment is discussed in detail.

It has been established that microorganisms in soil oxidize hydrogen molecules to water [4.93, 4.94] and convert it to methane [4.95, 4.96]. Laboratory experiments using tritiated hydrogen molecules confirm rapid oxidation of both tritiated and non-tritiated hydrogen molecules by microbial activity in soil [4.88, 4.97].

The migration of tritiated water and tritiated gas from the atmosphere to the soil is governed by physical processes. The transfer of tritium to the ground from the air occurs through wet (rainfall or other precipitation) or dry depositions. Wet deposition affects only HTO, owing to the poor solubility of methane and hydrogen in rainwater. Besides precipitation, another transfer route of tritium into soils may be through melting of snow or ice that had formerly incorporated tritium from nuclear device testing in the Northern Hemisphere [4.98, 4.99]. Owing to the low diffusion rate of tritium in cold dry snow and its even lower diffusion rate in ice, tritium remains locked in the snowpack. Following any later melting episodes, the tritium is released with a delay that may range from the annual snowmelt cycle to even years or decades. Such a process can occur for high latitude snowpacks or permafrost areas. Therefore, any such liberated historical tritium can retrospectively enrich surface waters, soils and crops with larger than expected activities.

Tritiated gases are transferred to the soil column mainly by dry deposition from the atmosphere to the soil. This process can be conveniently expressed by the deposition velocity. Specifically, the amount deposited per unit time is characterized by the deposition velocity and the tritium activity concentration in air, so the flux density of tritiated gas into the soil is expressed as follows:

$$F = V_{\mathrm{dep}} C_{\mathrm{a}}$$

where

F is the flux density of HT into the soil ($Bq \cdot m^{-2} \cdot s^{-1}$);
V_{dep} is the deposition velocity of HT (m/s);

and C_a is the tritium activity concentration of HT in air (Bq/m^3).

The hydrogen deposition velocities cited in the literature differ by two orders of magnitude, from 10^{-5} to 10^{-3} m/s [4.100] (see Table 4.6). More recent experiments yield values from 1.3×10^{-5} to 3.7×10^{-5} m/s [4.109]. The key parameter that determines the deposition velocity of HT is the soil water content, which may significantly vary due to a number of factors — for example, soil type, as reported by Murphy [4.111] and demonstrated again by Koarashi et al. [4.109].

During metabolic activities, a portion of the HTO released to the soil after oxidation of HT gas will become incorporated into organic molecules, including plant structural material and soil organic matter. This OBT has a significantly longer residence time in biological matrices than HTO. The formation of organic matter incorporating tritium (OBT) has been reported for soil incubated with HTO over a period of several days [4.112].

TABLE 4.6. ESTIMATES OF THE VELOCITY OF HYDROGEN DRY DEPOSITION FROM ATMOSPHERE TO SOIL

Deposition velocity, V_{dep} ($\times 10^{-4}$ m/s)	Evaluation site	Reference
3.5–8.1	Canada	[4.101]
0.1–3.0	Canada	[4.102]
2.7–7.2	Canada	[4.103]
0.24–5.1	Canada	[4.104]
1.1–7.1	Canada	[4.105]
1.0– 15	Canada	[4.74]
0.8–12	Canada	[4.106]
0.69–1.6	Canada	[4.107]
1.7–11	France	[4.101]

TABLE 4.6. ESTIMATES OF THE VELOCITY OF HYDROGEN DRY DEPOSITION FROM ATMOSPHERE TO SOIL (cont.)

Deposition velocity, V_{dep} ($\times 10^{-4}$ m/s)	Evaluation site	Reference
1.7–3.6	France	[4.108]
16	Japan	[4.89]
0.13–0.37	Japan	[4.109]
1.8–11	Germany	[4.97]
0.96–11	UK	[4.110]

4.6. CONCLUSIONS

Besides naturally produced cosmogenic tritium, the current tritium activities in the atmosphere reflect influences from past atmospheric nuclear weapon tests (now substantially reduced owing to radioactive decay) and from continuing tritium releases from nuclear industries. Activities of tritium in precipitation are now generally in the range 0.5–2 Bq/L in the Northern Hemisphere at sites without a specific local input. They are derived from peak emissions in the early 1960s from nuclear testing that resulted in tritium activities in precipitation of up to several hundred becquerel per litre. In the Southern Hemisphere, the maximum tritium levels reached in the early 1960s were substantially lower (around 10 Bq/L) and are now generally below 0.5 Bq/L. Variations observed between monitoring stations located in the same hemisphere are due to various seasonal trends, continental or marine influence, latitude, and other local and less obvious phenomena, such as the amount of precipitation or altitude. Tritium levels in precipitation in the Northern Hemisphere are highest in spring and summer, with typical ratios of approximately 2 to 4 between the maximum and minimum tritium content, depending on the site.

With respect to atmospheric background levels, worldwide values of tritium activity concentration in air are currently found to be less than 0.7 Bq/L for tritium as HTO in water vapour, which result in volumetric activity concentrations in air lower than 9 mBq/m^3. For example, recent measurements of tritium activity concentrations away from the influence of local sources in France indicated HTO levels of 0.18 ± 0.01 Bq/L (corresponding to 1.7 ± 0.2 mBq/m^3 in air) and HT and CH_3T activity concentrations in air lower than 10 mBq/m^3 and 3 mBq/m^3, respectively. Even though these values cannot be directly extrapolated worldwide, they provide information on the magnitude of reference background levels.

Tritium emissions produced by nuclear facilities influence at local scales to an extent that depends on prevailing weather conditions. Examples of such facilities are NPPs, nuclear reprocessing plants, civil and militaries research sites and laboratories, and industrial sites producing tritium powered luminescent items or tritium labelled compounds. The impact of these emissions (many gaseous, but also in liquids) is variable but detectable in the air, soil and rainwater.

Compared with the release of other radionuclides, tritium releases from major accidents at NPPs (Three Mile Island, Chernobyl and Fukushima Daiichi) were relatively low, with estimated releases ranging from 4 to 700 TBq. At the same time, large tritium releases are known to be associated with accidents at research sites, such as LLNL, where two major accidents in 1965 and 1970, were estimated to yield 29 000 TBq of tritium; SRNL, with estimated tritium releases of 38 000 TBq during 1974–1988, following 13 incidents; and CRL (Canada) in 1952 and 1956, when approximately 500 TBq of tritium was released.

The IAEA GNIP programme continues to provide effective global monitoring of the isotopic composition of precipitation. Persistent monitoring activities implemented by the Member States around their nuclear sites and ongoing extensive international research ensure good knowledge of the behaviour of tritium in different forms across the atmospheric compartment.

Completion of fusion research projects and the eventual operation of future fusion power reactors and the accompanying systematic monitoring around these facilities will result in additional knowledge on the behaviour of tritium in the atmosphere.

Contemporary tritium science is being challenged by a need to better understand transitional, non-equilibrium processes and exchanges of tritium at the air–ground, air–sea and air–river environmental interfaces. These investigations are important because in many cases, tritium releases to aqueous ecosystems, primarily the marine environment, are much higher than gaseous releases. Currently, few studies exist on the transfer of tritium from water to air.

REFERENCES TO CHAPTER 4

[4.1] MASARIK, J., BEER, J., An updated simulation of particle fluxes and cosmogenic nuclide production in the Earth's atmosphere, J. Geophys. Res. **114** (2009) D11103, https://doi.org/10.1029/2008JD010557

[4.2] OMS, P.-E., et al., Inventory and distribution of tritium in the oceans in 2016, Sci. Total Environ. **656** (2019) 1289,
https://doi.org/10.1016/j.scitotenv.2018.11.448

[4.3] NIE, B., et al., Anthropogenic tritium: Inventory, discharge, environmental behavior and health effects, Renew. Sustain. Energy Rev. **135** (2021) 110188,
https://doi.org/10.1016/j.rser.2020.110188

[4.4] MICHEL, R.L., Tritium inventories of the world oceans and their implications, Nature **263** 5573 (1976) 103,
https://doi.org/10.1038/263103a0

[4.5] UNITED NATIONS, Biological Effects of Selected Internal Emitters–Tritium, 2016 Report of the United Nations Scientific Committee on the Effects of Atomic Radiation (UNSCEAR) to the General Assembly, Annex C, UN, New York (2017).

[4.6] JACOBS, D.G., Sources of Tritium and Its Behavior Upon Release to the Environment. U.S, Atomic Energy Commission, Division of Technical Information, US Atomic Energy Commission, Oak Ridge, TN (1968),
https://doi.org/10.2172/4799828

[4.7] NATIONAL COUNCIL ON RADIATION PROTECTION AND MEASUREMENTS, Tritium and Other Radionuclide Labeled Organic Compounds Incorporated in Genetic Material, Report No. 63, NCRP, Bethesda, MD (1979).

[4.8] POLUIANOV, S., KOVALTSOV, G.A., USOSKIN, I., A new full 3-D model of cosmogenic tritium ^{3}H production in the atmosphere (CRAC:3H), J. Geophys. Res. **125** 18 (2020) e2020JD033147,
https://doi.org/10.1029/2020JD033147

[4.9] KAUFMAN, S., LIBBY, W.F., The natural distribution of tritium, Phys. Rev. **93** 6 (1954) 1337,
https://doi.org/10.1103/PhysRev.93.1337

[4.10] FLORKOWSKI, T., KUC, T., ROZANSKI, K., Influence of the Chernobyl accident on the natural levels of tritium and radiocarbon, Int. J. Rad. Appl. Instr. A **39** 1 (1988) 77,
https://doi.org/10.1016/0883-2889(88)90095-0

[4.11] UNITED NATIONS, Sources and Effects of Ionizing Radiation, Vol. I: Sources, 2000 Report of the United Nations Scientific Committee on the Effects of Atomic Radiation (UNSCEAR) to the General Assembly, UN, New York (2000).

[4.12] ANDRYUSHIN, I., et al., Catalog of Worldwide Nuclear Testing, V.N. Mikhailov., Begell House, Danbury, CT (1999).

[4.13] INSTITUT DE RADIOPROTECTION ET DE SÛRETÉ NUCLÉAIRE, Les Essais Atmosphériques d'Armes Nucléaires: Des Retombées Radioactives à l'Échelle Planétaire, IRSN, Fontenay-aux-Roses (2006).

[4.14] GUÉTAT, P., DOUCHE, C., HUBINOIS, J.C., Le Tritium et l'Environnement: Sources, Mesures et Transferts, Radioprotection **43** 4 (2008) 547,
https://doi.org/10.1051/radiopro:2008027

[4.15] INTERNATIONAL ATOMIC ENERGY AGENCY, Global Network of Isotopes in Precipitation (GNIP), IAEA, Vienna (2019),
https://www.iaea.org/services/networks/gnip

[4.16] PALOMO, M., PEÑALVER, A., AGUILAR, C., BORRULL, F., Tritium activity levels in environmental water samples from different origins, Appl. Radiat. Isot. **65** 9 (2007) 1048, https://doi.org/10.1016/j.apradiso.2007.03.013

[4.17] RADWAN, I., PIETRZAK-FLIS, Z., WARDASZKO, T., Tritium in surface waters, tap water and in precipitation in Poland during the 1994–1999 period, J. Radioanal. Nucl. Chem. **247** 1 (2001) 71, https://doi.org/10.1023/A:1006775300812

[4.18] INSTITUT DE RADIOPROTECTION ET DE SÛRETÉ NUCLÉAIRE, Le Tritium Dans l'Environnement: Synthèse Critique des Connaissances et Perspectives de Recherches, IRSN/DEI/2009-05, Vol. IRSN/DEI/2009-05, IRSN, Fontenay-aux-Roses (2009).

[4.19] AUTORITÉ DE SÛRETÉ NUCLÉAIRE, Livre Blanc du Tritium, ASN, Montrouge (2010).

[4.20] INSTITUT DE RADIOPROTECTION ET DE SÛRETÉ NUCLÉAIRE, Bilan de l'État Radiologique de l'Environnement Français de 2015 à 2017, IRSN, Fontenay-aux-Roses (2018).

[4.21] PERRAULT, D., Nuclear safety aspects on the road towards fusion energy, Fusion Eng. Des. **146** (2019) 130, https://doi.org/10.1016/j.fusengdes.2018.11.053

[4.22] NIE, B., JIANG, M., NI, M., LI, F., Preliminary environmental radiation considerations for CFETR, Fusion Eng. Des. **140** (2019) 16, https://doi.org/10.1016/j.fusengdes.2019.01.119

[4.23] WEI, S., et al., Preliminary safety analysis of tritium source term for the CFETR tritium plant, Fusion Sci. Technol. **76** 7 (2020) 869, https://doi.org/10.1080/15361055.2020.1777668

[4.24] BELL, A., BRENNAN, P., JET tritium inventory-control and measurements during DT, shutdown and DD operational phases, Fusion Eng. Des. **54** 3–4 (2001) 337–348, https://doi.org/10.1016/S0920-3796(00)00552-4

[4.25] HORTON, L., et al., JET experiments with tritium and deuterium–tritium mixtures, Fusion Eng. Des. **109–111** (2016) 925–936, https://doi.org/10.1016/j.fusengdes.2016.01.051

[4.26] FOURRÉ, E., et al., Past and recent tritium levels in Arctic and Antarctic polar caps, Earth Planet. Sci. Lett. **245** 1 (2006) 56–64, https://doi.org/10.1016/j.epsl.2006.03.003

[4.27] CAUQUOIN, A., et al., The global distribution of natural tritium in precipitation simulated with an Atmospheric General Circulation Model and comparison with observations, Earth Planet. Sci. Lett. **427** (2015) 160–170, https://doi.org/10.1016/j.epsl.2015.06.043

[4.28] GAT, J.R., MOOK, W.G., MEIJER, H.A.J., "Tritium in the atmosphere", Environmental Isotopes in the Hydrological Cycle, Principles and Applications, Vol. II, UNESCO/IAEA, Paris (2001) 63–74.

[4.29] EHHALT, D.H., Vertical profiles and transport of HTO in the troposphere, J. Geophys. Res. **76** (1971) 7351–7367,
https://doi.org/10.1029/JC076i030p07351

[4.30] KIM, S.B., et al., Organically bound tritium (OBT) activity concentrations in surface soil at the Chalk River Laboratories, Canada, J. Environ. Radioact. **208–209** (2019) 105999,
https://doi.org/10.1016/j.jenvrad.2019.105999

[4.31] OKAI, T., MOMOSHIMA, N., TAKASHIMA, Y., Variation of atmospheric tritium concentrations in three different chemical forms in Fukuoka, Japan, J. Radioanal. Nucl. Chem. **239** 3 (1999) 527–531,
https://doi.org/10.1007/BF02349063

[4.32] UDA, T., SUGIYAMA, T., ASAKURA, Y., MUNAKATA, K., TANAKA, M., Development of high performance catalyst for oxidation of tritiated hydrogen and methane gases, Fusion Sci. Technol. **48** 1 (2005) 480–483,
https://doi.org/10.13182/FST05-A970

[4.33] UDA, T., SUGIYAMA, T., TANAKA, M., MUNAKATA, K., MOMOSHIMA, N., Developments of gaseous water, hydrogen and methane sampling system for environmental tritium monitoring, Fusion Eng. Des. **81** 8 (2006) 1385–1390,
https://doi.org/10.1016/j.fusengdes.2005.08.077

[4.34] UDA, T., TANAKA, M., MUNAKATA, K., Characteristics of honeycomb catalysts for oxidation of tritiated hydrogen and methane gases, Fusion Eng. Des. **83** 10 (2008) 1715–1720,
https://doi.org/10.1016/j.fusengdes.2008.06.037

[4.35] MUNAKATA, K., et al., Oxidation of hydrogen isotopes over honeycomb catalysts, J. Nucl. Mater. **417** 1 (2011) 1170–1174,
https://doi.org/10.1016/j.jnucmat.2011.02.013

[4.36] CONNAN, O., et al., Atmospheric tritium concentrations under influence of AREVA NC La Hague reprocessing plant (France) and background levels, J. Environ. Radioact. **177** (2017) 184–193,
https://doi.org/10.1016/j.jenvrad.2017.06.015

[4.37] UDA, T., TANAKA, M., SUGIYAMA, T., YAMAGUCHI, T., MOMOSHIMA, N., The background levels of atmospheric tritium concentration in the environment at NIFS Toki site of Japan, Fusion Sci. Technol. **54** 1 (2008) 281–284,
https://doi.org/10.13182/FST08-A1813

[4.38] MOMOSHIMA, N., et al., Tritium in the atmospheric environment, J. Radioanal. Nucl. Chem. **8** 2 (2007) 117–120,
https://doi.org/10.14494/jnrs2000.8.117

[4.39] AKATA, N., et al., Tritium concentrations in the atmospheric environment at Rokkasho, Japan before the final testing of the spent nuclear fuel reprocessing plant, J. Environ. Radioact. **102** 9 (2011) 837–842,
https://doi.org/10.1016/j.jenvrad.2011.05.005

[4.40] KOARASHI, J., MIKAMI, S., AKIYAMA, K., ASANO, T., Determination of chemical forms of ^{3}H released to the atmosphere from the Tokai Reprocessing Plant, J. Radioanal. Nucl. Chem. **262** 3 (2004) 569–572, https://doi.org/10.1007/s10967-004-0477-8

[4.41] MIHOK, S., et al., Tritium dynamics in soils and plants grown under three irrigation regimes at a tritium processing facility in Canada, J. Environ. Radioact. **153** (2016) 176–187, https://doi.org/10.1016/j.jenvrad.2015.12.025

[4.42] SAEY, L., HÉMIDY, P.Y., GONTIER, G., Étude des Niveaux de Radioactivité Atmosphérique à Proximité du CNPE de Cruas-Meysse, IRSN, Paris (2017).

[4.43] CHAE, J.-S., KIM, G., Dispersion and removal characteristics of tritium originated from nuclear power plants in the atmosphere, J. Environ. Radioact. **192** (2018) 524–531, https://doi.org/10.1016/j.jenvrad.2018.08.007

[4.44] MILJEVIĆ, N., ŠIPKA, V., ŽUJIĆ, A., GOLOBOČANIN, D., Tritium around the Vinča Institute of Nuclear Sciences, J. Environ. Radioact. **48** 3 (2000) 303–315, https://doi.org/10.1016/S0265-931X(99)00082-X

[4.45] WOOD, M.J., WORKMAN, W.J.G., Environmental monitoring of tritium in air with passive diffusion samplers, Fusion Technol. **21** 2P2 (1992) 529–535, https://doi.org/10.13182/FST92-A29801

[4.46] KIM, C.-K., HAN, M.-J., Dose assessment and behavior of tritium in environmental samples around Wolsong nuclear power plant, Appl. Radiat. Isot. **50** 4 (1999) 783–791, https://doi.org/10.1016/S0969-8043(98)00075-X

[4.47] INTERNATIONAL ATOMIC ENERGY AGENCY, Global Network of Isotopes in Precipitation (GNIP), https://www.iaea.org/services/networks/gnip

[4.48] CARTWRIGHT, I., MORGENSTERN, U., Constraining groundwater recharge and the rate of geochemical processes using tritium and major ion geochemistry: Ovens catchment, southeast Australia, J. Hydrol. **475** (2012) 137–149, https://doi.org/10.1016/j.jhydrol.2012.09.037

[4.49] TADROS, C.V., HUGHES, C.E., CRAWFORD, J., HOLLINS, S.E., CHISARI, R., Tritium in Australian precipitation: A 50 year record, J. Hydrol. **513** (2014) 262–273, https://doi.org/10.1016/j.jhydrol.2014.03.031

[4.50] TERZER-WASSMUTH, S., ARAGUÁS-ARAGUÁS, L.J., COPIA, L., WASSENAAR, L.I., High spatial resolution prediction of tritium (^{3}H) in contemporary global precipitation, Sci. Rep. **12** 1 (2022) 10271, https://doi.org/10.1038/s41598-022-14227-5

[4.51] EHHALT, D.H., ROHRER, F., SCHAUFFLER, S., POLLOCK, W., Tritiated water vapor in the stratosphere: Vertical profiles and residence time, J. Geophys. Res. **107** D24 (2002) 4757, https://doi.org/10.1029/2001JD001343

[4.52] CRAWFORD, J., HUGHES, C.E., PARKES, S.D., Is the isotopic composition of event based precipitation driven by moisture source or synoptic scale weather in the Sydney Basin, Australia? J. Hydrol. **507** (2013) 213–226, https://doi.org/10.1016/j.jhydrol.2013.10.031

[4.53] DULIU, O.G., VARLAM, C., SHNAWAW, M.D., 18 years of continuous observation of tritium and atmospheric precipitations in Ramnicu Valcea (Romania): A time series analysis, J. Environ. Radioact. **190–191** (2018) 105–110, https://doi.org/10.1016/j.jenvrad.2018.05.011

[4.54] NEWELL, R.E., Transfer through the tropopause and within the stratosphere, Q. J. R. Meteorol. Soc. **89** 380 (1963) 167–204, https://doi.org/10.1002/qj.49708938002

[4.55] GUSYEV, M.A., et al., Application of tritium in precipitation and baseflow in Japan: A case study of groundwater transit times and storage in Hokkaido watersheds, Hydrol. Earth Syst. Sci. **20** 7 (2016) 3043–3058, https://doi.org/10.5194/hess-20-3043-2016

[4.56] HARMS, P.A., VISSER, A., MORAN, J.E., ESSER, B.K., Distribution of tritium in precipitation and surface water in California, J. Hydrol. **534** (2016) 63–72, https://doi.org/10.1016/j.jhydrol.2015.12.046

[4.57] BOLIN, B., On the use of tritium as a tracer for water in the nature, Proceedings of the Second United Nations International Conference on the Peaceful Uses of Atomic Energy, UN, Geneva (1958).

[4.58] STEWART, G.L., FARNSWORTH, R.K., United States tritium rainout and its hydrologie implications, Water Resour. Res. **4** 2 (1968) 273–289, https://doi.org/10.1029/WR004i002p00273

[4.59] MOMOSHIMA, N., OKAI, T., KAJI, T., TAKASHIMA, Y., Distribution and transformation of various chemical forms of tritium in the environment, Radiochim. Acta **54** 3 (1991) 129–132, https://doi.org/10.1524/ract.1991.54.3.129

[4.60] ROZANSKI, K., GONFIANTINI, R., ARAGUAS-ARAGUAS, L., Tritium in the global atmosphere: Distribution patterns and recent trends, J. Phys. G**17** S (1991) S523, https://doi.org/10.1088/0954-3899/17/S/053

[4.61] INSTITUT DE RADIOPROTECTION ET DE SÛRETÉ NUCLÉAIRE, Actualisation des Connaissances Acquises sur le Tritium dans l'environnement "7 Ans Après le Livre Blanc de l'ASN: Que Sait-on de plus et que Reste-il à Faire ?", Report PRP-ENV/SERIS/2017-004, IRSN, Fontenay-aux-Roses (2017).

[4.62] CONNAN, O., et al., Tritium in precipitation on 5 sites in North-West France during the 2017–2019 period, J. Environ. Radioact. **212** (2020) 106129, https://doi.org/10.1016/j.jenvrad.2019.106129

[4.63] MATSUURA, K., SASA, Y., NAKAMURA, C., KATAGIRI, H., Levels of tritium concentration in the environmental samples around JAERI TOKAI, J. Radioanal. Nucl. Chem. **197** 2 (1995) 295–307, https://doi.org/10.1007/BF02036007

[4.64] SRB TECHNOLOGIES, 2019 Annual Compliance and Performance Report, SRBT, Ontario (2019),
http://srbt.com/ACR2019.pdf

[4.65] SAVANNAH RIVER SITE, Environmental Report SRNS-RP-2020-000, SRNS-RP-2020-00064, SRS, Aiken (2019).

[4.66] LAWRENCE LIVERMORE NATIONAL LABORATORY, Environmental Report 2019, UCRL-TR-50027-19, LLNL, Livermore, CA (2019).

[4.67] VOSTROTIN, V.V., YANOV, A.Y., FINASHOV, L.V., Tritium in environmental objects in the area affected by FSUE Mayak Production Association in 2014–15, J. Radiol. Prot. **41** 2 (2021) S56,
https://doi.org/10.1088/1361-6498/abe8c8

[4.68] MINISTRY OF ECONOMY, TRADE AND INDUSTRY OF JAPAN, Tritiated Water Task Force Report, Tokyo (2016).

[4.69] PETERSON, S.R., RASKOB, W., Historical doses to the public from routine and accidental releases of tritium, Lawrence Livermore National Laboratory, 1953–2005, Fusion Sci. Technol. **54** 1 (2008) 277–280,
https://doi.org/10.13182/FST08-A1812

[4.70] MURPHY, C.E., Jr., et al., Tritium in the Savannah River Site Environment, WSRC-RP-90-424-1, Vol. WSRC-RP-90-424-1, SRS, Aiken (1991).

[4.71] BROWN, R.M., HT and HTO in the environment at Chalk River, Fusion Technol. **8** 2P2 (1985) 2539–2543,
https://doi.org/10.13182/FST85-A24661

[4.72] BROWN, R.M., OGRAM, G.L., SPENCER, F.S., Oxidation and dispersion of HT in the environment: The August 1986 field experiment at Chalk River, Health Phys. **58** 2 (1990) 171–181,
https://doi.org/10.1097/00004032-199002000-00004

[4.73] DAVIS, P.A., et al., Overview of the 1994 chronic HT release experiment at Chalk River, Fusion Technol. **28** 3P1 (1995) 840–845,
https://doi.org/10.13182/FST95-A30509

[4.74] DJERRASSI, H., GULDEN, W., Overview of the tritium release experiment in France, Fusion Technol. **14** 2P2B (1988) 1216–1221,
https://doi.org/10.13182/FST88-A25305

[4.75] GARLAND, J.A., WAKEFORD, R., Atmospheric emissions from the Windscale accident of October 1957, Atmos. Environ. **41** 18 (2007) 3904–3920,
https://doi.org/10.1016/j.atmosenv.2006.12.049

[4.76] CRICK, M.J., LINSLEY, G.S., An assessment of the radiological impact of the Windscale reactor fire, October 1957, Int. J. Radiat. Biol. Relat. Stud. Phys. Chem. Med. **46** 5 (1984) 479–506,
https://doi.org/10.1080/09553008414551711

[4.77] KOGA, T., MORISHIMA, H., NIWA, T., KAWAI, H., Tritium precipitation in European cities and in Osaka, Japan owing to the Chernobyl nuclear accident, J. Radiat. Res. **32** 3 (1991) 267–276,
https://doi.org/10.1269/jrr.32.267

[4.78] INTERNATIONAL ATOMIC ENERGY AGENCY, Environmental Consequences of the Chernobyl Accident and Their Remediation, Twenty Years of Experience, Radiological Assessment Reports Series No. 8, IAEA, Vienna (2006).

[4.79] POVINEC, P.P., et al., Cesium, iodine and tritium in NW Pacific waters — a comparison of the Fukushima impact with global fallout, Biogeosciences **10** 8 (2013) 5481–5496,
https://doi.org/10.5194/bg-10-5481-2013

[4.80] KOTOVENKO, O., MIROSHNYCHENKO, O., MARCHENKO, I., Tritium and its place in the overall radiation Ukraine technogenesis, Environmental problems **1** 1 (2016) 23.

[4.81] SCHÖNHOFER, F., WEISZ, J., Measurement by ultra low level liquid scintillation counting following the Chernobyl accident, J. Radioanal. Nucl. Chem. **115** 1 (1987) 125–140,
https://doi.org/10.1007/BF02041981

[4.82] SOYFER, V., GORYACHEV, V., VAKULOVSKY, S., KATRICH, I., Some results of tritium researches of natural waters in Russia (1960–2010), Radioprotection **46** 6 (2011) S355–S361,
https://doi.org/10.1051/radiopro/20116467s

[4.83] NISHIHARA, K., et al., Radionuclide release to stagnant water in Fukushima-1 nuclear power plant, Trans. Atom. Energy Soc. **11** (2012) 13–19,
https://doi.org/10.3327/taesj.J11.040

[4.84] NERH NUCLEAR EMERGENCY RESPONSE HEADQUARTERS, Disaster Provision Main Office of Fukushima Prefecture (Nuclear Power Team), Readings 2012 (2012).

[4.85] KAKIUCHI, H., et al., Concentration of ^{3}H in plants around Fukushima Dai-ichi Nuclear Power Station, Sci. Rep. **2** 1 (2012) 947,
https://doi.org/10.1038/srep00947

[4.86] UEDA, S., et al., Nuclear accident-derived ^{3}H in river water of Fukushima Prefecture during 2011–2014, J. Environ. Radioact. **146** (2015) 102–109,
https://doi.org/10.1016/j.jenvrad.2015.04.014

[4.87] MATSUMOTO, T., et al., Tritium in Japanese precipitation following the March 2011 Fukushima Daiichi Nuclear Plant accident, Sci. Total Environ. **445–446** (2013) 365–370,
https://doi.org/10.1016/j.scitotenv.2012.12.069

[4.88] McFARLANE, J.C., ROGERS, R.D., BRADLEY, D.V., Jr., Tritium oxidation in surface soils: A survey of soils near five nuclear fuel reprocessing plants, Environ. Sci. Technol. **13** 5 (1979) 607–608,
https://doi.org/10.1021/es60153a001

[4.89] ICHIMASA, M., et al., Oxidation of atmospheric molecular tritium in plant leaves, lichens and mosses, J. Radiat. Res. **30** 4 (1989) 323–329,
https://doi.org/10.1269/jrr.30.323

[4.90] YOKOYAMA, S., NOGUCHI, H., ICHIMASA, Y., ICHIMASA, M., Re-emission of heavy water vapour from soil to the atmosphere, J. Environ. Radioact. **71** 3 (2004) 201–213,
https://doi.org/10.1016/S0265-931X(03)00169-3

[4.91] OTA, M., YAMAZAWA, H., MORIIZUMI, J., IIDA, T., Measurement and modeling of oxidation rate of hydrogen isotopic gases by soil, J. Environ. Radioact. **97** 2 (2007) 103–115,
https://doi.org/10.1016/j.jenvrad.2007.03.006

[4.92] BAGLAN, N., ALANIC, G., LE MEIGNEN, R., POINTURIER, F., A follow up of the decrease of non exchangeable organically bound tritium levels in the surroundings of a nuclear research center, J. Environ. Radioact. **102** 7 (2011) 695–702,
https://doi.org/10.1016/j.jenvrad.2011.03.014

[4.93] CONRAD, R., SEILER, W., The role of hydrogen bacteria during the decomposition of hydrogen by soil, FEMS Microbiol. Lett. **6** 3 (1979) 143–145,
https://doi.org/10.1111/j.1574-6968.1979.tb04296.x

[4.94] FALLON, R.D., Influences of pH, temperature and moisture on gaseous tritium uptake in surface soils, Appl. Environ. Microbiol **44** (1982) 171–178,
https://doi.org/10.1128/aem.44.1.171-178.1982

[4.95] BENDER, M., CONRAD, R., Effect of CH_4 concentrations and soil conditions on the induction of CH_4 oxidation activity, Soil Biology and Biochemistry **27** 12 (1995) 1517–1527,
https://doi.org/10.1016/0038-0717(95)00104-M

[4.96] CZEPIEL, P., CRILL, P., HARRISS, R., Environmental factors influencing the variability of methane oxidation in temperate zone soils, J. Geophys. Res. **100** (D5) (1995) 9359–9364,
https://doi.org/10.1029/95JD00542

[4.97] FOERSTEL, H., LEPA, K., TRIERWEILER, H., Re-emission of HTO into the atmosphere after HT/HTO conversion in the soil, Fusion Technol. **14** 2P2B (1988) 1203–1208,
https://doi.org/10.13182/FST88-A25303

[4.98] GALERIU, D., DAVIS, P., WORKMAN, W., Tritium profiles in snowpacks, J. Environ. Radioact. **101** 10 (2010) 869–874,
https://doi.org/10.1016/j.jenvrad.2010.05.013

[4.99] BOND, M.J., CARR, J., Permafrost thaw and implications for the fate and transport of tritium in the Canadian north, J. Environ. Radioact. **192** (2018) 295–311,
https://doi.org/10.1016/j.jenvrad.2018.07.006

[4.100] BELOT, Y., ROY, M., MÉTIVIER, H., Le Tritium de l'Environnement à l'Homme, Les Editions de Physique, Paris (1996).

[4.101] FOERSTEL, H., HT to HTO conversion in the soil and subsequent tritium pathway, Fusion Technol. **14** 2 (1988) 1241–1246,
https://doi.org/10.13182/FST88-A25310

[4.102] NOGUCHI, H., MATSUI, T., MURATA, M., Tritium behavior observed in the Canadian HT Release Study, Fusion Technol. **14** 2P2B (1988) 1187–1192,
https://doi.org/10.13182/FST88-A25300

[4.103] OGRAM, G.L., SPENCER, F.S., BROWN, R.M., Field studies of HT behaviour in the environment: 2. The interaction with soil, Fusion Technol. **14** 2P2B (1988) 1170–1175,
https://doi.org/10.13182/FST88-A25297

[4.104] SPENCER, F., OGRAM, G., BROWN, R., Field studies of HT behaviour in the environment: 3. Tritium deposition and dynamics in vegetation, Fusion Technol. **14** 2P2B (1988) 1176–1181,
https://doi.org/10.13182/FST88-A25298

[4.105] TASCHNER, M., WIENER, B., BUNNENBERG, C., HT dispersion and deposition in soil after experimental releases of tritiated hydrogen, Fusion Technol. **14** 2 (1988) 1264–1269,
https://doi.org/10.13182/FST88-A25314

[4.106] BURNHAM, C.D., BROWN, R.M., OGRAM, G.L., SPENCER, F.S., An overview of experiments at Chalk River on HT dispersion in the environment, Fusion Technol. **14** 2P2B (1988) 1159–1164,
https://doi.org/10.13182/FST88-A25295

[4.107] KIM, S.B., STUART, M., BREDLAW, M., FESTARINI, A., BEATON, D., HT to HTO conversion and field experiments near Darlington Nuclear Power Generating Station (DNPGS) site, J. Environ. Radioact. **132** (2014) 73–80,
https://doi.org/10.1016/j.jenvrad.2014.02.001

[4.108] PAILLARD, P., CALANDO, J.P., CLERC, H., GROS, R., BELOT, Y., Tritium release experiment in France results concerning HT/HTO conversion in the air and soil, Fusion Technol. **14** 2P2B (1988) 1226–1230,
https://doi.org/10.13182/FST88-A25307

[4.109] KOARASHI, J., AMANO, H., ANDOH, M., LIDA, T., KOMA, T., A new method for determining HT deposition velocity and dependence of the velocity on environmental factors, Radiat. Prot. Dosimetry **93** (2001) 237–243,
https://doi.org/10.1093/oxfordjournals.rpd.a006435

[4.110] GARLAND, J.A., The absorption and evaporation of tritiated water vapor by soil and grassland, Water Air Soil Pollut. **13** 3 (1980) 317–333,
https://doi.org/10.1007/BF02145476

[4.111] MURPHY, C.E., Jr., The Transport, Dispersion and Cycling of Tritium in the Environment, Report WSRC-RP-90-462 Westinghouse Savannah River Company, Aiken, SC (1990).

[4.112] PAPKE, H., FÖRSTEL, H., Formation rate of nonexchangeable organically bound tritium from tritiated soil water., Health Phys. **60** 6 (1991) 773–779,
https://doi.org/10.1097/00004032-199106000-00002

Chapter 5

TRITIUM IN MARINE AND BRACKISH ENVIRONMENTS

I.W. CROUDACE, A. ULANOWSKI

5.1. INTRODUCTION

Following the discovery of cosmogenic tritium in 1950 (see Section 5.2.2), it was quickly recognized that tritium, oxidized to its species HTO, would serve as the perfect, conservative, hydrological tracer. However, the availability of suitably sensitive and precise isotope ratio mass spectrometers and techniques capable of measuring the minute concentrations of tritium in ocean and rainwaters were not readily available until the mid-1970s. The pioneering developments of these specialized gas-source mass spectrometers were made by Clarke et al. [5.1] and other notable contributors (see Section 9.5.1). These instruments evolved to enable routine ultraprecise measurements of tritium, ^{3}He, ^{4}He and other noble gases.

Although there was an extensive terrestrial monitoring network for tritium from the mid-1950s, there were few measurements taken in ocean settings in the early years. Over the past 50 years, however, a substantial volume of high precision validated data has been acquired from samples from the global oceans and enclosed seas (e.g. Refs [5.2–5.6]). The acquisition of these tritium data has been made through a series of international oceanographic programmes that led to dramatically improved understandings of ocean circulation, ventilation, mixing and air–sea exchange. The continued significant improvements in measurement sensitivity even offered the potential for filling gaps in the early depositional record through sampling of ocean island tree rings [5.7] (see additional examples in Section 5.3.1 and in Chapter 7). In principle, such data could allow the reconstruction of annually averaged records of tritium precipitation added to the oceans in the early 1950s, when few actual data were acquired.

The contemporary knowledge on tritium presence in the principal environmental compartments indicates that the oceans and seas are the main reservoir of tritium at the global scale (see Table 5.1).

TABLE 5.1. DISTRIBUTION OF TRITIUM IN THE MAIN ENVIRONMENTAL COMPARTMENTS [5.8]

Environmental compartment	Form or compound	Fraction of global tritium (%)
Hydrosphere (mostly in the oceans)	Liquid water	~90
Atmosphere, troposphere	Water vapour	0.1
	Molecular hydrogen gas	0.02–0.2
	Methane gas	<0.04
Atmosphere, stratosphere	Water	~10
	Molecular hydrogen	0.004–0.007

5.1.1. The role of tritium in understanding ocean processes

The understanding of physical and chemical processes operating in the immense volume of the oceans (ventilation, inter-gyre mixing, thermohaline circulation) has undoubtedly developed through the acquisition and interpretation of copious high precision data on tritium and its progeny ^{3}He in the last 50 years. These data explain how tritium, as the ideal transient hydrological tracer, disperses in marine systems. They also contributed to understanding the mechanisms of atmospheric exchange with the oceans, as well as rates of ocean circulation horizontally and vertically, and terrestrial transfer. These discoveries have been crucial, for example, in deepening an understanding of the role and impacts of CO_2 on global climate.

Ocean currents (see Fig. 5.1) drive mixing and are in turn driven by solar energy fluxes, winds, tides, differences in temperature and salinity, and the rotation of the Earth around its axis. Global winds drag on the surfaces of oceans, causing them to move and build up in the direction that the wind is blowing. The Coriolis effect deflects winds to the right in the Northern Hemisphere and to the left in the Southern Hemisphere, and this phenomenon results in the deflection of major surface ocean currents in a clockwise spiral in the Northern Hemisphere and in a counterclockwise spiral in the Southern Hemisphere. These major spirals of oceanic currents, known as ‘gyres’, occur north and south of the Equator. There are five major ocean-wide gyres: the North Atlantic, South Atlantic, North Pacific, South Pacific and Indian Ocean gyres (see Fig. 5.1). Each is flanked by

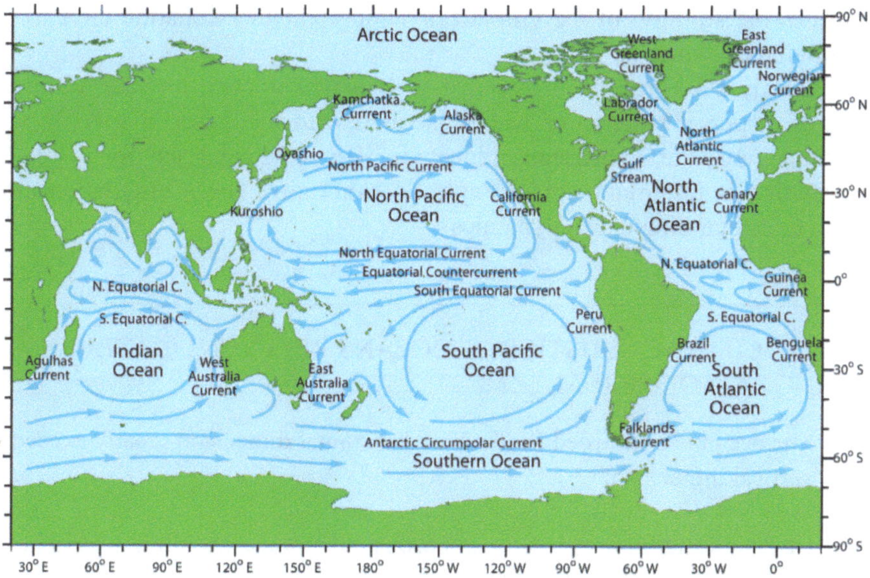

FIG. 5.1. Summary of major ocean currents and gyres existing in the North Atlantic, South Atlantic, North Pacific, South Pacific and Indian Ocean. (Image credit: Science Education Through Earth Observation for High Schools project, chapter "Ocean currents"; https://seos-project.eu/oceancurrents/oceancurrents-c02-p01.html; reproduced with permission.)

a strong and narrow western boundary current (e.g. the Gulf Stream) and a weak and broad eastern boundary current (e.g. the Canary Current).

5.1.2. Tritium as a transient tracer of ocean processes

Transient tracers allow major perturbations, such as ocean ventilation and impacts from climate change, to be investigated. They are new to the oceanic environment and evolve with time and can therefore be used to study physical, chemical and biological processes such as ocean ventilation and circulation [5.9]. Transient tracers are analogous to gigantic dye experiments and provide powerful insights into perturbations occurring in the interior of the ocean, along with information on process timescales. Examples of transient tracers in the oceans are chlorofluorocarbons, tritium, tritiugenic ^{3}He and anthropogenic ^{14}C. Tritium is the perfect conservative tracer of water masses such as oceans (see Box 5.1 for details). Transient tracers such as tritium are uniquely suited to determining the amount of interior exchange of ocean waters through mixing and dilution and downward propagation rates in such areas as the Atlantic. Studies in this region have suggested that newly ventilated North Atlantic Deep Water indicates the strength of the climatically important thermohaline circulation or meridional overturning.

Steady state tracers, in contrast to transient tracers, provide a smoothed view of dispersal patterns of waters labelled with the tracer at specific source locations (e.g. the mid-ocean ridges for ^{3}He or glacial ice for ^{4}He). Such information is important, for example, in calibrating ocean circulation models and in understanding ocean dynamics.

BOX 5.1. CHARACTERISTICS OF OCEANS RELEVANT TO TRITIUM MIXING AND DISPERSAL

(a) The ocean is a stratified fluid whose horizontal dimension greatly exceeds its vertical dimension.

(b) Oceanic water masses can travel horizontally thousands of kilometres before being mixed away vertically.

(c) Mixing (ventilation) is controlled by buoyancy forces and is affected by salinity, temperature, freshwater fluxes and pressure effects; for example, thermohaline circulation.

(d) The upper oceanic layer is mixed with a distinct thermal, density and compositional structure. The upper part includes the thermocline and the pycnocline. In the thermocline, the temperature decreases rapidly from the mixed upper layer of the ocean to much colder deep water in the thermocline. From a depth of 1000 m to approximately 4000 m, the water temperature remains constant. At layers deeper than 4000 m, the temperature varies very little, ranging from near freezing to just above the freezing point of water, as depth increases.

(e) The depth and strength of the thermocline varies with location, season and from year to year. It is semi-permanent in the tropics, variable in temperate regions (often deepest during the summer) and shallow or non-existent in the polar regions, where the water column is cold from the surface to the bottom.

(f) Ocean ventilation is the physical process whereby 'young' surface waters that have been in contact with the atmosphere are transported to the ocean interior. Thus, ocean ventilation brings oxygenated water to where oxygen is being consumed by biogeochemical and other remineralization processes. Ventilation also transports surface changes, such as heat and natural and anthropogenic substances (e.g. CO_2 from fossil fuel burning, radionuclides such as tritium and ^{14}C from nuclear weapon tests) to the interior ocean.

(g) Thermohaline circulation (global conveyor system) is part of a process whereby large scale ocean circulation is driven by global density gradients created by surface heat and freshwater fluxes (see Meridional Overturning Circulation below).

(h) The Meridional Overturning Circulation refers to large scale observable phenomena that incorporate thermohaline circulation, wind and tidal effects and describe a global system of ocean currents. In the Atlantic Meridional Overturning Circulation, warm water flows northwards (Gulf Stream) where it cools, and evaporation occurs, leading to increases in the salt content. The lower water temperature and higher salt content make the water denser and it then sinks into the ocean (deep water formation). The colder, denser water slowly spreads southwards, several kilometres below the surface. Eventually, it returns to the surface and warms in a process called 'upwelling' and the circulation is complete. This global process provides a mixing mechanism for the world oceans, transferring heat and substances, and plays a pivotal role in climate.

5.1.3. Measurement of tritium and ^3He in ocean waters

5.1.3.1. ^3He ingrowth technique

Measuring tritium indirectly via its radiogenic, inert progeny ^3He provides the most sensitive, precise and accurate means of determining tritium in waters [5.1, 5.10, 5.11]. ^3He is differentiated as tritiugenic if derived from the decay of tritium or terrigenic if it is primordial in origin and derived from outgassing from rocks. Measurement of marine and other waters for tritium and ^{3}He is generally a process comprising three steps:

(1) Degassing of a volume of water to remove all dissolved helium.
(2) Storage of the outgassed water in a helium leaktight container (usually low He permeability glass or metal) for an ingrowth period of several weeks to a year or more. Shielding the stored samples from cosmic rays is important to prevent cosmogenic ^{3}He production in the sample.
(3) Extraction of ^{3}He from the water sample and analysis using ^{3}He ingrowth mass spectrometry.

This method offers the highest sensitivity, precision and accuracy, typically better than 1%, and a lower ultimate detection limit than other methods.

5.1.3.2. Radioactivity counting

Radioactivity counting is significantly less sensitive than helium ingrowth mass spectrometry and usually involves isotopic enrichment of tritium by electrolysis, followed by low level radioactive decay counting using liquid scintillation [5.12] (see also Chapter 9) or gas flow proportional counting. Measurements are made relative to prepared standards and accuracy is limited by the reproducibility of the enrichment process to 3–10%.

5.1.3.3. Tritium–helium dating

A logical complement to tritium data is data from its stable decay product ^{3}He (tritiugenic ^{3}He). Helium-3 forms through beta decay of ^{3}H with a half-life of 12.3 years. The variations in this isotopic pair form the basis of a powerful radiometric clock. The ratio $^{3}H/^{3}He$ [5.13] leads to a timescale that is independent of tritium input. So long as a water parcel is in contact with the atmosphere, the ^{3}He formed by tritium decay is exchanged with any ^{3}He in the atmosphere, and the tritium–helium radiometric clock is constantly reset to zero, as helium is able to outgas from the water. However, if atmospheric exchange is lost (e.g. through ventilation and circulation of ^{3}H in the oceans), the tritiugenic ^{3}He is able to accumulate and the $^{3}H/^{3}He$ ratio changes according to the tritium content and to the time the water parcel is isolated from the atmosphere. The $^{3}H/^{3}He$ age, also called water mass age, records the duration of the isolation of a water parcel from the atmosphere, and timescales from months to years can be inferred [5.14]. The age is defined as follows:

$$\tau = \frac{T_{1/2}}{\ln 2} \ln\left(1 + \frac{\left[{}^{3}\mathrm{He}_{\mathrm{trit}}\right]}{[{}^{3}\mathrm{H}]}\right) \tag{5.1}$$

where $T_{1/2}$ is the ^{3}H half-life (s), and [^{3}H] and [$^{3}He_{trit}$] are the concentrations of ^{3}H and tritiugenic ^{3}He, respectively, expressed in TU (the number of tritium atoms per 10^{18} atoms of hydrogen; see Section 2.3.2 for additional details).

5.1.4. Tritium data collection in the oceans

Insights into the distribution and dispersion and/or transport of tritium in the global oceans are largely derived from a series of international oceanographic expeditions. The most significant programmes were GEOSECS (Geochemical Ocean Sections Survey; 1970–1980), TTO (Transient Tracers in the Ocean; 1981–1983), SAVE (South Atlantic Ventilation Experiment; 1980–1985), WOCE (World Ocean Circulation Experiment; 1990–2002) and GEOTRACES (2010–2020). Some smaller programmes (European Phycemed and Topogulf) also contributed significant tritium and ^{3}He data [5.5, 5.6, 5.15]. Over the past 50 years, data from more than 60 000 high precision measurements for tritium, 63 000 for He isotopes, 57 000 dissolved He and approximately 34 000 dissolved Ne concentrations have been accumulated following scientific expeditions aimed at better understanding oceanic processes. More details on these expeditions are given in Table 5.2.

TABLE 5.2. MAIN INTERNATIONAL OCEANOGRAPHIC PROGRAMMES CONTRIBUTING TRITIUM AND $^{3}He/^{4}He$ DATA

Programme	Place and period	Description
GEOSECS	Atlantic Ocean (1972–1973) Pacific Ocean (1973–1974) Indian Ocean (1977–1978)	GEOSECS ran from 1970 to 1980 and was a global survey of the 3-D distribution of chemical, isotopic and radiochemical tracers in the ocean. A key objective was to investigate the pathway of the deep thermohaline circulation of the ocean, using chemical and radionuclide tracers. It provided the first comprehensive data set for the distribution of chemical species (nutrients, radioisotopes, stable isotopes, carbon, gases, trace metals and particles) in the world ocean.

TABLE 5.2. MAIN INTERNATIONAL OCEANOGRAPHIC PROGRAMMES CONTRIBUTING TRITIUM AND $^{3}He/^{4}He$ DATA (cont.)

Programme	Place and period	Description
TTO	North Atlantic Ocean (1981) Tropical Atlantic Ocean (1983)	The TTO programme was a series of major geochemical and physical oceanographic expeditions, sponsored by the US National Science Foundation and the US Department of Energy, investigating the movement of anthropogenic geochemical tracers into the interior of the ocean. The programme was based on the GEOSECS programme and involved intensive sampling in the northern two-thirds of the Atlantic. It investigated for the first time the penetration of human made tracers derived from military and industrial activities into the thermocline and deep ocean. Two of the main targets were tritium and ^{14}C derived from the atmospheric testing of nuclear weapons up to 1962.
GEOTRACES	Global (2010–2020)	The international programme GEOTRACES aimed to improve understanding of biogeochemical cycles and large scale distribution of trace elements and their isotopes in the marine environment. Scientists from 35 nations were involved in the programme, which was designed to study all major ocean basins. The GEOTRACES Science Plan was published in 2006 and the GEOTRACES programme formally launched its seagoing effort in January 2010.
SAVE	Atlantic Ocean (1980–1985)	SAVE had two campaigns (TTO/NAS and TTO/Equatorial) to measure transient tracers in the Atlantic Ocean and was funded by the US National Science Foundation. A crucial aim was to provide large scale distributions of tritium and ^{3}He to assess ventilation mechanisms in the Southern Atlantic. About 2400 samples were collected from 140 stations (surface, thermocline, intermediate and bottom waters), with the highest sampling intensity in the upper layers. Dating of water masses using $^{3}H/^{3}He$ ratios was made and interpreted along with other hydrographic and anthropogenic data [5.16].

TABLE 5.2. MAIN INTERNATIONAL OCEANOGRAPHIC PROGRAMMES CONTRIBUTING TRITIUM AND $^{3}He/^{4}He$ DATA (cont.)

Programme	Place and period	Description
CLIVAR	Global (1995–ongoing)	CLIVAR is one of six core projects of the World Climate Research Programme. CLIVAR's mission is to understand the dynamics, the interaction and the predictability of the climate system, with emphasis on ocean–atmosphere interactions. The concept of CLIVAR arose from the recognition that observed climate variations result from natural variability superimposed on long term trends that could be induced by anthropogenic modifications of the global environment and other external forcing factors.
GO-SHIP	2007–ongoing	GO-SHIP brought together scientists with interests in physical oceanography, the carbon cycle, marine biogeochemistry and ecosystems, and other users and collectors of ocean interior data. The programme was sponsored by US CLIVAR and OCB (Ocean Carbon and Biogeochemistry Program) and was funded by the US National Science Foundation and the National Oceanic and Atmospheric Administration of the USA.
PHYCEMED	1981	A cruise focusing on the chemical oceanography of pollutants in the general framework of the physics and chemistry of the Mediterranean Sea.
TOPOGULF	1983	TOPOGULF was a joint programme initiated by IFREMER (Brest, France) and IFM (Kiel, Germany). Tritium and ^{3}He, along with other tracers, were measured in the north-eastern Atlantic Ocean following the 1983 TOPOGULF cruise. The sampled area was located near the Azores Islands on both sides of the mid-Atlantic ridge.
WOMARS	1998–2001	The project WOMARS: Radionuclide Levels in Oceans and Seas operated through international collaborations and multi-national funding. The final report was published in 2012 [5.17]. Data collected were included in the Global Marine Radioactivity Database [5.18].

TABLE 5.2. MAIN INTERNATIONAL OCEANOGRAPHIC PROGRAMMES CONTRIBUTING TRITIUM AND $^3He/^4He$ DATA (cont.)

Programme	Place and period	Description
WOCE	Global (1990–2002)	WOCE was a component of the international World Climate Research Program. WOCE's field phase ran between 1990 and 1998 and was followed by an analysis and modelling phase that ran until 2002. The main purpose of WOCE was to develop models to describe the present state of the ocean and to predict future changes in relation to long term climate changes. The measurements considered essential for the WOCE Hydrographic Program included the collection of transient tracer data (chlorofluorocarbons CFC 11 and CFC 12, tritium, tritiugenic ^3He and anthropogenic ^{14}C), as well as the steady state tracers ^3He, ^4He and stable isotopes of water. The programme provided the most comprehensive sampling of the South Atlantic in 1990–1995 [5.19].

Note: CLIVAR: Climate and Ocean: Variability, Predictability and Change; GEOSECS: Geochemical Ocean Sections Survey; GO-SHIP: Global Ocean Ship-based Hydrographic Investigations Program; SAVE: South Atlantic Ventilation Experiment; TTO: Transient Tracers in the Ocean; WOCE: World Ocean Circulation Experiment; WOMARS: Worldwide Marine Radioactivity Studies.

5.2. TRITIUM SOURCE TERMS

5.2.1. Tritium measurement units and activities in natural systems

Tritium concentrations in natural waters are traditionally reported in TU, defined as a $^3H/^1H$ atomic fraction of 10^{-18}. For liquid water at normal conditions, TU is equivalent to an activity concentration of 0.119 Bq/L (see Chapter 2). Historically, the definition of TU arose from the estimate [5.20] of the natural abundance of tritium in Norwegian surface water, which was taken to approximate secular equilibrium between cosmogenic production and radioactive decay. Tritium concentrations in natural waters are very low, of the order of 1 TU, while the mean tritium content in continental precipitation is currently about 0.2–10 TU [5.21–5.23] (see also Section 2.4.2).

The results of oceanographic studies are also often converted to estimates of local tritium inventory — namely, the relative concentration of tritium integrated from the ocean surface to the ocean floor — and reported in units of TU · m.

5.2.2. Natural (cosmogenic) tritium

The first convincing empirical evidence for cosmogenic tritium production was obtained in 1950 and confirmed later [5.24–5.26]. This quickly led to its proposed use as a powerful hydrological tracer (e.g. Ref. [5.27]). Natural tritium is mainly produced from the interaction of cosmic rays with nitrogen (and oxygen) in the upper atmosphere, with a production rate varying as a function of geographical latitude and of solar activity cycle, with higher generation in subpolar (c. 60° latitude) areas and during lower activity of the Sun. Cosmic ray interactions also result in the production of secondary neutrons, and those with energy exceeding a threshold of 4 MeV can react with atmospheric nitrogen to produce ^{12}C and ^{3}H in the reaction $^{14}N(n, {}^{3}H)^{12}C$ [5.28]. Other reactions, such as the (p, ^{3}H) reaction on nitrogen and oxygen nuclei, also contribute to cosmogenic tritium production. The tritium atoms produced in these reactions are rapidly converted to tritiated water (HTO), which then becomes part of the hydrological cycle (Fig. 5.2; see also Chapter 3).

The production rate of cosmogenic radionuclides (e.g. ^{3}H, ^{14}C) in the upper atmosphere changes with time because the incoming galactic cosmic ray flux that reaches the Earth's atmosphere depends on the general level of solar activity [5.29]. During intervals of low sunspot activity, the magnetic shielding properties of the solar wind are such that a larger GCR flux arrives in the Earth's upper atmosphere, whereas cosmic ray fluxes are lower during periods when sunspot activity is higher.

Available estimates of the mean production rates of cosmogenic tritium span the range of 0.14–0.36 atoms · cm^{-2} · s^{-1} [5.21, 5.30–5.34]. Recent Monte Carlo modelling [5.35] estimated the global average tritium production rate at moderate solar activity value to be 0.345 atoms · cm^{-2} · s^{-1} and a range from 0.09 atoms · cm^{-2} · s^{-1} in the troposphere at high solar activity to 0.92 atoms · cm−2 · s−1 for the entire atmosphere in polar regions at low solar activity. Stratospheric production of tritium was estimated to be about 50% by Ehhalt et al. [5.36] and 67–71% by Poluianov et al. [5.35].

5.2.3. Tritium transfer to the oceans from atmospheric nuclear testing

From 1952 until 1963, significant quantities of tritium were introduced to the atmosphere following high yield nuclear and thermonuclear weapon tests (see Fig. 5.3). By the mid-1960s, the natural background of tritium in precipitation was surpassed by this human made tritium by over two orders of magnitude (e.g. Ref. [5.37]). In the summer of 1963, the tritium in precipitation increased to

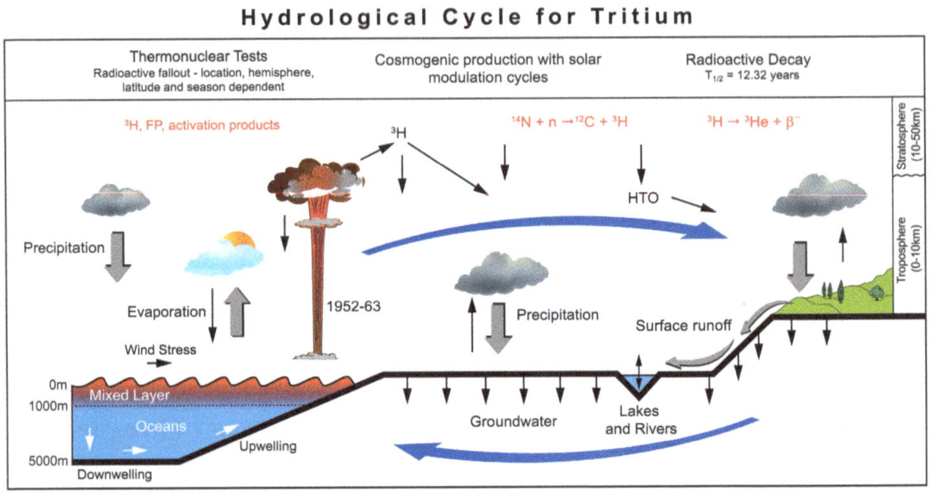

FIG. 5.2. Schematic of the hydrological cycle, indicating tritium production, dispersion and decay.

6000 TU in temperate continental regions in the Northern Hemisphere. This newly introduced anthropogenic tritium was quickly exploited as a valuable transient tracer to study processes in the oceans. The testing of atmospheric nuclear weapons that could have impacted the environment covered the period 1945–1998, with most radioactivity entering the environment during the main tests, from 1952–1962. The last atmospheric test occurred in 1980 [5.38] (see also Section 3.2.1).

The approximate concentrations of tritium in marine precipitation for the Northern and Southern Hemispheres have varied with time and global latitude (see Fig. 5.4). Most of the nuclear weapon tests were carried out in the Northern Hemisphere and, since atmospheric circulation in each hemisphere is relatively independent, the tritium distribution is strongly asymmetric, with higher concentrations existing in the Northern Hemisphere compared with the Southern Hemisphere. The tritium from nuclear weapon tests, in common with all associated radioactive fallout, was mainly released in three temporal pulses: 1954, 1958–1959 and, predominantly, 1963. Most of the tritium from thermonuclear weapon tests was injected into the stratosphere above the hemisphere where the test occurred, from where it then mixed down into the troposphere and subsequently participated in the hydrological cycle (see also Chapter 4).

The amounts of tritium delivered to the global oceans and enclosed seas (e.g. Mediterranean Sea) have been evaluated following the work of many international groups over several decades (e.g. Refs [5.6, 5.15, 5.39–5.41]). Transfer to the surface waters of the oceans, which are the largest sink for tritium, occurs by water vapour exchange, precipitation and continental runoff

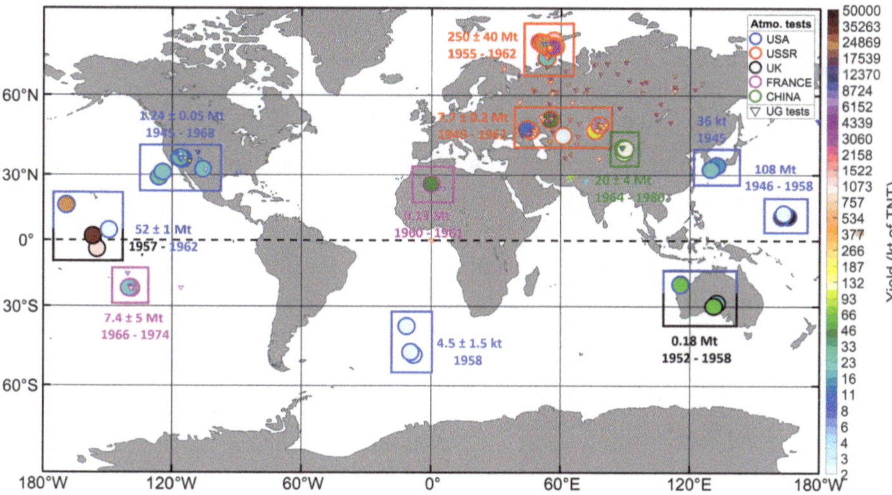

FIG. 5.3. Locations of nuclear explosions in 1945–1998, along with the total weapon yield in kilotonnes or megatonnes of TNT. Atmospheric tests are shown with a circle and underground tests with a triangle. The line colour of circles corresponds to the State that carried out the test. The rectangles represent the sum of yields from atmospheric tests for the indicated years. Data from the Swedish National Defence Research Institute [5.38]. Figure reproduced from Ref. [5.15] under a Creative Commons Attribution 4.0 International (CC BY 4.0) licence.

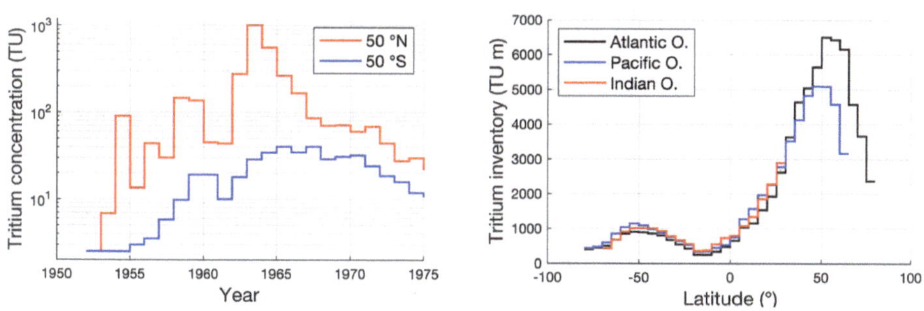

(a) (b)

FIG. 5.4. Temporal (a) and latitudinal (b) distributions of tritium in oceans in 1972, as estimated by Weiss and Roether [5.3]. (a) Year averaged tritium concentrations in marine precipitation along circles of 50° N and 50° S latitudes. (b) Estimated local inventory of tritium in oceans as a function of latitude.

[5.3, 5.37]. Tritium concentrations are currently of the order of 10 TU in continental precipitation. In the oceans, the maximum tritium concentrations observed in the Northern Hemisphere surface water are of the order of 2 TU. Regions with high runoff components, such as the Arctic Ocean and the enclosed Mediterranean Sea basin, typically have tritium concentrations above those of the open oceans. Tritium concentrations in Southern Ocean surface waters are currently of the order of 0.15 TU. In the Mediterranean Sea, tritium in surface waters rose to 20–30 TU in 1964, which was about 100 times the natural level, and then declined 30 times by 2011 through radioactive decay (Section 5.4.7).

5.3. TRANSPORT AND CIRCULATION OF TRITIUM IN THE MARINE ENVIRONMENT

5.3.1. Atmospheric dynamics and tritium in the oceans

The testing of thermonuclear weapons in the atmosphere from 1952 onwards led to the injection of several hundred kilograms of tritium into the troposphere and the stratosphere. Since most testing occurred north of the Equator, the global controls on atmospheric circulation effectively quarantined the fallout in the Northern Hemisphere (cf. Fig. 5.4). Exchanges between the two hemispheric air masses were found to take a year or two, and the residence time for tritium injected into the North Hemisphere stratosphere was in the range 1–2 years. The residence time of any tritium in the troposphere is generally in the range 1–30 days, with precipitation and vapour exchange being the main removal processes. The amount and distribution of tritium from nuclear weapon tests in the atmosphere at any time depended on the yield of the nuclear devices, and it has been stated that much of the tritium was derived directly from tritium components in the pre-detonated nuclear device rather than from tritium generated by nuclear reactions. Mass estimations suggest that approximately 2 kg of tritium was injected per megatonne of explosive force [5.42, 5.43]. After deposition, tritium, as HTO, enters the terrestrial and marine hydrological cycles.

5.3.2. Significant sources of tritium entering the oceans

Since the beginning of the nuclear era in the 1940s, the contribution of the different paths for tritium entry into the oceans varied significantly. From the late 1950s to the early 1970s, the dominating tritium input sources into oceans were those related to tritium in the atmosphere, namely, water vapour exchange and precipitation. At that time, tritium deposition to the North Atlantic via vapour exchange and precipitation was more than an order of magnitude (c. 30–50 times)

BOX 5.2. SUMMARY OF TRITIUM TRANSFER FROM THE ATMOSPHERE TO THE HYDROSPHERE
(*adapted from Ref. [5.44]*)

(a) Tritium in the water in the troposphere is removed rapidly, and typically vapour in the atmosphere has a residence time of 10–30 days. Most tritium is removed by molecular exchange, primarily to the ocean.

(b) Tritium (and other substances and/or radionuclides) in the stratosphere is mixed into the troposphere by two processes in the Northern Hemisphere: (a) during thunderstorms, thunderheads penetrate into the tropopause, resulting in rain that brings down tritium or other nuclear debris from the stratosphere; (b) the spring leak, which prominently occurs in the Northern Hemisphere. During the spring leak, the tropopause breaks down between latitudes of 30° N and 60° N, resulting in the mixing of stratospheric air with higher tritium concentrations and air in the troposphere. As a result, the tritium concentration in precipitation tends to increase in the spring of each year.

(c) The spring leak is more influential in the Northern Hemisphere because it is created by heating up of the land masses at those latitudes in the spring and summer. This seasonal effect is seen for tritium, anthropogenic radionuclides and cosmogenic radioisotope data.

(d) There is a marked latitude dependence where fallout occurs from the stratosphere. Most nuclear debris will ultimately be deposited in the 30–60° N latitude band.

(e) The mean residence time for tritium and other aerosols in the stratosphere is 1–2 years.

(f) The spring leak is hardly seen in the Southern Hemisphere, because fewer land masses are present at the latitude band 30–60° S and the breakup of the tropopause is inhibited. Penetration of thunderheads into the stratosphere does occur in the Southern Hemisphere, and the stratosphere can come close to the land surface in the high plateaus of Antarctica, but these processes to have lower potential to introduce nuclear debris into the troposphere.

higher than that from runoff or inflow from the southern Atlantic or from the Arctic. Later, after the early 1970s, the influx of fresh water from the Arctic played a more prominent role [5.13].

The dominant tritium transfer processes from the atmosphere to the hydrosphere (summarized in the review of Michel et al. [5.44]) are as follows (see also Box 5.2):

— Vapour exchange. This is considered to be the dominant mechanism for extraction of atmospheric tritium and exceeds precipitation inputs by a factor of three [5.40]. It involves the diffusive extraction of HTO vapour by water vapour in the troposphere and stratosphere.
— Direct precipitation. Deposition occurs mainly through rainfall and snow.
— Continental runoff and river flow. Tritium deposited to the continents flows to the oceans from sources such as lakes, rivers and groundwater flow. Its presence in these dynamic terrestrial reservoirs leads to a delayed entry into the oceans, which may take several years. For the northeastern Atlantic, it is estimated that river runoff contributes no more than 10% of the tritium entering the ocean [5.39].
— Industrial inputs. These are largely represented by discharges from nuclear reprocessing plants (Sellafield and La Hague into the northeastern Atlantic Ocean; see Section 3.2.3); nuclear sites (e.g. Marcoule nuclear industrial site into the Mediterranean Sea); nuclear accident discharges (Fukushima Daiichi NPP into the northwestern Pacific); and Russian nuclear legacy sites (Arctic). All of these have impacts that depend on the magnitude and frequency of discharges and marine processes (e.g. currents, storms).

5.4. TRITIUM IN THE OCEANS

5.4.1. Inventories of tritium in the ocean

Tritium is distributed non-uniformly across the world ocean waters. One study [5.3] summarized outcomes of various oceanographic studies in 1952–1975 and quantified global inventories of tritium in the oceans. Shown in Table 5.3 are the total activity values of tritium in the oceans from the two hemispheres. As seen, the global tritium inventory in 1972 in the Northern Hemisphere was approximately three times higher than in the Southern Hemisphere, thus largely reflecting the contribution from the nuclear weapon tests that had occurred a decade before. Similarly, Östlund and Fine [5.45] estimated the global tritium inventory to be equal to 164 kg (decay corrected to 1 January 1972) and evaluated its distribution between the major oceanic pools (Table 5.4).

TABLE 5.3. ESTIMATES OF TOTAL TRITIUM ACTIVITY INPUTS TO THE WORLD OCEANS UP TO 1972
(data from the GEOSECS programme; table adapted from Ref. [5.3])

Ocean	Tritium activity in 1972 (PBq)				
	Northern Hemisphere		Southern Hemisphere		Total
	Direct input	Continental input	Direct input	Continental input	
Atlantic	19 200	6 400	3 400	400	29 400
Pacific	25 000	2 200	7 700	300	35 200
Indian	2 000	1 200	5 700	100	9 100
Total	46 200	9 900	16 800	800	73 700
Hemisphere total	56 100		17 600		73 700
Hemisphere total (1963 equiv.)	93 100		29 300		122 300

TABLE 5.4. ESTIMATED OCEANIC TRITIUM INVENTORY IN 1972 [5.45]

Ocean	Tritium mass in 1972 (kg)
North Atlantic	66
South Atlantic	7
Arctic	6
North Pacific	59
South Pacific	14
Antarctic	6
Indian	6
Total oceanic	164

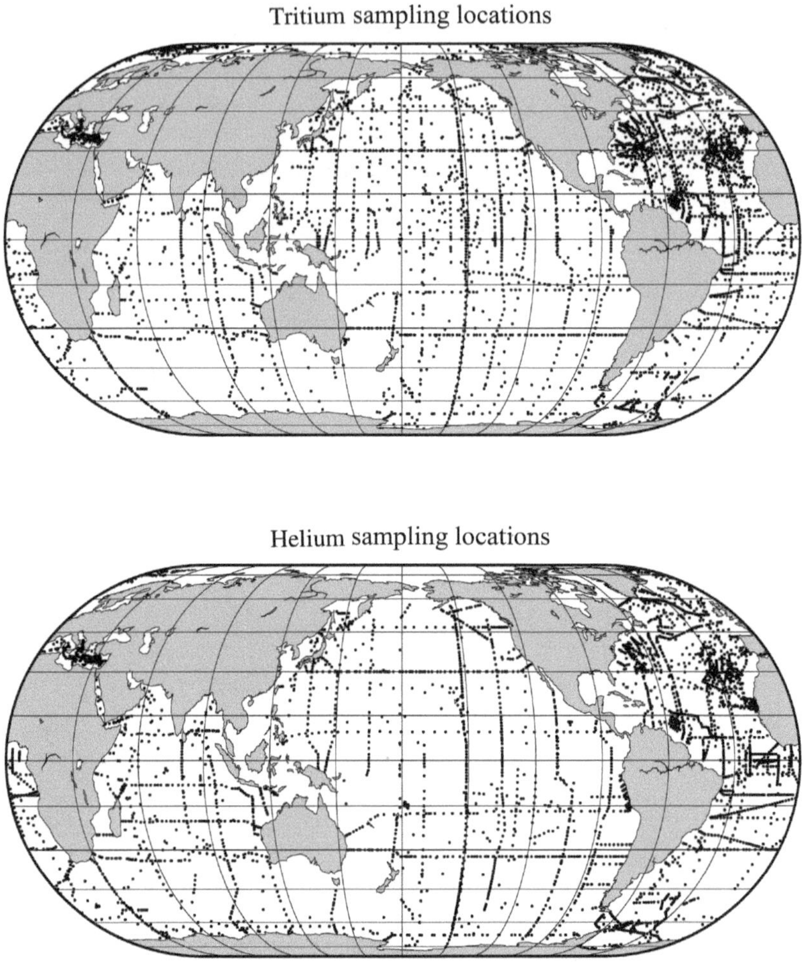

FIG. 5.5. Locations (solid dots) of international scientific sampling of the oceans to study processes of ventilation, circulation and mixing using tritium and ^{3}He data. Reproduced from Ref. [5.6] under a Creative Commons Attribution 4.0 International (CC BY 4.0) licence.

High precision measurements of tritium and its progeny ^{3}He have provided critical insights into ocean circulation and mixing (ventilation) and ocean–atmosphere interactions. The large number of data currently existing have been generated by a small number of expert laboratories using noble gas (helium ingrowth) mass spectrometric techniques to determine tritium and ^{3}He/^{4}He ratios [5.1, 5.6, 5.10, 5.46, 5.47]. The largest number of datasets have been collected for the Atlantic Ocean, although the Pacific Ocean and the Mediterranean Sea are also well represented (see Fig. 5.5).

The study of Jenkins et al. [5.6] summarized data on the variation of tritium concentrations in near-surface ocean waters (up to 50 m depth) as a function of time for several latitude bands. The most distinctive feature is the increase northwards in the concentration ranges. The time–latitude trends reflect the globally asymmetric delivery of tritium. Some of the structure is caused by regional variations in atmospheric input and ocean circulation, also including data from different major ocean basins. The summary of tritium in the oceans is given in Box 5.3.

The highest recorded values of tritium concentration in surface ocean water, exceeding 50 TU, were observed in the Arctic area (60–90° N) in the 1970s. In the latitude band 45–60° N, the highest values (20–60 TU) were recorded in the 1960s. Starting as early as the 1950s, many datasets were recorded in the latitude band 15–45° N, showing tritium concentrations in the range 10–30 TU during 1955–1975. In the equatorial band, from 5° S to 15° N, tritium concentrations peaked at 20 TU in the early 1960s, being mostly below 10 TU. In the Southern Hemisphere, within the latitude band 90°–5° S, the recorded values of tritium concentrations were substantially lower, with maximal values measured in the mid-1960s not exceeding 5 TU.

After 1980, the global tritium concentrations substantially declined to values well below 1 TU in polar areas of the Southern Hemisphere (90–45° S), while being notably higher in other areas of the world oceans, systematically increasing in the northern direction.

A recent review by Oms et al. [5.15] has provided a valuable summary of tritium distribution in the oceans. The study was based on a compilation of 41 000 measurements held in online databases from large scale international oceanographic observation programmes over the period 1967–2016. It also included 610 measurements acquired from French oceanographic campaigns from 1996–2016. The authors made rigorous efforts to evaluate data quality and assert that, even with potential uncertainties in some data, their study provided an estimation of the inventories of tritium at a global scale. The total inventory of tritium in the global ocean, holding approximately 90% of the Earth inventory, is currently estimated to be 26.8 ± 14 kg (referenced to 2016). The study [5.15] attributes 3.8 kg of this amount to cosmogenic origin and consider the rest 23.0 ± 14 kg as being of anthropogenic origin. This compares well to other 2016 inventory estimates of 22–23.5 kg remaining after radioactive decay of the 520–550 kg originating from atmospheric nuclear tests since 1960 [5.2, 5.48].

Alternative estimates (see Chapter 2) can result in a natural steady state inventory of 4.5–4.9 kg, given a natural input of 258–278 g/a. The steady state inventory from industrial inputs is estimated to be 1.3 kg (from an input of approximately 78 g/a). Therefore, the total tritium inventory in the Earth (referenced to 2016) from natural, industrial and weapon test sources is 27.8–29.3 kg, as estimated from the literature.

BOX 5.3. TRITIUM IN THE OCEANS [5.6, 5.16, 5.22, 5.40]

(a) Natural tritium was dominated by inputs from the atmospheric thermonuclear tests in the 1950s, and especially in the early 1960s; the global tritium inventory increased from its natural value of ~1.6 EBq to ~186 EBq. Much of the tritium originating from nuclear weapon tests has now decayed.

(b) Tritium enters the hydrological cycle in the form of tritiated water (HTO) and ends up in the oceans, where it is slowly dispersed into the immense volume of overturning ocean water.

(c) The delivery and timing of tritium from nuclear weapon tests to the oceans was dependent on the location of weapon test sites and occurred as a strong pulse-like injection in the Northern Hemisphere and a smaller, more 'diffuse' input into the Southern Hemisphere.

(d) Tritium, as a transient tracer, was ideally suited to study the ventilation of deep waters on decade to century timescales. Variations in water column tritium concentrations with time (based on the GEOSECS Atlantic Expedition data) reveal North Atlantic deep water formation. The evolution of the tritium distributions is also valuable, as 'packets' of tritiated water penetrate the subtropical thermocline and intermediate and deep waters.

(e) The maximal tritium content in water columns in the South Atlantic was found in the subtropics, where by 1993 the peak nuclear weapon test signal had penetrated to ~1000 m depth.

(f) In 1993, the total water column tritium inventory between 0° S and 50° S according to WOCE data amounted to 2.88×10^{16} TU_{81} m^3 (where TU_{81} indicates TU decay corrected to 1981; in activity terms, it equals 3.44 EBq with uncertainty ±10%) with 80% residing in the upper 800 m [5.16].

(g) Tritium initially disperses in the oceans by mostly lateral flow along contours of constant density before mixing downwards.

(h) Surface ocean water data for 1959–1973 indicate a strong north–south gradient in tritium concentration as a result of the input from the atmosphere. An east–west gradient is also present, possibly resulting from oceanic current movements.

The difference between tritium content in the Northern and the Southern Hemispheres is significant, with higher artificial inventories in the Northern Hemisphere. The artificial inventory present in the Indian Ocean is higher than in similar latitudes in the Atlantic and Pacific oceans and is likely due to the transport of tritium labelled water masses from the northwestern Pacific via the Indonesian Seas to the Indian Ocean [5.49]. The highest artificial inventories per unit area occur for the Arctic Ocean, the North Atlantic Ocean and the Mediterranean Sea, being more than three times higher than the inventories per unit area in the Pacific Ocean.

TABLE 5.5. INDICATIVE TRITIUM CONCENTRATIONS IN THE OCEANS [5.15]

Area	Tritium content (TU_{16})[a]	Comment
North Atlantic	0.6–0.8	Dispersing under the mixing influence of North Atlantic subtropical anti-cyclonic gyre.
North Pacific	0.2–0.8	For sub-Equatorial, subtropical and North Pacific waters. Tritium content in the northwestern Pacific after the Fukushima Daiichi NPP accident was reported as 0.4–1.3 TU (decay corrected to 2011) compared with pre-accident values of 0.4 ± 0.1 TU [5.50].
South Pacific	0.15–0.17	
Indian Ocean	0.18–0.29	
Arctic Ocean	1.0 ± 0.9	The Arctic Ocean is found to contain the highest amount of tritium, which is inferred to result from nuclear reprocessing plant releases located in north-western Europe. It is estimated that 9% of this concentration is of natural origin, 16% comes from European reprocessing facilities and 75% is due to tritium from nuclear weapon tests and unidentified sources.
Southern Ocean	0.05 ± 0.04	In 1973, tritium showed concentrations of 1–2 TU, compared with 5–20 TU in the Northern Hemisphere ocean waters, reflecting the main nuclear weapon testing inputs in that hemisphere.
Mediterranean Sea	0.71 ± 0.2	Mediterranean tritiated waters outflow through the Straits of Gibraltar and then sink to 1000 m before spreading out into the North Atlantic.

[a] TU_{16} indicates tritium concentration decay corrected to 2016.

According to Oms et al. [5.15], the latitudinal distribution of tritium in the global ocean can be characterized as follows (TU data are arbitrarily decay corrected to 2016; see also Table 5.5):

- — The average tritium concentration in the Northern Hemisphere is 0.51 ± 0.50 TU_{16} (n = 8711; TU_{16} indicates tritium concentration decay corrected to 2016) and approximately three times higher than that in the Southern Hemisphere, 0.15 ± 0.11 TU_{16} (n = 8176), with respective inventories of 16 ± 8 kg and 7 ± 6 kg;
- — In the Northern Hemisphere, the tritium concentrations increase with latitude;
- — The highest tritium concentrations, 1.00 ± 0.91 TU_{16} (n = 1610), occur in the Arctic Ocean;
- — Tritium variations have a similar latitudinal distribution in the Pacific, Atlantic and Indian oceans;
- — The lowest tritium concentrations, 0.05 ± 0.04 TU_{16} (n = 2280), are found in the Southern Ocean;
- — The mean tritium concentration in the Mediterranean Sea is equal to 0.71 ± 0.19 TU_{16} (n = 312), which is significantly higher than the concentration in North Atlantic waters at the same latitude, 0.55 ± 0.13 TU_{16} (n = 852).

5.4.2. Atlantic Ocean

The most intensive sampling and analysis of tritium and helium isotopes in the oceans took place in the Atlantic within the frameworks of the GEOSECS and WOCE oceanographic programmes. These studies made major contributions to the understanding of ocean circulation and rates of lateral and vertical mixing [5.37, 5.51, 5.52].

5.4.2.1. Thermocline ventilation

Ocean ventilation is the physical process with which surface waters carrying heat and natural and anthropogenic substances such as tritium, ^{14}C and CO_2 are transported to the ocean's interior. The process transports oxygenated waters to where oxygen is being consumed by biogeochemical and other remineralization processes. Tritium, with its 12.3 year half-life, is ideally suited to study oceanographic processes such as lateral dispersion and ventilation, which are known to vary over annual through decadal timescales. Thermocline

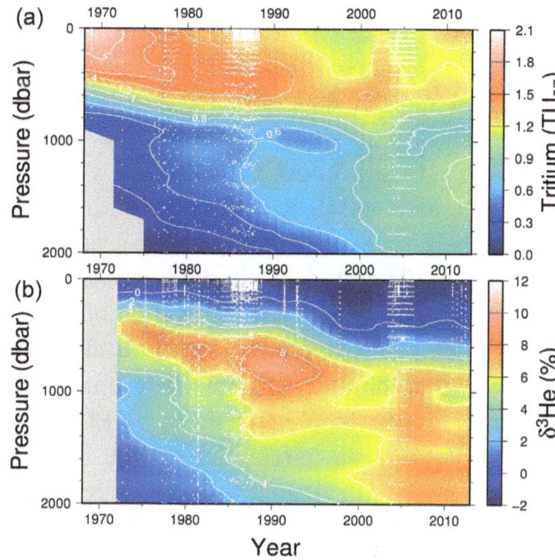

FIG. 5.6. Time series of tritium and ^3He measurements at various depths in the proximity of Bermuda (North Atlantic). (a) Tritium data shown as relative concentration decay corrected to 1997. (b) ^3He data shown as excess of ^3He isotope. White dots indicate (time–pressure) coordinates of the sampling points. Reproduced from Ref. [5.6] under a Creative Commons Attribution 4.0 International (CC BY 4.0) licence.

ventilation provides important information for integrated ocean–atmosphere models. The rates of downward mixing and tritium distributions in the North Atlantic show that the main thermocline is found to ventilate at a rate faster than predicted by the so-called Ekman pumping [5.14, 5.53]. The WOCE project was able to compare the ventilation rates of several subtropical gyres and oceans. Differences in wind stress patterns and thermohaline stratification, combined with the tracer observations, provided insights into the mechanisms of subduction and ventilation (see Fig. 5.6 for $^3H/^3He$ data in a subtropical area of the northern Atlantic [5.6]).

5.4.2.2. Intergyral exchange

Tritium, with its strong interhemispheric concentration difference, is well suited to trace interhemispheric and intergyral exchange of waters. Tritium and its decay product ^3He are used to understand the rates of exchange and mixing. The extensive datasets accumulated from numerous long time research programmes (see Table 5.2) have greatly improved models of oceanographic circulation, to quantify their parameters and to increase their descriptive power.

5.4.2.3. Deep and intermediate water formation

The large body of ^3H and ^3He data obtained during the WOCE programme has been analysed to develop a visualization of the penetration of tritium (in effect, a surface dye) into the interior ocean (see Fig. 5.6). For example, recently ventilated intermediate and deep waters were identified through tritium and $^3He/^4He$ studies in the Greenland and Labrador seas, where deep water is formed by deep convection. The studies allowed estimates of formation rates and variability of deep water and enabled an evaluation of the penetration of anthropogenic CO_2 or other surface perturbations into the deep Atlantic [5.54]. Additionally, tritium water column inventories determined for the mid-1990s could be compared with those obtained from the GEOSECS and TTO data to reveal features of the large scale circulation of the ocean averaged over a decadal timescale. These studies evaluated the exchange of surface water with deeper levels and the scale of interhemispheric exchange (exploiting the large north–south asymmetry of tritium distribution).

5.4.3. Pacific Ocean

The concentrations of selected human made radionuclides in the Pacific Ocean up to the early 2000s were summarized by Povinec et al. [5.49]. Tritium and radiocarbon data were largely acquired as parts of international oceanographic sampling (e.g. WOCE) to understand oceanographic processes in the Pacific Ocean (e.g. oceanic circulation and ventilation, climate change impacts). In 2005, the IAEA project Worldwide Marine Radioactivity Studies (WOMARS) provided an invaluable updated data compilation describing the distribution and behaviour of anthropogenic radionuclides in the world ocean and filling gaps in the WOCE data compilation. The major source of tritium in the Pacific Ocean and its marginal seas up to the early 2000s was from global and regional fallout from nuclear weapon testing in the 1950s and early 1960s. The dominant input was into the Northern Hemisphere (Fig. 5.2), where many large atmospheric nuclear tests took place.

The northwestern Pacific Ocean is well known as the area with the highest deposition of global fallout radionuclides following atmospheric nuclear weapon tests in 1952–1963. The dominant current in the northwestern Pacific is the Kuroshio Current, which brings equatorial waters to the north. Further, surface water masses from the Pacific Ocean flow via the Indonesian Seas to the northeastern Indian Ocean and then become part of the Indian Ocean subtropical gyre, occupying the region between 10° S and 40° S. Recent estimations show that the flow between the Pacific and Indian oceans is significant and might play an important role in the El Niño–Southern Oscillation and Asian monsoon climate phenomena.

North Pacific Intermediate Water (NPIW) ventilation is generally accepted to occur in the subpolar western Pacific, specifically in the Sea of Okhotsk, where outcropping of NPIW dense waters, as well as deep convection, have been observed. The northeastern Pacific is also a region where ventilation of the intermediate layers takes place. Tritium data indicate that the Alaska Gyre is a site of NPIW ventilation.

An evaluation of data for the Pacific Ocean from 1959–1973 [5.22] is summarized as follows:

- Surface water data indicate a strong north–south gradient in tritium concentration derived from atmospheric weapon test inputs. An east–west gradient is also present, possibly resulting from ocean current movements. A model indicates that the average residence time of water in the mixed layer of the North Pacific is 9–15 years.
- In the North Pacific Gyre, the tritium distribution below the winter mixed layer is essentially controlled by lateral transport down to the salinity value of 26.1 g/L. Below this level, tritium concentrations drop monotonically with depth. Water with a salinity greater than 26.1 g/L outcrops only in the northwestern Pacific. This is possibly the reason why penetration of tritium into waters of such high densities is slow.
- In the equatorial Pacific — in particular, near the North Equatorial Current — tritium concentrations show a distinct maximum under the mixed layer, coinciding with a salinity maximum indicative of water that flows under the mixed layer. In the equatorial zone, strong stratification of the water column is evident. As tritium concentrations are higher in the north than in the south, it is possible to identify the sources of the various water layers.
- In the South Pacific Gyre, tritium concentrations are much lower than in the North Pacific. However, tritium is present there well below the mixed layer owing to the sinking of surface water at the Antarctic Convergence. South of the Convergence, the temperature minimum is an effective barrier to downward transport of tritium.
- For most locations, the main control on tritium distribution as a function of depth is horizontal water movement.

Following the Fukushima Daiichi NPP accident in March 2011, the immediate and later impact on the western Pacific attracted broad scientific attention to determine the dispersion and fate of the released radionuclides [5.55, 5.56]. Measurements of radionuclides (particularly, tritium, radiocarbon and radiocaesium) in the western part of the North Pacific Ocean confirmed that their distribution has been influenced by the accident.

The 2011–2012 Ka'imikai-o-Kanaloa scientific cruise collected water column samples from coastal sea water off the shore of the Fukushima Daiichi NPP and in the other parts of the northwestern Pacific Ocean and measured activity concentrations of ^{3}H, ^{14}C and Cs radionuclides [5.50, 5.57]. The highest activity levels in surface sea water were observed for ^{134}Cs and ^{137}Cs in seawater samples collected off the shore of the Fukushima Daiichi NPP. The results were up to 1.1 Bq/L at 10 m water depth, which represented an increase of about 30 times compared with the pre-accident values. In comparable samples, tritium levels were approximately six times higher than those observed before the accident, which were about 0.15 Bq/L or 0.4–1.3 TU [5.50]. The levels of ^{14}C were 9% above the global fallout background. Tritium seawater profiles generally followed ^{137}Cs distributions, with an average $^{3}H/^{137}Cs$ activity ratio of 0.5 ± 0.3. The estimated total inventory of ^{137}Cs and ^{134}Cs in the water column of the investigated western North Pacific region was 2.7 ± 0.4 PBq and for ^{3}H 0.3 ± 0.2 PBq. Direct releases of contaminated water from the damaged Fukushima Daiichi NPP into the ocean, as well as dry and wet depositions of the radionuclides released in air over the western North Pacific, considerably changed the radionuclide distribution patterns in sea water. The presence of the shorter lived ^{134}Cs (half-life 2.07 years) enabled the identification of waters labelled by releases from the Fukushima Daiichi NPP accident. Following the decay of ^{134}Cs, new distribution patterns of ^{3}H, ^{14}C and ^{137}Cs in the Pacific Ocean need to be established for future oceanographic studies.

In another study, Kaizer et al. [5.58] evaluated and compared data from three oceanographic expeditions conducted along the 149° E meridian in 1993, 2005 and 2012 to evaluate temporal changes in the concentrations of tritium and other radionuclides in the western waters of the North Pacific Ocean. They found that in 1993, the distribution of tritium concentration data in surface sea water and in the water column was quite uniform. These data turned out to be an important comparator when they resampled the waters in the winter of 2012 after the Fukushima Daiichi NPP accident. These post-accident samples, collected from the mid-latitudinal and northern part of the transect (10–40° N), showed tritium concentrations that were approximately 60% above the expected background values according to the previous studies. In 2012, these measured concentrations peaked at approximately 2.0 TU (240 Bq/m^3). The 2012 sampling expedition was conducted 10 months after the accident, allowing time for diffusion of tritium into the water columns. This resulted in tritium concentrations exceeding the 1993 data by 2–7 times (see Table 5.6). The ^{3}H inventory at the sampling station near 1.2° N was found to be 2–4 times higher than at the other two stations.

TABLE 5.6. WATER COLUMN INVENTORIES FOR TRITIUM FROM THE WESTERN PACIFIC [5.58]

Latitude	Longitude	Tritium inventory[a] (kBq/m^2)	
		1993	2012
1.20° N	146.50° E	10.30 ± 0.05	69.00 ± 10.1
2.00° S	145.60° E	10.30 ± 0.03	17.70 ± 1.4
3.50° S	144.90° E	11.40 ± 0.05	28.50 ± 1.9

[a] Mean ± standard deviation.

The background tritium level (decay corrected to June 2011) in the northwestern Pacific Ocean was estimated as 47 Bq/m^3 (0.4 ± 0.1 TU) [5.50]. Reportedly, 0.7 ± 0.3 PBq of the ^3H derived from the Fukushima Daiichi NPP accident was released in 2012 into the North Pacific and influenced surface and vertical ^3H profiles in sea water in the region 145–149° E, 10–40° N [5.56, 5.58, 5.59].

5.4.4. Indian Ocean

The IAEA project WOMARS [5.17] studied the distribution and behaviour of anthropogenic radionuclides in the world ocean. A series of expeditions in 1997–1999 in the Pacific and Indian oceans, Arabian Sea and Antarctic contributed data on the distribution of ^3H, ^{14}C, ^{90}Sr and ^{129}I in the surface waters of the Indian Ocean [5.49].

During the WOMARS project, a sampling campaign was carried out in March–April 1998 along a track from New Zealand to the Mediterranean Sea. The aim of the study was to determine the concentrations of anthropogenic radionuclides, including ^3H, in surface water as a function of latitude. The latitudinal radionuclide concentration showed the same trend as that found in the 1978 GEOSECS tritium data.

Comparing radionuclide concentrations in the northwestern sector of the Pacific Ocean with those observed in the northern and southern parts of the Indian Ocean led Povinec et al. [5.49] to conclude the following:

— The surface distributions of ^3H in the Indian Ocean are higher than expected from decay corrected values from the GEOSECS programme.

— Observed ^{3}H and ^{90}Sr concentrations are comparable with surface values found in the northwestern Pacific Ocean.
— The distribution of radionuclide concentrations in surface waters reveals a strong latitudinal variation defined by the water fronts.
— The observed ^{3}H and ^{90}Sr levels do not follow the global deposition pattern and are influenced by oceanographic circulation fronts.
— Surface ^{3}H and ^{14}C concentrations observed in the northwestern Pacific in 1997 are comparable to the Subtropical Gyre data (at 20–40° S) in the southern Indian Ocean. The higher radionuclide concentrations observed there are associated with the Subtropical Gyre, which acts as a reservoir of radionuclides transported from the northwestern Pacific Ocean via the Indonesian Seas to the Indian Ocean.

5.4.5. Arctic Ocean

A recent comprehensive review by Oms et al. [5.15] indicated that elevated tritium concentrations, on average 1.00 ± 0.91 TU_{16}, were found in the Arctic Ocean in the surface layer above 500 m depth. It was assumed that between 1952 and 2016, the total release of anthropogenic (artificial) radionuclides from the European nuclear reprocessing plants of Sellafield and La Hague entered the Arctic Ocean through marine transport. By 2016, this total contribution amounted to 203 PBq (570 g), taking into account radioactive decay.

For comparison, the mean annual release of European nuclear facilities in the English Channel from 1984 to 2016 was 11.5 PBq/a (32 g/a), which is twice as high as the natural flux in the Arctic Ocean. The tritium inventory in the Arctic Ocean can be apportioned as coming from the natural production (9%), from European nuclear reprocessing plants (16%) and from nuclear weapon tests (75%).

The relatively highly tritiated waters of the upper Arctic Ocean are transported out of the enclosed basin back into the North Atlantic Ocean via the East Greenland Current [5.60] and sink into the deep North Atlantic Ocean at convection sites [5.54]. By 2021, the residual component of atmospheric nuclear weapon test tritium can be considered as low relative to annual inputs and represents approximately 10% of the natural input.

5.4.6. Southern Ocean

Measurements of tritium in waters of the Southern Ocean were used to estimate formation rates and pathways of the bottom waters of the Southern Ocean; particularly, the Weddell Sea Bottom Water and the Antarctic Bottom Water [5.37, 5.53].

The lowest tritium concentrations, of 0.05 ± 0.04 TU_{16}, were found in the Southern Ocean according to 2280 data values [5.15]. It is hypothesized that the tritium concentration in the Southern Ocean is close to an apparent steady state natural tritium input and represents a close estimation of the natural tritium inputs in the upper ocean layer (0–500 m). The total inventory of tritium in the Southern Ocean shows a high uncertainty (748 ± 886 PBq) and, given these large uncertainty limits, encompasses the estimated natural inventory of 233 PBq.

5.4.7. Mediterranean Sea

The Mediterranean Sea has a surface area of about 2.5×10^6 km^2 and a mean depth of 1500 m. It is a mid-latitude, land-locked evaporative sea with many characteristics and processes that are similar to those observed in the world oceans; for example, exchanges of water, salt and heat. It is composed of two nearly equally sized basins connected by the Strait of Sicily. Its relatively small size and the interplay between freshwater input from rivers and the Black Sea and the narrow connection to the Atlantic Ocean through the Strait of Gibraltar make it a particularly sensitive natural laboratory to investigate climatic, environmental and anthropogenic changes. Tritium variations in waters of the western Mediterranean Sea, collected in 1981 as part of the Phycemed Cruise, were reported by Andrie and Merlivat [5.61]. Saline water of Mediterranean origin may affect water formation processes, variabilities and the stability of the global ocean thermohaline equilibrium state [5.62].

The tritium concentrations in the surface water of the Mediterranean Sea increased from approximately 0.3 TU in 1952 (before the atmospheric nuclear weapon test inputs) [5.41] to 20–30 TU in 1964 (at the peak time of atmospheric nuclear weapon tests), which represented a factor of about 100 above the natural levels. From 1964, the tritium concentrations declined systematically mostly owing to reduced anthropogenic inputs after the atmospheric nuclear test spike and subsequent radioactive decay [5.41].

Studies by Ayache et al. [5.62] modelled the tritium transient of the 1950s–1960s nuclear weapon tests using ^3He to develop a high resolution model. The formulation used for the tritium boundary conditions led to results of sufficient quality to be used for evaluating a Mediterranean circulation model. Observations of $^3H/^3He$ ratios available from a number of oceanographic cruises enabled the development of a decadal scale variability model for the thermohaline circulation in the Mediterranean Sea. The model correctly simulated the main features of the thermohaline circulation in the Mediterranean Sea, with mechanisms having a realistic timescale compared with evaluations from observations. The model estimates indicated similar values and time trends of tritium concentrations in both the western and eastern basins of the Mediterranean Sea from 1970 to 2010.

However, in the period of peak atmospheric nuclear testing in the early 1960s, the tritium concentrations in the western basin were approximately 1.5 times higher than in the eastern basin. In the western basin, the thermohaline circulation cell is driven by deep-water formation.

Tritium inputs to the Mediterranean Sea have been derived from various release sources that included species (HTO, OBT) from nuclear test fallout, discharges from several French nuclear sites (e.g. Marcoule) and industrial inputs from watchmaking workshops in France and Switzerland from 1962–2008. The latter are generally considered to exist as micrometre sized particulate organic matter (POM) variably locked up in refractory and/or resilient particulate residual matter [5.63, 5.64].

Tritium fluxes that flow from French non-nuclearized rivers towards the Mediterranean Sea ranged from 0.85 to 4.34 TBq/a, representing only 0.29–1.5% of total tritium fluxes delivered in this area. At the same time, these rivers contribute more than 10% of the water input to the sea. The Rhône river overall exports up to 293 TBq/a of tritium (i.e. 98.5–99.7%), which originated from France's various nuclear facilities and represents the main terrestrial source of tritium for the northwestern Mediterranean Sea. The main source terms in the Rhône valley are the Marcoule nuclear industrial site and the Bugey, Saint-Alban, Cruas and Tricastin NPPs. The Pierrelatte site and the Creys-Malville site have a lesser impact and are both currently undergoing decommissioning [5.65, 5.66].

5.5. TRITIUM DISPERSION FROM COASTAL NUCLEAR FACILITIES

5.5.1. Tritium dispersion from nuclear facilities

Nuclear facilities (NPPs, SFRPs, radiopharmaceutical and industrial facilities) are often located on coastal margins and estuaries and contribute directly to the marine inventory of radionuclides, including tritium. Some facilities located on shores or banks of freshwater bodies (lakes, rivers) or brackish waters (river estuaries, tidal marshes) contribute to marine pollution indirectly in a more complex way. Indicative releases of some nuclear facilities [5.67–5.74] in Europe and Japan are given in this section for illustrative and comparative purposes and are shown in Table 5.7. Other specific data on tritium liquid releases to the environment from various nuclear facilities can be found in Chapters 3, 4, 6 and 7 (e.g. Fig. 3.7 and Table 3.4).

TABLE 5.7. INDICATIVE LIQUID RELEASES (ROUTINE AND ACCIDENTAL) OF TRITIUM FOR NUCLEAR FACILITIES

Site	Tritium liquid release or release rate	Comment	Reference
Sellafield SFRP (UK)	3 PBq/a	Approximately 8 g of ^3H (2.8 PBq) has been released annually but the amount varies with time; see Fig. 5.7.	[5.69][a]
Severn Estuary NPPs (UK)	~11 PBq over 30 years	In general terms, liquid discharges of ^3H from the former Magnox sites at Oldbury, Berkeley and Hinkley B were similar in magnitude to those at the Cardiff site. For example, in 1997 releases from the Hinkley B NPPs was 620 TBq compared with 473 TBq from the Cardiff plant.	[5.66, 5.67, 5.69][a]
Radiopharmaceutical plant, Cardiff (UK)	9.3 PBq over 18 years	Maximum discharge of 820 TBq in 1986–1987. Technogenic organic compounds made up 20–30% of the total tritium releases until 1998 and about 80% after 1998. This radiopharmaceutical site ceased manufacturing in 2010.	[5.66, 5.67, 5.69, 5.70][a]
NPPs (France)	<0.3 PBq/a	Releases into rivers and the Mediterranean Sea (see Chapter 6).	[5.68][a]
La Hague SFRP (France)	10 PBq/a	Approximately 30 g of ^3H per year per 1600 t of nuclear fuel, but the amount varies with time; see Fig. 5.7.	[5.63, 5.65][a]
Fukushima Daiichi NPP (Japan)	0.3 ± 0.2 PBq Range 0.1–0.5 PBq	During normal operation, the Fukushima Daiichi NPP discharged tritium at a rate of ~22 TBq/a from its six units. At the time of the accident in March 2011, the total inventory of tritium, including the 3 damaged units, was estimated to be ~3.4 PBq. Only part of this inventory was released in the accident, while part of the inventory remained on the site. The NPP operator (TEPCO) estimated that 20–40 TBq of ^3H leaked into the sea in 2011–2013.	[5.50]

TABLE 5.7. INDICATIVE LIQUID RELEASES (ROUTINE AND ACCIDENTAL) OF TRITIUM FOR NUCLEAR FACILITIES (cont.)

Site	Tritium liquid release or release rate	Comment	Reference
Fukushima Daiichi NPP (Japan)	Inventory ~0.7 PBq (March 2023) Release rate ~0.2 PBq/a	After treatment in the ALPS system for removal of radionuclides, water with residual (unremovable) tritium is currently stored on site. Its gradual controlled discharges started in 2023 and will continue for 30 years.	TEPCO[b]
Rokkasho SFRP (Japan)	~1.8 PBq	The facility is not yet in full operation (start of operation provisionally planned for 2027)[c]. In 2006–2008, controlled releases of tritium were performed as part of the facility's pre-commissioning testing.	[5.70]
Tokai SFRP (Japan)	~4.5 PBq	The facility is currently closed, but in 1977–2007 tritium was released into the ocean, with 90% occurring before 1997.	[5.71]

[a] See also OSPAR reports and data at https://www.ospar.org/about/publications

[b] https://www.tepco.co.jp/en/hd/decommission/information/newsrelease/reference/pdf/2023/reference_20230928_01-e.pdf

[c] https://www.jnfl.co.jp/en/release/press/2024/detail/20241213-1.html

Note: ALPS: Advanced Liquid Processing System; NPP: nuclear power plant; SFRP: spent fuel reprocessing plant; TEPCO: Tokyo Electric Power Company Holding, Inc.

5.5.2. Tritium from spent fuel reprocessing plants in Europe

In Europe, the routine operations of SFRPs result in the most significant releases of anthropogenic tritium. The La Hague SFRP makes liquid releases into the English Channel; the Sellafield SFRP, before its closure in 2022, released into the Irish Sea; and the SFRP in the Marcoule nuclear industrial site (France) releases into the Rhône river and the Mediterranean Sea.

Tritium is mostly released when irradiated nuclear fuel is reprocessed and is then released in liquid effluents into coastal waters. For example, at the La Hague SFRP, every 600 t of reprocessed nuclear fuel material generates approximately

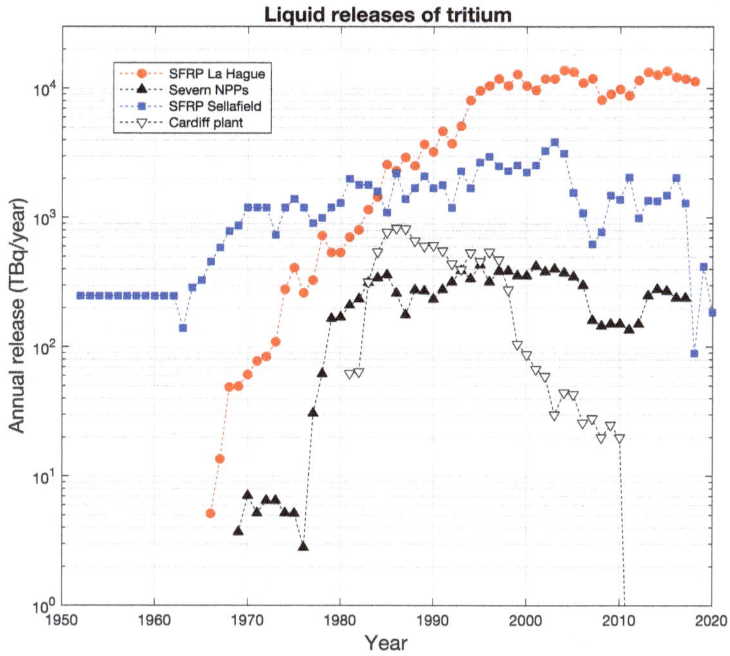

FIG. 5.7. Tritium liquid releases into northwestern European seas and estuaries from the Sellafield SFRP, La Hague SFRP, the Severn Estuary NPPs and from the former radiopharmaceutical plant in Cardiff. Data from Refs [5.69–5.71, 5.73, 5.74].

30 g/a (or 10 PBq/a) of tritium. By contrast, the Sellafield SFRP released about 8 g/a (or 2.8 PBq/a).

The two most significant SFRPs affecting coastal and oceanic waters of the northeastern Atlantic and Arctic oceans are the La Hague SFRP (1976–now) and the Sellafield site (1950–2022). The main current source of controlled tritium liquid discharges into the Atlantic marine environment is the La Hague SFRP on the Cotentin Peninsula, with a capacity to process approximately 1700 t of spent fuel per year. Its liquid discharge rate of tritium is estimated at approximately 10 PBq/a, which represents about 100 times the input from an NPP, with other sources contributing significantly less (Fig. 5.7) [5.67].

The dispersal behaviour and nature of the discharges from European reprocessing plants and other significant sources of marine tritium discharges have been studied through hydrodynamic modelling and monitoring [5.68, 5.69], which indicated a dominant eastward transfer through the English Channel and the Northern Sea along the northwestern coastal areas of Europe.

5.5.3. Tritium from spent fuel reprocessing plants in Japan

There are two SFRPs on the east coast of Japan, in Tokai (now closed) and in Rokkasho. By 2022, these facilities were sources of anthropogenic tritium and cumulatively discharged about 6.3 PBq into the western North Pacific Ocean [5.70, 5.71].

The Tokai SFRP of the Japan Atomic Energy Agency started trial operations in 1977 and was the first nuclear reprocessing plant in Japan. The plant had annual reprocessing capacity of 210 t of spent fuel, and from 1977 to 2008 it processed 1140 t of spent fuel. The annual reprocessing capacity was reportedly approximately 10% of that of the La Hague SFRP and Sellafield SFRPs. During its operation, a total of 4.5 PBq of tritium liquid effluent was discharged into the sea. Annual releases of tritium from the Tokai plant closely followed the amount of the processed spent fuel and varied in the range 0.01–0.5 PBq/a, with a peak value in 1994. Most of the tritium (87%) was released from this plant before 1997, while the remaining 13% was released over the period 2000–2007, with the maximum annual amount not exceeding 0.15 PBq [5.71]. The Tokai reprocessing plant ceased operations in 2007.

The Rokkasho SFRP of Japan Nuclear Fuel Limited in Aomori Prefecture lies at the northern end of Honshu Island (see Fig. 5.8). It was built as a successor to the smaller Tokai reprocessing plant. Construction began in 1993, and the plant was designed with an annual handling of 800 t of uranium or 8 t of plutonium. During the testing phase using spent nuclear fuel, small amounts of radionuclides were released into the atmosphere and the ocean episodically from 2006.

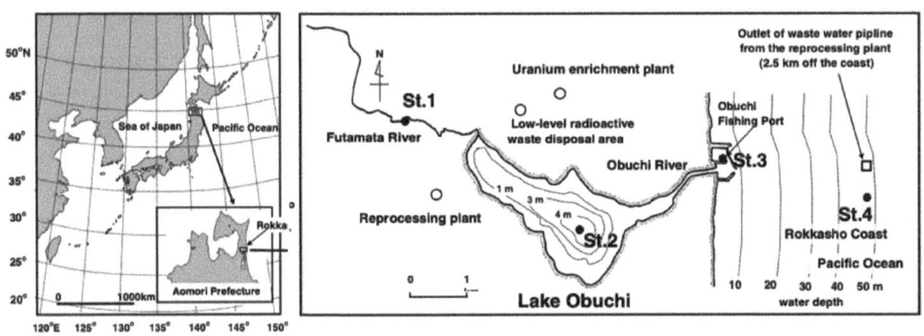

FIG. 5.8. Location of the Rokkasho reprocessing plant, freshwater and brackish lakes and the release point of radioactive waste. The numbered black circles (St.1–St.4) indicate locations where water samples were collected for determination of ^{3}H concentrations during test releases from Japan Nuclear Fuel Ltd. Copyright © American Nuclear Society; reprinted from Ref. [5.75] by permission of Taylor & Francis Ltd, http://www.tandfonline.com on behalf of American Nuclear Society.

Following the testing phase, the plant was originally intended to begin operating from 2013, but a series of delays have so far failed to allow the plant to operate.[1]

During the testing phase for Rokkasho SFPR, in 2005–2008, systematic releases of tritium occurred, varying from 1 to 500 TBq per month (see Fig. 5.9) [5.75]. Field sampling was systematically implemented along the coast, both upstream and downstream of the discharge pipe, as well on Lake Obuchi, which has brackish waters [5.75, 5.76]. Sample collection was coordinated with the release times of tritiated wastewater to understand dispersion processes (Fig. 5.9). The input of ^3H released from the Rokkasho SFRP to Lake Obuchi was evaluated, following sampling and hydrodynamic modelling, and certain amounts of ^3H were considered as produced by the reprocessing plant discharges into the Pacific Ocean during testing. As seen in Fig. 5.9, the activity concentration of tritium in the brackish lake Obuchi correlated with that on the coast of the Obuchi fishing port, thus indicating a likely input of tritium into the lake with tidal waters from the ocean.

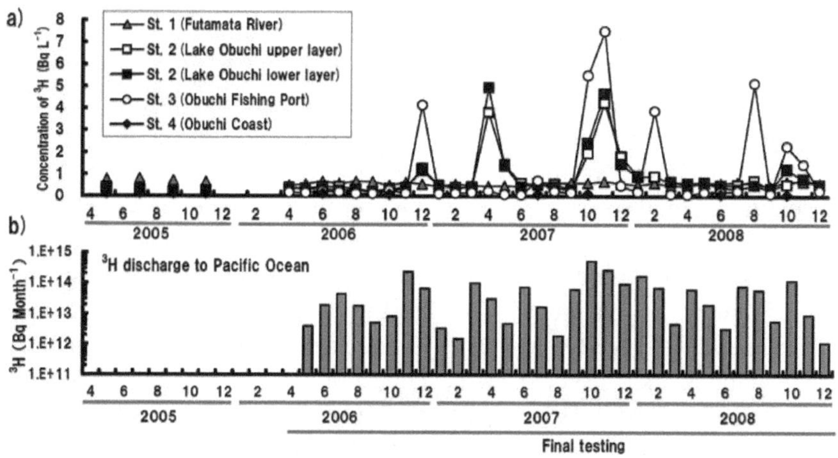

FIG. 5.9. Activity concentration of tritium in water (a) and tritium releases (b) from the Rokkasho spent fuel reprocessing plant during the testing phase in 2006–2008. The sampling point notations are shown in Fig. 5.8. Copyright © American Nuclear Society; reprinted from Ref. [5.75] by permission of Taylor & Francis Ltd, http://www.tandfonline.com on behalf of American Nuclear Society.

[1] https://www.world-nuclear-news.org/Articles/Japanese-utilities-revise-MOX-utilisation-plan

Muranaka et al. [5.77] investigated the effect of the amount and the timing of episodic releases of radioactive wastewater into the coastal waters of the eastern Aomori Peninsula during the early testing phase in 2006–2009. During testing, about 600 t of tritium-contaminated wastewater were released via the discharge point 2.5 km from the seashore. Most discharges occurred with an estimated inventory of the order of 10–100 TBq, with seawater samples being collected one day after release of tritium-contaminated wastewater. The activity concentrations of tritium in seawater samples collected a day after discharge were found to vary between the baseline levels of 0.2–0.3 Bq/L and 4 Bq/L.

A study reported by Shirotani et al. [5.78] showed that the Fukushima Daiichi NPP accident made virtually no impact on the activity concentrations of ^{3}H in surface ocean waters around Rokkasho.

5.5.4. Tritium from nuclear power plants

The normal operation of nuclear power reactors leads to the production of tritium in the fuel (ternary fission), control rods, burnable absorbers and coolant. However, releases of tritium due to the normal operation of NPPs are dwarfed by the releases from reprocessing plants (e.g. La Hague, Sellafield). For a 900 MW(e) LWR, the tritium release rate is about 0.01 g/a (0.02–0.03 PBq/a). Owing to use of heavy water as a moderator and coolant, the release of tritium from HWRs is substantially higher than that from reactors with light water design. HWRs are significant producers of tritium, which is generated by the neutron activation of the deuterium moderator. Depending on the design, a 900 MW(e) HWR can release up to 2.3 g/a (0.8 PBq/a) of tritium (see also Chapter 3 for more details).

Accidents at NPPs and subsequent remedial activities may result in significant releases of radioactive materials, including tritium. For example, the accident at the Fukushima Daiichi NPP in March 2011, followed by intensive remediation and decontamination works, resulted in the collection of approximately one million cubic metres of radioactively contaminated wastewater on the site. Successful decontamination of this water using the ALPS allowed the removal of many radionuclides, but not tritium. Following a request by the Japanese Government, the IAEA coordinated a special international task force to provide an independent expert advice and safety assessment of the intended disposal of the ALPS treated water into the Pacific Ocean [5.79]. The IAEA coordinated an independent sampling and interlaboratory comparison to corroborate the source and environmental monitoring data reported by the operator and regulator [5.80]. The planned gradual disposal of approximately 0.7 PBq of tritium in the ALPS treated water from the Fukushima Daiichi NPP can be put into context by comparing this activity with other natural and anthropogenic sources of tritium. For example, it is estimated that 8000 PBq of tritium from global atmospheric nuclear testing conducted in 1950–1960s remains in the oceans. Approximately 100 PBq of tritium is naturally produced annually by interaction of GCRs with nitrogen and oxygen in the atmosphere. NPPs emit tritium at rates that, depending on plant design, can reach values of several petabecquerels per year. Large coastal nuclear reprocessing plants, such as La Hague and Sellafield, can routinely produce liquid marine releases at rates of the order of 10 PBq/a (see Chapter 3 and Section 5.6.3).

5.5.5. Tritium from other coastal facilities

The former radiopharmaceutical complex in Cardiff (UK) can serve as an example of a facility that has released tritium-containing compounds into an estuarine environment. This coastal facility was a producer of radiolabelled pharmaceutical compounds for the life science sector from 1982 until 2010, and during that period released significant amounts of technogenic tritiated organic molecules (T-TOM), such as hydrocarbons, alcohols and amino acids, into the estuary [5.81]. Most releases (see Figs 5.7 and 5.10) occurred from 1982 to 1998, with a maximum of 820 TBq in 1986–1987. In the early years, releases occurred directly into macerated untreated sewage, which was then released into the estuary at high tide. This practice ceased in the late 1990s, when the Cardiff East Sewage Treatment Works was opened. It was estimated that 20–30% of the tritium releases up to 1998 may have been in the form of T-TOM, whereas it was closer to 80% T-TOM from 1998.

The environmental behaviour of T-TOM differs from that of OBT, where the latter is generally formed through biogenic processes [5.82]. At Cardiff, from 1998, it became recognized that at least some of the T-TOM released into the estuary were not biodegrading as had been expected and that bioaccumulation in marine biota was occurring. The observation that fish (flounder) were significantly accumulating tritium led the UK Environment Agency to demand a significant reduction and holdback of specific classes of components in the releases. They also directed the

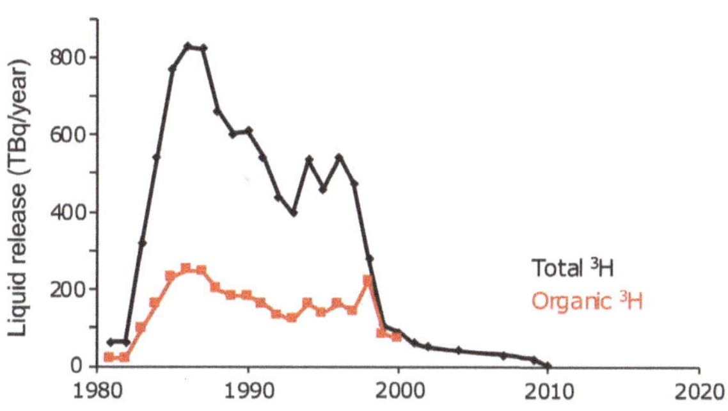

FIG. 5.10. Annual releases of total tritium and technogenic tritiated organic molecules from the Cardiff Maynard Centre in 1982–2010. The data are from Refs [5.73, 5.74].

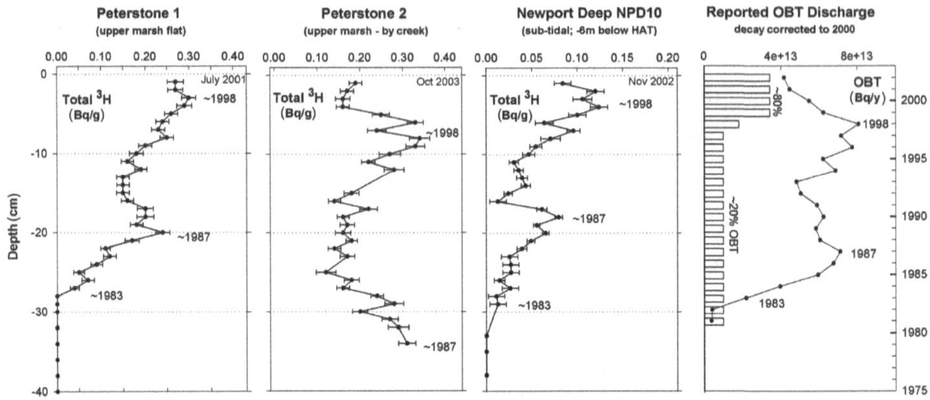

FIG. 5.11. Tritium profiles in Severn Estuary saltmarsh sediment cores from Peterstone saltmarsh and from a subtidal sediment core (Newport Deep NPD10); the OBT (T-TOM) discharge data and the relative proportion of OBT (T-TOM) to total tritium are shown as bars. All data are decay corrected to 2000. HAT: highest astronomical tide. Reprinted with permission from Ref. [5.74], Copyright 2012, American Chemical Society.

site management to initiate targeted scientific research to better understand the scale and radiological implications of the problem [5.81, 5.83–5.86]. One of the studies showed that accumulating estuarine sediment in intertidal and subtidal settings in the Severn Estuary retained a faithful record of the reported release pattern, indicating that biodegradation was likely to be limited (Fig. 5.11) [5.74]. By 2010, the owners of the Cardiff radiopharmaceutical plant decided to terminate manufacturing and began decommissioning the nuclear licensed site. By 2020, the site was formally delicensed.

5.6. BIOAVAILABILITY

5.6.1. Tritium in marine biota of the English Channel and the North Sea

In the English Channel, the magnitude of the La Hague SFRP releases dominates over those from the other nearby NPPs in northern France (Gravelines, Paluel, Penly and Flamanville). The tritium released from the La Hague SFRP is rapidly dispersed and diluted in sea waters around the Cotentin Peninsula and transported to the North Sea, so the concentrations of HTO and OBT in the English Channel marine environment are generally lower than 10 Bq/L. Such activity concentrations are not considered to pose any consequential hazard to biodiversity or human health (for comparison, the WHO limit for tritium in drinking water is equal to 10 000 Bq/L [5.87]).

The impact of tritium discharges from the La Hague SFRP on marine biota has been investigated over several years [5.67]. The observed relationships between spikes in SFRP discharges and tritium activity concentrations and exchange in sea water and seaweed provided helpful insights into the biokinetics of rapid uptake. The data [5.67] demonstrated that HTO in sea water and in seaweed quickly equilibrate, so rapid alterations of tritium activity concentrations in sea water are closely followed by similar changes in seaweed HTO content. By contrast, the OBT component was shown to be more stable. The OBT did not follow rapid environmental changes, but better represented the average tritium content in the environment. The data also indicated that HTO conversion to OBT by organisms is a slow process. The radiological impact to marine biota due to rapid and short pulse releases is more strongly created by HTO. These data provide important insights for all regions where HTO is discharged into the marine environment.

The activity concentrations of tritium in various marine organisms and ratios of OBT to HTO, as found off the coast of the Cotentin Peninsula [5.67], are given in Table 5.8.

The ICRP (International Commission on Radiological Protection) Derived Consideration Reference Level (DCRL) band for most marine biota ranges from 40 to 400 μSv/h [5.88]. The currently observed tritium activity concentration levels in the English Channel marine environment correspond to radiation doses 10^4–10^5 times lower than the biota reference levels.

TABLE 5.8. RATIOS OBT/HTO AND RANGES OF TRITIUM ACTIVITY CONCENTRATION IN MARINE BIOTA OFF THE COAST OF THE NORTHERN COTENTIN PENINSULA (NORMANDY, FRANCE) [5.67]

Biota	Sample size	Ratio OBT/HTO (mean ± SD)	Activity concentration of ^3H (Bq/kg fresh mass)
Seaweed	94	1.1 ± 0.4	1.4–26.4
Mollusc	19	1.2 ± 0.6	2.5–16.1
Crustacean	7	1.3 ± 0.7	1.8–10.4
Fish	7	1.5 ± 1.2	4–12.6
All	127	1.2 ± 0.5	1.4–26.4

5.6.2. Tritium in marine biota of the western Mediterranean Sea

In the western Mediterranean Sea, OBT measurements in filter-feeding organisms show that away from the influence of the releases from nuclear facilities (e.g. Rhône river deliveries), concentrations are uniform, at ~0.3–1.4 Bq/L of sample combustion water [5.72]. Levels are 3–10 times higher than HTO concentrations in the surface sea water of the western Mediterranean Sea (~0.1 Bq/L) and this leads to significant apparent OBT/HTO disequilibria. The OBT levels in filter-feeders are of a similar magnitude to those recorded in water and sediments or suspended matter in estuaries: 0.08 ± 0.01 to 0.50 ± 0.01 Bq/L for HTO and 1 ± 1 to 29 ± 17 Bq/L for OBT in sediments.

Near the Rhône river, OBT levels in filter-feeders are more variable, with values up to 6 Bq/L of sample combustion water [5.65]. Those levels are explained by both HTO concentrations in estuarine waters at the river mouth (0.2–2 Bq/L) and OBT concentrations in suspended particles transferred in the lower reaches of the Rhône river (about 40 Bq/L of combustion water). These results suggest that where HTO concentration in surface sea water is low (~0.1 Bq/L), the higher tritium inputs from the Rhône river and other small coastal rivers explain the elevated OBT in various marine organisms in coastal areas. It is recognized that further research is needed to establish the transfer pathways along the food chain. Tritium concentrations in the marine biota studied suggest significant transfer of tritium in an organically bound form either from suspended particles delivered by rivers or by autogenic sources (phytoplankton) produced near the river mouths. At present, it cannot be excluded that tritium associated with dissolved organic molecules (particularly nutrients) or submicrometric particles (technogenic tritium) exported by the Rhône river could also be transferred to these organisms [5.72].

OBT levels in Mediterranean Sea fish show concentrations that vary from 0.4 to 3.5 Bq/L of sample combustion water, which are similar to concentrations recorded in filter-feeder organisms collected away from the influence of nuclear facilities. Higher concentrations are recorded for samples taken in the Gulf of Lion, which is affected by deliveries from the Rhône river. This is different from the situation observed in the English Channel previously discussed, where OBT concentrations in biota are equilibrated with HTO concentrations in sea water [5.67]. Biota in the Mediterranean Sea (an oligotrophic sea) seem to be influenced by terrigenous nutrient inputs. In the English Channel, HTO concentrations are approximately 10 Bq/L, which is nearly 100 times higher than those in the Mediterranean sea water, and 10 times higher than those of the nutrients provided by river deliveries and those of fresh river waters [5.66, 5.72]. Inputs of allogenic tritium from rivers into the marine environment need to be taken into consideration, particularly in coastal areas, to credibly explain high OBT concentrations in living organisms.

5.6.3. Tritium in marine biota of the Irish Sea

The Sellafield nuclear site is located on the eastern margin of the Irish Sea on the northwest UK coastline. The results of routine regulatory monitoring around the Sellafield site presented in annual reports by UK governmental agencies [5.73] show an influence of tritium liquid discharges from the site on marine biota. The data (see Fig. 5.12 and Table 5.9) are reported as total tritium ($^3H_{Tot}$) (fresh weight), which does not specifically distinguish HTO, and OBT. It is highlighted that the reported tritium data strongly depend on the specific analytical procedure used [5.89] and may not be directly comparable with data produced by more widely used thermal desorption methods. At present, the applied methodology is distinctive and determines total tritium (tritiated water plus any OBT) in biota using chromic acid wet oxidation, distillation and liquid scintillation counting (LSC). Samples specifically analysed for OBT are dried at 40°C, then wet-oxidized with chromic acid, followed by distillation. Seawater samples (filtered) are analysed by double distillation followed by LSC of the distillate.

To investigate any enhancement of tritium concentrations by marine biota, Hunt et al. [5.89] compared regulatory monitoring data for the Sellafield site

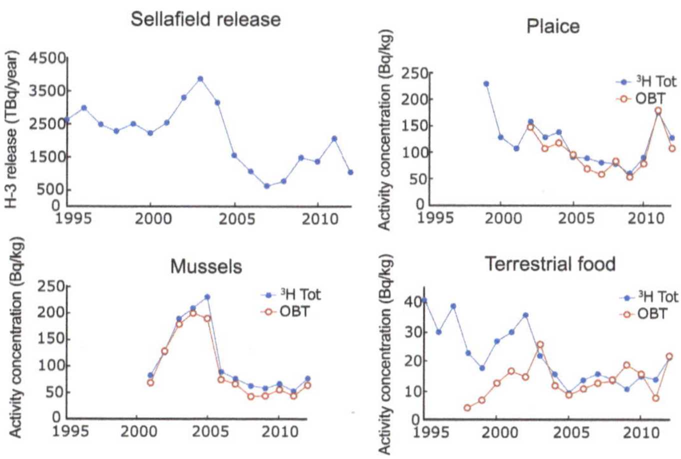

FIG. 5.12. Relationship between liquid total tritium discharges from the Sellafield site (UK) over time and concentrations of organic and total tritium in marine and terrestrial biota. Data from Ref. [5.73].

TABLE 5.9. ACTIVITY OF TRITIUM RELEASED FROM THE SELLAFIELD SITE AND FOUND IN SEA WATER AND MARINE BIOTA EITHER AS TOTAL OR ORGANICALLY BOUND TRITIUM *(adapted from Ref. [5.89])*

Year	$^3H_{Tot}$ release (TBq)	$^3H_{Tot}$ activity concentration in sea water (Bq/L)	Normalized activity concentration (Bq/L per TBq/a)	Plaice (*Pleuronectes platessa*) Activity concentration (Bq/kg fresh mass)		Plaice CR[a] (L/kg)	Mussels Activity concentration (Bq/kg fresh mass)		Mussels CR (L/kg)
				$^3H_{Tot}$	OBT	$^3H_{Tot}$	$^3H_{Tot}$	OBT	$^3H_{Tot}$
1999	2520	12	0.0048	212	—[b]	18	—	—	—
2000	2260	17	0.0075	140	—	8.2	—	—	—
2001	2560	17	0.0065	108	—	6.5	85	71	5.1
2002	3320	23	0.0069	168	145	7.3	130	130	5.7
2003	3900	21	0.0054	132	103	6.3	191	182	9.1
2004	3170	23	0.0073	205	153	8.9	210	202	9.1
2005	1570	19	0.0121	161	153	8.5	230	190	12
2006	1090	9	0.0083	122	105	14	91	78	10

[a] CR: concentration ratio, defined as a quotient of activity concentration in an organism (Bq/kg) and activity concentration in water (Bq/L).

[b] —: no data.

using concentration ratios (CRs; L/kg), of tritium activity concentrations in biota (Bq/kg fresh mass) and in water (Bq/L). If tritium exists as HTO, then at equilibrium conditions CR = 1 for marine biota and ambient sea water, unless other forms of tritium exist [5.90].

The close correlation between $^3H_{Tot}$ and OBT (Fig. 5.12) indicates that the marine biota sampled quickly accommodates to the changing tritium activity in water. The same observation applies less clearly to terrestrial foodstuffs, which seem to indicate a time lag. The activity concentrations for $^3H_{Tot}$ and OBT are

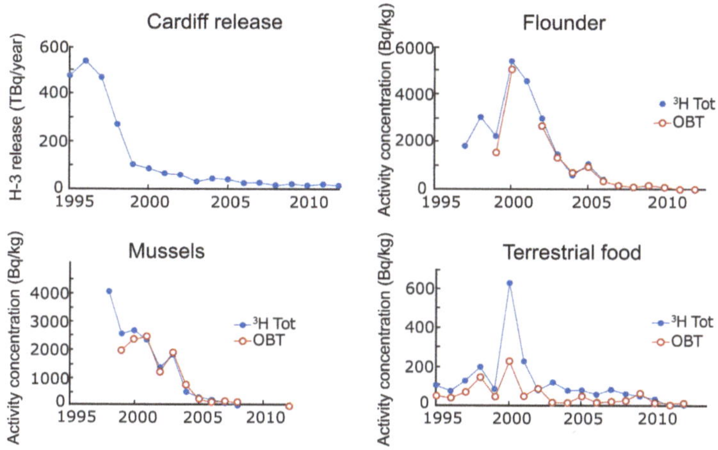

FIG. 5.13. Relationship between liquid tritium releases from the former Cardiff Maynard Centre (UK) over time and concentrations of total tritium and OBT in marine and terrestrial biota. Data from Refs [5.73, 5.91].

elevated — 100–200 Bq/kg fresh mass for plaice, 50–200 Bq/kg fresh mass for mussels and 10–20 Bq/kg fresh mass for terrestrial foodstuff — and reflect the high tritium (HTO) concentration in the liquid effluent released into sea water.

The data [5.73] show a small enhancement of CRs, reflecting a significant fraction of OBT relative to $^{3}H_{Tot}$ (Table 5.9). In plaice, for example, the fraction of OBT in total tritium ranges from 0.75 to 0.95 and CRs range from 6 to 18. This could be due to formation of organically bound complexes of tritium either in the effluents prior to their release or in the marine environment due to natural processes. Characterization of the OBT composition of the liquid effluents is required to identify whether organic complexation and sediment interactions might explain enhancement pathways.

Tritium activity concentrations in marine biota around the former radiopharmaceutical production site in Cardiff were substantially higher than those seen in similar biota investigated around the Sellafield site, which has a significantly higher tritium release (cf. Section 5.5.6 and Fig. 5.7; see Fig. 5.13 and Table 5.10). The Cardiff site ceased manufacturing in 2010 and was releasing T-TOM into the Severn Estuary between 1982 and 2010. The impact of T-TOM and their environmental fate became the subject of several studies [5.74, 5.81, 5.83–5.86, 5.89].

A significantly enhanced tritium uptake in the Cardiff marine environment was reported and assessed by Hunt et al. [5.89]. Large CRs in marine biota have been explained by the incorporation of a complex mixture of tritiated organic compounds in the liquid effluent. It is notable that up to the early 2000s, this

TABLE 5.10. ACTIVITY OF TRITIUM RELEASED FROM CARDIFF AREA AND FOUND IN SEA WATER AND MARINE BIOTA EITHER AS TOTAL OR ORGANICALLY BOUND TRITIUM
(adapted from Ref. [5.89])

Year	$^3H_{Tot}$ release (TBq)	$^3H_{Tot}$ activity concentration in sea water (Bq/L)	Normalized activity concentration (Bq/L per TBq/a)	Flounder (*Platichthys flesus*)			Mussels		
				Activity concentration (kBq/kg fresh mass)		CR[a] (L/kg)	Activity concentration (kBq/kg fresh mass)		CR (L/kg)
				$^3H_{Tot}$	OBT	$^3H_{Tot}$	$^3H_{Tot}$	OBT	$^3H_{Tot}$
1997	473	53	0.11	19	—[b]	358	—	—	—
1998	277	28	0.1	31	—	1107	41	—	1464
1999	105	9.2	0.09	23	16	2500	26	20	2826
2000	87.2	8.1	0.09	54	51	6667	27	24	3333
2001	67.2	6	0.09	46	n.d.	7667	24	25	4000
2002	59.5	4.9	0.08	30	27	6122	14	12	2857
2003	30.2	5.9	0.2	15	14	2542	19	19	3220
2004	44.4	5.3	0.12	6.6	7.6	1245	5.7	7.9	1075
2005	40.4	10	0.25	11	9.7	1100	3.3	3.0	330
2006	24.8	5	0.2	4.4	4.0	880	2.5	2.3	500

[a] CR: concentration ratio, defined as a quotient of activity concentration in an organism (Bq/kg) and activity concentration in water (Bq/L).

[b] —: no data.

liquid effluent was simply mixed with untreated macerated sewage before being released into the Severn Estuary at high tides (see Table 5.10) [5.89]. Regulatory monitoring of tritium in flounder and mussels for 1997–2006 [5.73] shows that CRs increased significantly around 2000 by a factor ranging from 20 to 7700, followed by a sharp decrease in 2004 by a factor of approximately 10. This was explained to be the result of a change in the chemical form of the organic tritium (T-TOM) that was released over that period. The data were consistent with the presence in the effluent up to 2001 of organic components more readily taken up by biota, followed by a decrease in such components.

The doses of radiation exposure due to consumption of seafood containing tritium from plant releases in 1995 were estimated to be approximately 0.01 mSv for a member of the critical population group [5.91, 5.92], thus being substantially below regulatory dose constraints [5.93].

The observational differences between OBT and HTO in biota seen around Sellafield, La Hague and Cardiff are provisionally explained by the respective composition of the liquid releases (HTO, OBT, T-TOM) into the coastal seas. However, other factors may have played a role, such as the specific local hydrodynamic and sedimentological characteristics of the three sites. For example, the liquid releases to sea water from La Hague are rapidly transported away in an easterly direction [5.94], whereas at Sellafield the releases may have longer residence time in the Irish Sea basin before being transported away. Locally, the composition of settled and suspended sediment may also have played a role. For example, around the Sellafield site, there is a distinct opportunity for the releases to interact with the accumulations of muddy sediments (Eastern Irish Sea Mud Patch), in contrast to the coarser sandy sediment found in the English Channel closer to the La Hague site.

5.6.4. Tritium in marine biota of the northwestern Pacific

The accident at the Fukushima Daiichi NPP, the continuing remedial activities and Japan's national plans for the safe disposal of residual tritium-containing water to the ocean have all resulted in increased attention to the possible radiological consequences of tritium. Povinec et al. [5.95], based on measurements of tritium in sea water and marine organisms in proximity to the Fukushima Daiichi NPP, reported the tissue-free water tritium (TFWT) levels in fish muscle to be within a range of 5.8–5.9 mBq/L and the tritium activity concentrations in sea water around Japan to be 70–98 mBq/L. OBT was found to be less than 0.28 Bq/kg wet weight. In the ocean, within the 20 km radius area from the Fukushima Daiichi NPP, the TFWT levels in 2019 were reportedly as 91–96 mBq/L in muscle of olive flounder, which was consistent with the tritium activity concentration levels of 75–100 mBq/L in sea water [5.95].

In 2023, the IAEA presented a comprehensive report [5.79] summarizing the results of an international safety review of the Japan national plans to gradually discharge ALPS treated water into the ocean. On the basis of the corroborated results of source and environmental monitoring [5.80] and the comprehensive review, the IAEA concluded that the radiological impact of the planned controlled releases would be very low, with conservatively estimated doses to public from all radionuclides, including tritium, being 100–1000 lower than the existing dose constraints. The IAEA recommended continued long term monitoring to verify the radiological impact estimates for people and the environment.

5.7. CONCLUSIONS

The global oceans are the major tritium reservoir, receiving natural (cosmogenic) and anthropogenic inputs through the hydrological cycle, which includes delayed transfers from the continents (weathering, rivers, lakes, meltwaters). The predominant tritium input component to the oceans was the injection of tritium from atmospheric nuclear weapon testing (from the mid-1950s to early 1960s), in the range 520–550 kg. This has now reduced, after five radioactive half-lives, to approximately 20 kg. Smaller ongoing and significant contributions also occur from civil nuclear, military and industrial discharges into rivers, estuaries and marginal and semi-enclosed seas.

The tritium introduced as a by-product of atmospheric nuclear weapon testing, along with its radiogenic noble gas progeny ^{3}He, provided two ideal hydrological tracers to investigate diverse oceanic processes. Exploitation of the immense potential of these tracers was accelerated and realized by the timely development of ultrasensitive He ingrowth mass spectrometers in the 1970s to measure the very low oceanic concentrations at high precision. International oceanographic campaigns rapidly incorporated and applied the new measurement technology to generate a database of tens of thousands of tritium and ^{3}He measurements to study the scale of horizontal and vertical mixing through the world ocean. These studies refined the understanding of ocean circulation and ventilation, biogeochemical processes, deep-ocean hydrothermal inputs, and the scale of CO_2 absorption and excess heat. These have contributed crucial insights into the role that the oceans play in climate regulation.

From the 2000s onwards, research focused on forms of anthropogenic tritium other than HTO that were inferred to arise in the environment from biological and technogenic processes. Additional data on these species (E-OBT and NE-OBT) can improve the understanding of the radiological impact of tritium on biological systems. Additionally, the release of other forms, such as technogenic tritium particles and tritiated organic molecules, was found to demonstrate persistence and longer residence in some marine and riverine sediment systems close to industrial discharge sources.

5.8. ACKNOWLEDGEMENTS

The technical artwork support provided by K. Davis (UK) is gratefully acknowledged.

REFERENCES TO CHAPTER 5

[5.1] CLARKE, W.B., JENKINS, W., TOP, Z., Determination of tritium by mass spectrometric measurement of ^{3}He, Int. J. Appl. Radiat. Isot. **27** 9 (1976) 515–522, https://doi.org/10.1016/0020-708X(76)90082-X

[5.2] MICHEL, R.L., Tritium inventories of the world oceans and their implications, Nature **263** (1976) 103–106, https://doi.org/10.1038/263103a0

[5.3] WEISS, W., ROETHER, W., The rates of tritium input to the world oceans, Earth Planet. Sci. Lett. **49** 2 (1980) 435–446, https://doi.org/10.1016/0012-821X(80)90084-9

[5.4] SARMIENTO, J.L., ROOTH, C., ROETHER, W., The North Atlantic tritium distribution in 1972, J. Geophys. Res. Oceans **87** C10 (1982) 8047–8056, https://doi.org/10.1029/JC087iC10p08047

[5.5] OMS, P.-E., Tritium in oceans: A compilation [dataset], PANGAEA (2018), https://doi.pangaea.de/10.1594/PANGAEA.892125

[5.6] JENKINS, W.J., et al., A comprehensive global oceanic dataset of helium isotope and tritium measurements, Earth Syst. Sci. Data **11** 2 (2019) 441–454, https://doi.org/10.5194/essd-11-441-2019

[5.7] STARK, S., STATHAM, P., STANLEY, R., JENKINS, W., Using tree ring cellulose as a tool to estimate past tritium inputs to the ocean, Earth Planet. Sci. Lett. **237** 3–4 (2005) 341–353, https://doi.org/10.1016/j.epsl.2005.06.040

[5.8] CAUQUOIN, A., JEAN-BAPTISTE, P., RISI, C., FOURRÉ, É., LANDAIS, A., Modeling the global bomb tritium transient signal with the AGCM LMDZ-iso: A method to evaluate aspects of the hydrological cycle, J. Geophys. Res. Atmos. **121** 21 (2016) 12612–12629, https://doi.org/10.1002/2016JD025484

[5.9] JENKINS, W.J., SMETHIE, W.M., Transient tracers track ocean climate signals, Oceanus (1 Dec. 1996), https://www.whoi.edu/oceanus/feature/transient-tracers-track-ocean-cimate-signals/

[5.10] BAYER, R., et al., "'Ultra low level' tritium measurements", Performance and Blank Components of a Mass Spectrometric System for Routine Measurement of Helium Isotopes and Tritium by the ^{3}He Ingrowth Method, Springer-Verlag, Berlin (1989) 39–40.

[5.11] LOTT, D., JENKINS, W., Advances in analysis and shipboard processing of tritium and helium samples, Int. WOCE Newsl. **30** (1998) 27–30.

[5.12] ÖSTLUND, H., WERNER, E., "The electrolytic enrichment of tritium and deuterium for natural tritium measurements", Tritium in the Physical and Biological Sciences, Vol. 1, IAEA, Vienna (1962) pp. 95–104.

[5.13] JENKINS, W.J., "Tritium–helium dating", Marine Chemistry and Geochemistry: A Derivative of the Encyclopedia of Ocean Sciences, 2nd edn (STEELE, J.H., THORPE, S.A., TUREKIAN, K.K., Eds), Academic Press, London (2001) 139–146.

[5.14] JENKINS, W.J., Studying subtropical thermocline ventilation and circulation using tritium and ^{3}He, J. Geophys. Res. Oceans **103** C8 (1998) 15817–15831, https://doi.org/10.1029/98JC00141

[5.15] OMS, P.-E., et al., Inventory and distribution of tritium in the oceans in 2016, Sci. Total Environ. **656** (2019) 1289–1303, https://doi.org/10.1016/j.scitotenv.2018.11.448

[5.16] BUTZIN, M., ROETHER, W., Tritium budget of the South Atlantic, J. Geophys. Res. Oceans **109** (2004) C03036, https://doi.org/10.1029/2002JC001561

[5.17] INTERNATIONAL ATOMIC ENERGY AGENCY, Worldwide Marine Radioactivity Studies (WOMARS): Radionuclide Levels in Oceans and Seas, IAEA-TECDOC-1429, IAEA, Vienna (2005).

[5.18] INTERNATIONAL ATOMIC ENERGY AGENCY, Global Marine Radioactivity Database (GLOMARD), IAEA-TECDOC-1146, IAEA, Vienna (2000).

[5.19] CHAPMAN, P., The world ocean circulation experiment (WOCE), Mar. Technol. Soc. J. **32** 3 (1998) 23–36.

[5.20] LIBBY, W., The potential usefulness of natural tritium, Proc. Natl Acad. Sci. **39** 4 (1953) 245–247, https://doi.org/10.1073/pnas.39.4.245

[5.21] CRAIG, H., LAL, D., The production rate of natural tritium, Tellus A **13** 1 (1961) 85–105, https://doi.org/10.3402/tellusa.v13i1.9430

[5.22] MICHEL, R., SUESS, H., Bomb tritium in the Pacific Ocean, J. Geophys. Res. **80** 30 (1975) 4139–4152, https://doi.org/10.1029/JC080i030p04139

[5.23] INTERNATIONAL ATOMIC ENERGY AGENCY, Global Network of Isotopes in Precipitation (GNIP), https://www.iaea.org/services/networks/gnip

[5.24] NIR, A., KRUGER, S.T., LINGENFELTER, R.E., FLAMM, E.J., Natural tritium, Rev. Geophys. **4** 4 (1966) 441–456, https://doi.org/10.1029/RG004i004p00441

[5.25] VON FALTINGS, V., HARTECK, P., Der Tritiumgehalt der Atmosphäre, Z. Naturforsch. A **5** 8 (1950) 438, https://doi.org/10.1515/zna-1950-0804

[5.26] GROSSE, A.V., JOHNSTON, W.M., WOLFGANG, R.L., LIBBY, W.F., Tritium in nature, Science **113** (1951) 2923,
https://doi.org/10.1126/science.113.2923.1

[5.27] KAUFMAN, S., LIBBY, W.F., The natural distribution of tritium, Phys. Rev. **93** 6 (1954) 1337–1344,
https://doi.org/10.1103/PhysRev.93.1337

[5.28] PALCSU, L., et al., Modulation of cosmogenic tritium in meteoric precipitation by the 11-year cycle of solar magnetic field activity, Sci. Rep. **8** 1 (2018) 12813,
https://doi.org/10.1038/s41598-018-31208-9

[5.29] STUIVER, M., QUAY, P.D., Changes in atmospheric carbon-14 attributed to a variable sun, Science **207** 4426 (1980) 11–19,
https://doi.org/10.1126/science.207.4426.11

[5.30] UNITED NATIONS, Sources and Effects of Ionizing Radiation, 1977 Report of the United Nations Scientific Committee on the Effects of Atomic Radiation (UNSCEAR) to the General Assembly, with Scientific Annexes, UN, New York (1977).

[5.31] MASARIK, J., REEDY, R.C., Terrestrial cosmogenic-nuclide production systematics calculated from numerical simulations, Earth Planet. Sci. Lett. **136** 3 (1995) 381–395,
https://doi.org/10.1016/0012-821X(95)00169-D

[5.32] O'BRIEN, K., Secular variations in the production of cosmogenic isotopes in the Earth's atmosphere, J. Geophys. Res. Space Phys. **84** A2 (1979) 423–431,
https://doi.org/10.1029/JA084iA02p00423

[5.33] MASARIK, J., BEER, J., An updated simulation of particle fluxes and cosmogenic nuclide production in the Earth's atmosphere, J. Geophys. Res. Atmos. **114** (2009) D11103,
https://doi.org/10.1029/2008JD010557

[5.34] MASARIK, J., BEER, J., Simulation of particle fluxes and cosmogenic nuclide production in the Earth's atmosphere, J. Geophys. Res. Atmos. **104** D10 (1999) 12099–12111,
https://doi.org/10.1029/1998JD200091

[5.35] POLUIANOV, S., KOVALTSOV, G.A., USOSKIN, I., A new full 3-D model of cosmogenic tritium ^{3}H production in the atmosphere (CRAC: 3H), J. Geophys. Res. Atmos. **125** 18 (2020) e2020JD033147,
https://doi.org/10.1029/2020JD033147

[5.36] EHHALT, D.H., ROHRER, F., SCHAUFFLER, S., POLLOCK, W., Tritiated water vapor in the stratosphere: Vertical profiles and residence time, J. Geophys. Res. Atmos. **107** D24 (2002) 4757,
https://doi.org/10.1029/2001JD001343

[5.37] WEISS, W., BULLACHER, J., ROETHER, W., "Evidence of pulsed discharges of tritium from nuclear energy installations in Central European precipitation", Proc. Behaviour of Tritium in the Environment, San Francisco, 1978, IAEA, Vienna (1979) 17–39,
https://inis.iaea.org/records/y4hp7-13p66

[5.38] BERGKVIST, N.-O., FERM, R., Nuclear Explosions 1945–1998, Defence Research Establishment, Stockholm (2000).

[5.39] DREISIGACKER, E., ROETHER, W., Tritium and ^{90}Sr in North Atlantic surface water, Earth Planet. Sci. Lett. **38** 2 (1978) 301–312, https://doi.org/10.1016/0012-821X(78)90104-8

[5.40] DONEY, S.C., GLOVER, D.M., JENKINS, W.J., A model function of the global bomb tritium distribution in precipitation, 1960–1986, J. Geophys. Res. Oceans **97** C4 (1992) 5481–5492, https://doi.org/10.1029/92JC00015

[5.41] ROETHER, W., JEAN-BAPTISTE, P., FOURRÉ, E., SÜLTENFUSS, J., The transient distributions of nuclear weapon-generated tritium and its decay product ^{3}He in the Mediterranean Sea, 1952–2011, and their oceanographic potential, Ocean Sci. **9** 5 (2013) 837–854, https://doi.org/10.5194/os-9-837-2013

[5.42] MISKEL, J.A., "Production of tritium by nuclear weapons", Tritium (MOGHISSI, A.A., CARTER, M.W., Eds), Messenger Graphics, Las Vegas (1973) 79–85.

[5.43] MASON, A., HUT, G., TELEGADAS, K., Stratospheric HTO and ^{95}Zr residence times, Tellus **34** 4 (1982) 369–375, https://doi.org/10.3402/tellusa.v34i4.10823

[5.44] MICHEL, R.L., JURGENS, B.C., YOUNG, M.B., Tritium Deposition in Precipitation in the United States, 1953–2012, Rep. 2018–5086, US Geological Survey, Reston, VA (2018), https://doi.org/10.3133/sir20185086

[5.45] ÖSTLUND, H., FINE, R., "Oceanic distribution and transport of tritium", Proc. Behaviour of Tritium in the Environment, San Francisco, 1978, IAEA, Vienna (1979).

[5.46] JEAN-BAPTISTE, P., et al., ^{3}He mass spectrometry for very low-level measurement of organic tritium in environmental samples, J. Environ. Radioact. **101** 2 (2010) 185–190, https://doi.org/10.1016/j.jenvrad.2009.10.005

[5.47] SÜLTENFUSS, J., ROETHER, W., RHEIN, M., The Bremen mass spectrometric facility for the measurement of helium isotopes, neon, and tritium in water, Isot. Environ. Health Stud. **45** 2 (2009) 83–95, https://doi.org/10.1080/10256010902871929

[5.48] UNITED NATIONS, Exposures from natural radiation sources, 2000 Report of the United Nations Scientific Committee on the Effects of Atomic Radiation (UNSCEAR) to the General Assembly, with Scientific Annexes, Annex B, UN, New York (2000).

[5.49] POVINEC, P. et al., Tritium, radiocarbon, ^{90}Sr and ^{129}I in the Pacific and Indian Oceans, Nucl. Instrum. Methods Phys. Res. B **268** 7–8 (2010) 1214–1218, https://doi.org/10.1016/j.nimb.2009.10.136

[5.50] POVINEC, P., et al., Impact of the Fukushima accident on tritium, radiocarbon and radiocesium levels in seawater of the western North Pacific Ocean: A comparison with pre-Fukushima situation, J. Environ. Radioact. **166** (2017) 56–66, https://doi.org/10.1016/j.jenvrad.2016.02.027

[5.51] ÖSTLUND, H.G., DORSEY, H.G., ROOTH, C.G., GEOSECS North Atlantic radiocarbon and tritium results, Earth Planet. Sci. Lett. **23** 1 (1974) 69–86, https://doi.org/10.1016/0012-821X(74)90033-8

[5.52] JENKINS, W., CLARKE, W., The distribution of ^{3}He in the western Atlantic Ocean, Deep-Sea Res. Oceanogr. Abstr. **23** (1976) 481–494, https://doi.org/10.1016/0011-7471(76)90860-3

[5.53] JENKINS, W.J., ^{3}H and ^{3}He in the beta triangle: Observations of gyre ventilation and oxygen utilization rates, J. Phys. Oceanogr. **17** 6 (1987) 763–783, https://doi.org/10.1175/1520-0485(1987)017<0763:AITBTO>2.0.CO;2

[5.54] DONEY, S.C., JENKINS, W.J., Ventilation of the deep western boundary current and abyssal western North Atlantic: Estimates from tritium and ^{3}He distributions, J. Phys. Oceanogr. **24** 3 (1994) 638–659, https://doi.org/10.1175/1520-0485(1994)024<0638:VOTDWB>2.0.CO;2

[5.55] BUESSELER, K.O., Opening the floodgates at Fukushima, Science **369** 6504 (2020) 621–622, https://doi.org/10.1126/science.abc1507

[5.56] MEN, W., Discharge of contaminated water from the Fukushima Daiichi Nuclear Power Plant Accident into the Northwest Pacific: What is known and what needs to be known, Mar. Pollut. Bull. **173** (2021) 112984, https://doi.org/10.1016/j.marpolbul.2021.112984

[5.57] BUESSELER, K.O., et al., Fukushima-derived radionuclides in the ocean and biota off Japan, Proc. Natl. Acad. Sci. **109** 16 (2012) 5984–5988, https://doi.org/10.1073/pnas.1120794109

[5.58] KAIZER, J., KUMAMOTO, Y., MOLNÁR, M., PALCSU, L., POVINEC, P.P., Temporal changes in tritium and radiocarbon concentrations in the western North Pacific Ocean (1993–2012), J. Environ. Radioact. **218** (2020) 106238, https://doi.org/10.1016/j.jenvrad.2020.106238

[5.59] KAIZER, J., et al., Tritium and radiocarbon in the western North Pacific waters: Post-Fukushima situation, J. Environ. Radioact. **184–185** (2018) 83–94, https://doi.org/10.1016/j.jenvrad.2018.01.015

[5.60] DORSEY, H., PETERSON, W., Tritium in the Arctic Ocean and East Greenland Current, Earth Planet. Sci. Lett. **32** 2 (1976) 342–350, https://doi.org/10.1016/0012-821X(76)90074-1

[5.61] ANDRIE, C., MERLIVAT, L., Tritium in the western Mediterranean Sea during 1981 Phycemed cruise, Deep-Sea Res. A **35** 2 (1988) 247–267, https://doi.org/10.1016/0198-0149(88)90039-8

[5.62] AYACHE, M., et al., Modelling of the anthropogenic tritium transient and its decay product helium-3 in the Mediterranean Sea using a high-resolution regional model, Ocean Sci. **11** 3 (2015) 323–342, https://doi.org/10.5194/os-11-323-2015

[5.63] EYROLLE, F., et al., Evidence for tritium persistence as organically bound forms in river sediments since the past nuclear weapon tests, Sci. Rep. **9** 1 (2019) 11487, https://doi.org/10.1038/s41598-019-47821-1

[5.64] MOREREAU, A., et al., Trajectories of technogenic tritium in the Rhône River (France), J. Environ. Radioact. **223** (2020) 106370,
https://doi.org/10.1016/j.jenvrad.2020.106370

[5.65] EYROLLE-BOYER, F., ANTONELLI, C., RENAUD, P., TOURNIEUX, D., Origins and trend of radionuclides within the lower Rhône River over the last decades, Radioprotection **50** 1 (2015) 27–34,
https://doi.org/10.1051/radiopro/2014022

[5.66] DUCROS, L., et al., Tritium in river waters from French Mediterranean catchments: Background levels and variability, Sci. Total Environ. **612** (2018) 672–682,
https://doi.org/10.1016/j.scitotenv.2017.08.026

[5.67] FIÉVET, B., et al., Transfer of tritium released into the marine environment by French nuclear facilities bordering the English Channel, Environ. Sci. Technol. **47** 12 (2013) 6696–6703,
https://doi.org/10.1021/es400896t

[5.68] BAILLY DU BOIS, P., Dispersion of Radionuclides in the European North-Western Seas: Observations and Modelling, Université de Rouen, Rouen (2013).

[5.69] BAILLY DU BOIS, P., et al., Dissolved radiotracers and numerical modeling in North European continental shelf dispersion studies (1982–2016): Databases, methods and applications, Water **12** 6 (2020) 1667,
https://doi.org/10.3390/w12061667

[5.70] ANZAI, K., et al., "Radioactive effluent releases from Rokkasho reprocessing plant (2): liquid effluent", International Congress of the International Radiation Protection Association (IRPA): Strengthening Radiation Protection Worldwide (IRPA 12), 2008, Buenos Aires (2008),
https://inis.iaea.org/records/wedqz-sdx51

[5.71] KOKUBUN, Y., FUJITA, H., NAKANO, M., SUMIYA, S., Tritium concentration and diffusion in seawater discharged from Tokai reprocessing plant, Prog. Nucl. Sci. Technol. **1** (2011) 384–387,
https://doi.org/10.15669/pnst.1.384

[5.72] EYROLLE, F., et al., An updated review on tritium in the environment, J. Environ. Radioact. **181** (2018) 128–137,
https://doi.org/10.1016/j.jenvrad.2017.11.001

[5.73] ENVIRONMENT AGENCY, FOOD STANDARDS AGENCY, FOOD STANDARDS SCOTLAND, NATURAL RESOURCES WALES, NORTHERN IRELAND ENVIRONMENT AGENCY, SCOTTISH ENVIRONMENT PROTECTION AGENCY, Radioactivity in Food and the Environment, 2021: RIFE-27, CEFA, London (2022),
https://webarchive.nationalarchives.gov.uk/ukgwa/20231023170654mp_/https://assets.publishing.service.gov.uk/government/uploads/system/uploads/attachment_data/file/1114928/RIFE-27-Radioactivity-in-food-and-the-environment-2021.pdf

[5.74] CROUDACE, I.W., WARWICK, P.E., MORRIS, J.E., Evidence for the preservation of technogenic tritiated organic compounds in an estuarine sedimentary environment, Environ. Sci. Technol. **46** 11 (2012) 5704–5712,
https://doi.org/10.1021/es204247f

[5.75] UEDA, S., KAKIUCHI, H., HASEGAWA, H., HISAMATSU, S., Validation of a radionuclide transfer model in brackish lake, Fusion Sci. Technol. **60** 4 (2011) 1296–1299,
https://doi.org/10.13182/FST11-A12668

[5.76] UEDA, S., KONDO, K., INABA, J., KUTSUKAKE, H., NAKATA, K., Development and application of an eco-hydrodynamic model for radionuclides in a brackish lake: Case study of Lake Obuchi, Japan, bordered by nuclear fuel cycling facilities, J. Radioanal. Nucl. Chem. **268** 2 (2006) 261–273,
https://doi.org/10.1007/s10967-006-0163-0

[5.77] MURANAKA, T., YAMASHITA, J., SHIMA, N., Variation of tritium concentration in coastal seawater collected along the pacific coast in Aomori prefecture, Fusion Sci. Technol. **60** 4 (2011) 1264–1267,
https://doi.org/10.13182/FST11-A12660

[5.78] SHIROTANI, Y., INATOMI, N., MIYAMOTO, K., YAMADA, M., KUSAKABE, M., Distributions of tritium in the coastal waters off Aomori and Iwate Prefectures, Fusion Eng. Des. **172** (2021) 112738,
https://doi.org/10.1016/j.fusengdes.2021.112738

[5.79] INTERNATIONAL ATOMIC ENERGY AGENCY, IAEA Comprehensive Report on the Safety Review of the ALPS Treated Water at the Fukushima Daiichi Nuclear Power Station, IAEA, Vienna (2023),
https://www.iaea.org/sites/default/files/iaea_comprehensive_alps_report.pdf

[5.80] INTERNATIONAL ATOMIC ENERGY AGENCY, IAEA Review of Safety Related Aspects of Handling ALPS Treated Water at TEPCO's Fukushima Daiichi Nuclear Power Station, IAEA, Vienna (2023),
https://www.iaea.org/sites/default/files/first_interlaboratory_comparison_on_the_determination_of_radionuclides_in_alps_treated_water.pdf

[5.81] LAMBERT, B., Welsh tritium, J. Radiol. Prot. **21** 4 (2001) 333,
https://doi.org/10.1088/0952-4746/21/4/002

[5.82] KIM, S.B., BAGLAN, N., DAVIS, P.A., Current understanding of organically bound tritium (OBT) in the environment, J. Environ. Radioact. **126** (2013) 83–91,
https://doi.org/10.1016/j.jenvrad.2013.07.011

[5.83] McCUBBIN, D., LEONARD, K.S., BAILEY, T.A., WILLIAMS, J., TOSSELL, P., Incorporation of organic tritium (^{3}H) by marine organisms and sediment in the Severn Estuary/Bristol Channel (UK), Mar. Pollut. Bull. **42** 10 (2001) 852–863,
https://doi.org/10.1016/S0025-326X(01)00039-X

[5.84] WILLIAMS, J., RUSS, R., McCUBBIN, D., KNOWLES, J., An overview of tritium behaviour in the Severn Estuary (UK), J. Radiol. Prot. **21** 4 (2001) 337,
https://doi.org/10.1088/0952-4746/21/4/301

[5.85] HODGSON, A., SCOTT, J., FELL, T., HARRISON, J., Doses from the consumption of Cardiff Bay flounder containing organically bound tritium, J. Radiol. Prot. **25** 2 (2005) 149,
https://doi.org/10.1088/0952-4746/25/2/003

[5.86] HARRISON, J., Doses and risks from tritiated water and environmental organically bound tritium, J. Radiol. Prot. **29** 3 (2009) 335, https://doi.org/10.1088/0952-4746/29/3/S02

[5.87] WORLD HEALTH ORGANIZATION, Guidelines for Drinking-Water Quality, WHO, Geneva (2022).

[5.88] INTERNATIONAL COMMISSION ON RADIOLOGICAL PROTECTION, Environmental Protection – the Concept and Use of Reference Animals and Plants, Publication 108, Elsevier, Oxford (2008).

[5.89] HUNT, G., BAILEY, T., JENKINSON, S., LEONARD, K., Enhancement of tritium concentrations on uptake by marine biota: experience from UK coastal waters, J. Radiol. Prot. **30** 1 (2010) 73, https://doi.org/10.1088/0952-4746/30/1/N01

[5.90] INTERNATIONAL ATOMIC ENERGY AGENCY, Sediment Distribution Coefficients and Concentration Factors for Biota in the Marine Environment, Technical Reports Series No. 422, IAEA, Vienna (2004).

[5.91] ENVIRONMENT AGENCY, ENVIRONMENT AND HERITAGE SERVICE, FOOD STANDARDS AGENCY, SCOTTISH ENVIRONMENT PROTECTION AGENCY, Radioactivity in Food and the Environment, 2005: RIFE-11, CEFA, London (2006).

[5.92] MINISTRY OF AGRICULTURE, FISHERIES AND FOOD, Radioactivity in Food and the Environment, 1995: RIFE-1, Ministry of Agriculture, Fisheries and Food, London (1996).

[5.93] INTERNATIONAL ATOMIC ENERGY AGENCY, Radiation Protection and Safety of Radiation Sources: International Basic Safety Standards, IAEA Safety Standards Series No. GSR Part 3, IAEA, Vienna (2014), https://doi.org/10.61092/iaea.u2pu-60vm

[5.94] BAILLY DU BOIS, P., DUMAS, F., Fast hydrodynamic model for medium- and long-term dispersion in seawater in the English Channel and southern North Sea, qualitative and quantitative validation by radionuclide tracers, Ocean Model. **9** 2 (2005) 169–210, https://doi.org/10.1016/j.ocemod.2004.07.004

[5.95] POVINEC, P., HIROSE, K., TATEDA, Y., Fukushima Accident: 10 Years Later, Elsevier, Amsterdam (2021).

Chapter 6

TRITIUM IN FRESHWATER SYSTEMS

F. EYROLLE, V. VOSTROTIN, I.W. CROUDACE

6.1. INTRODUCTION

Among the waters of the Earth, 97.5% are saline oceanic and/or marine water and only 2.5% are fresh water, representing approximately 35.2×10^{15} m^3 (35.2×10^6 km^3). Fresh water is distributed between glaciers (68.7%), groundwater (30.1%), permafrost (0.8%), surface water (lakes and rivers) and the atmosphere (0.4%). Less than 1% of the water on the Earth is fresh water in liquid form. Lake Baikal in Siberia, with a volume of 23.6×10^{12} m^3, constitutes the largest reservoir of liquid fresh water on the Earth; in comparison, the combined volume of the Great Lakes in North America is smaller, at 22.7×10^{12} m^3. The distribution of water between the various environmental pools on the Earth is illustrated in Fig. 6.1, while the numerical values are given in Table 6.1 [6.1].

Fresh water is either locked up in ice caps and mountainous glaciers (68.7%) or found in the ground as underground fresh water (30.1%) or as surface or near-surface fresh waters (1.2%). Most of the surface or near-surface fresh water is

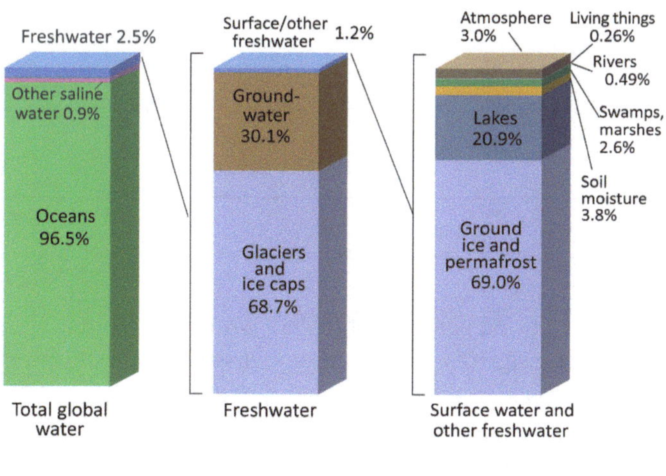

FIG. 6.1. Freshwater reservoirs on the Earth. Data from Ref. [6.1]. Credit: US Water Science School, https://www.usgs.gov/special-topic/water-science-school.

TABLE 6.1. GENERALIZED ESTIMATES OF GLOBAL WATER DISTRIBUTION [6.1]

Water type or location	Water volume (km^3)	Global water fraction (%)	Fresh water fraction (%)
Oceans, seas and bays	1 338 000 000	96.5	—[a]
Ice caps, glaciers and permanent snow	24 064 000	1.74	68.6
Groundwater	23 400 000	1.7	—
Fresh	10 530 000	0.76	30.1
Saline	12 870 000	0.93	—
Soil moisture	16 500	0.001	0.05
Ground, ice and permafrost	300 000	0.022	0.86
Lakes	176 400	0.013	—
Fresh	91 000	0.007	0.26
Saline	85 400	0.007	—
Atmosphere	12 900	0.001	0.04
Swamps and marshes	11 470	0.000 8	0.03
Rivers	2 120	0.000 2	0.006
Biological water	1 120	0.000 1	0.003

[a] —: not applicable.

locked up in ground ice and permafrost (69%), while approximately 21% is found in lakes. Of the remaining 10% of surface or near-surface fresh water, soil waters, swamps and marshes account for 6.4% and the atmosphere for 3.0%, while the residual is distributed between rivers (0.49%) and biota (0.26%).

Freshwater systems, including rivers, lakes and groundwater, are transitory sinks for meteoric waters (i.e. rainfall) before recycling through solar forcing. Until recently, these freshwater natural reservoirs were associated with residence times varying from days, for the atmosphere and rivers, to decades, for lakes. This compares with approximately 3 000 years in oceans and almost 10 000 years in glaciers and ice sheets.

Tritium produced in the upper atmosphere is almost totally oxidized to produce tritiated water and is delivered to the Earth's surface by precipitation. Freshwater reservoirs constitute intermediate repositories for tritium supplied from the atmosphere, with most of the tritium eventually being stored in the oceans. In freshwater systems, tritium mostly exists as free and/or mobile and poorly reactive HTO (>99%), with its mobility being controlled by hydrological exchanges and transport. Tritium levels in continental surface waters are generally higher than those in the ocean surfaces [6.2–6.8]. This is due to the relatively small dilution following mixing with surface water over land masses, compared with the potential for very large dilution in oceanic waters (see Chapter 5 for further information).

Most tritium in freshwater systems falls on the Earth's surface as precipitation, with inputs being influenced by anthropogenic releases, geographic latitude, altitude and season (see Chapter 4). Additionally, the residence of water masses in drainage basins, lakes and groundwaters, as well as water mass mixing and exchanges, leads to various dilution processes. In this context, tritium as HTO has been widely used as an ideal hydrological tracer to study water transfer from surface to subsurface, and as an age indicator for groundwater after the 1950s [6.9, 6.10].

Liquid releases of tritium from NPPs are defined by the design and construction of the plants. While most (>99%) tritium releases from PWRs are in the form of liquid effluents, only small amounts of tritium are released directly to fresh waters from BWRs and PHWRs, and most tritium from such facilities is first released to the atmosphere. Therefore, activity concentrations of tritium in rivers downstream of NPPs are often found to be low and show temporal variations mostly due to dispersion and dilution of the liquid releases along the water course.

Numerical models are often used to describe tritium transfer and exchanges with surrounding water bodies, including both interstitial waters in sediments and water vapour in the atmosphere. Exchanges of surface waters with the atmosphere are governed by evaporation and condensation processes, which may lead to significant tritium contamination at a local or regional scale. In freshwater systems, OBT is associated with naturally occurring fresh or detrital organic compounds in sediments, suspended solid particles and aquatic biota [6.11]. In certain cases, the presence of tritiated organic compounds can be caused by technogenic tritium originating from anthropogenic activities. For example, in the case of the Rhône river (France), such tritiated organic compounds are derived from comminuted tritiated residual particles from luminescent paints formerly used in the French and Swiss watchmaking industries [6.12].

In freshwater systems, different processes govern the mobility and activity levels of tritium in water (transport, mixing and dilution) and in organic matter in drainage basins (origin, transfer and recycling). These processes result in OBT/HTO disequilibria in the various environmental compartments, including sediments and aquatic biota. Proper characterization of the various physicochemical forms of tritium in the various environmental compartments that exchange with freshwater systems is fundamental for the understanding of the short to long term fate of tritium in rivers and other freshwater pools.

6.2. TRITIUM SOURCE TERMS

6.2.1. Natural production

The natural production of tritium in the upper atmosphere is 0.22–0.33 kg per year (of the order of 100 PBq/a) [6.13]. The natural production of tritium is balanced by its radioactive decay [6.14]. The widely adopted value of equilibrium inventory of natural tritium equals to approximately 3.5 kg (1.2×10^{18} Bq) [6.15, 6.16]; however, some alternative estimates of the equilibrium inventory yield the range 4.5−4.9 kg, given a natural input of 258−278 g/a (see Chapter 2).

Approximately 99% of the tritium produced in the upper atmosphere is oxidized to tritiated water and delivered to the Earth's surface by rainfall, becoming part of the hydrological cycle. A small fraction of natural tritium is geogenic and is formed via neutron capture by ^{6}Li contained in rocks ($^{6}Li + n \rightarrow {}^{3}H + {}^{4}He$). Neutrons involved in such reactions originate from the spontaneous fission of naturally occurring $^{235,238}U$ isotopes or from nuclear interactions of high energy protons and alpha particles from extraterrestrial radiation. Compared with tritium production in the atmosphere, such lithospheric production of tritium is negligible.

Apart from the oceans, which contain over 90% of the global tritium inventory (see Chapter 5), freshwater reservoirs (including rivers, lakes, groundwater, permafrost sinks and glaciers) constitute smaller repositories for tritium. Owing to tritium's relatively short half-life, tritium contents in the various reservoirs mostly depend on the residence time of the water masses in the continental compartment of the hydrological cycle. The rates of exchange of water masses with the atmosphere are powered by solar energy fluxes.

6.2.2. Nuclear tests

The atmospheric and ground testing of nuclear weapons from 1945 to 1980 resulted in the substantial release and dispersion of significant amounts of radioactive materials, including tritium [6.17]. The main contribution came from atmospheric tests of thermonuclear weapons from 1952 to 1963. Enhanced tritium concentrations in the environment, substantially exceeding the natural background levels, were observed globally but mostly in the Northern Hemisphere. In 1963, the USSR, USA and UK signed the Partial Test Ban Treaty to establish a definitive ban on nuclear weapon testing in the atmosphere, in space and under water. China and France were not signatories to this treaty, and they continued atmospheric nuclear tests until 1980.

Global testing of nuclear devices in the atmosphere resulted in a release of approximately 520–560 kg of tritium [6.15, 6.16]. Of this amount, Guétat et al. [6.16] estimated that approximately 420 kg was dispersed in the Northern Hemisphere and 140 kg in the Southern Hemisphere [6.15, 6.18].

By 2019, following radioactive decay, approximately 20 kg (7.1×10^{18} Bq) of the original amount of anthropogenic tritium from nuclear tests remained globally distributed between the atmosphere (~10%), the oceans (~90%) and continental waters (~1%) (Table 5.1). This amount still exceeds the equilibrium inventory of naturally produced cosmogenic tritium, which is approximately 3.5 kg (1.2×10^{18} Bq) [6.15, 6.16].

Over the nuclear era, tritium releases to the atmosphere led to significant changes in the natural levels of this nuclide within environmental compartments at the global scale. The atmospheric nuclear weapon tests carried out from 1945 to 1980 introduced approximately 200 times more tritium into the atmosphere than is present naturally at equilibrium on a global scale. During the peak releases in 1963, the tritium content in rainwater increased by a factor of almost 1000 throughout the Northern Hemisphere, where most of the nuclear tests were performed.

The significant input of anthropogenic tritium has declined over time owing to a combination of radioactive decay, natural recycling, water mass mixing and other mixing within the various environmental compartments (atmosphere, oceans, surface waters, groundwaters, biosphere). The annual tritium input rates to the environment and the estimated range of cumulative tritium yield resulting from atmospheric nuclear testing are summarized in Table 6.2.

By 2019, the tritium concentration in the atmosphere had decreased to levels close to that characterizing the period before atmospheric nuclear weapon testing. The reductions with time are due to radioactive decay and its progressive transfer to the oceans and freshwater systems. Without considering any natural dilution processes, the theoretical residual tritium levels from the peak emission in 1963

TABLE 6.2. ANNUAL TRITIUM INPUTS TO THE ENVIRONMENT

Tritium source	Activity (PBq/a)	Mass (g/a)
Natural cosmogenic production	54–117	150–330
Tritium from global nuclear reactors	28	79
Approximate production by HWRs	0.068	1.9
Production by PWR 1300 MW(e)	0.003 2	0.09
Production by PWR 900 MW(e)	0.001 1	0.03
La Hague SFRP in 2000	10	30
Sellafield SFRP in 2000	2.8	8
France (all nuclear)	15	45
Canada in 2006	3	9
Atmospheric nuclear weapon tests (total) [6.15, 6.16]	180 000–200 000 PBq	520 000–560 000 g

Note: HWR: heavy water reactor; PWR: pressurized water reactor; SFRP: spent fuel reprocessing plant.

would be expected to range from 10 to 30 Bq/L in 2020 because of radioactive decay only (Fig. 6.2). These values are significantly higher than current tritium contents measured in precipitation and river waters (<1 Bq/L). This theoretical approach helps to explain excess OBT contents (i.e. OBT/HTO disequilibria) observed in various natural environmental compartments where organic matter stores and isolates tritium as OBT [6.20] in riverine sedimentary reservoirs.

It is of particular importance to consider tritium repositories within the various biospheric compartments where tritium from the atmosphere is stored in weakly exchangeable forms of freshly synthetized organic matter. It is also important to evaluate tritium delivered from soil by leaching and erosion to river systems as OBT. OBT in its non-exchangeable form is not subject to dilution processes as is the case for its mobile forms in water masses. This implies that any tritium integrated into biomass during plant growth over the period of intensive atmospheric nuclear testing would be stored for the long term, if not

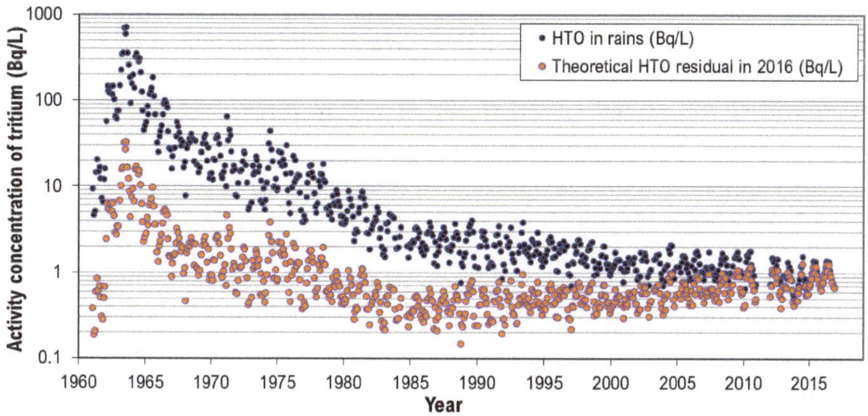

FIG. 6.2. Time series of tritium activity concentration measured in precipitation at the IAEA GNIP station in Vienna [6.19] since 1960 (blue symbols) and calculated residual expected in 2016 owing to radioactive decay only (red symbols).

biodegraded or recycled. This biogenic pool, created mostly in the 1950s and 1960s, constitutes a non-negligible source of OBT for river systems, depending on biomass density and recycling in the catchment.

Additional to any biomass repositories formed during the nuclear test era, permafrost and glaciers are also long term repositories for tritium inputs from nuclear tests. River waters, by comparison, reflect more closely the current tritium contents in the atmosphere. The tritium inventory in such freshwater systems is expected to be governed solely by radioactive decay.

6.2.3. Liquid releases of tritium from nuclear facilities

Nuclear industries that release tritium into the environment include PWRs, SFRPs and reactors dedicated to tritium production (see Table 6.2). In 2020, it was estimated that about 2.8×10^{16} Bq of tritium is released each year from nuclear facilities around the world, 1.17×10^{16} Bq to the atmosphere and 1.6×10^{16} Bq to fresh waters and marine waters [6.15, 6.21, 6.22]. Each year, these anthropogenic tritium emissions are of the same order as the natural tritium activity input at equilibrium $(5–12) \times 10^{16}$ Bq. Tritium emission rates from nuclear energy production are four orders of magnitude lower than the amount produced from atmospheric nuclear tests (about 200×10^{18} Bq or 200 EBq), even when considered cumulatively over several decades of input. Nevertheless, the environmental impact of these various tritium source terms cannot be directly compared owing to the heterogeneous dispersal from localized and distributed sources. Such nuclear facilities are non-uniformly distributed along large rivers, estuaries or coastal areas and produce different amounts of tritium releases depending on their type.

In HWRs, tritium is produced from neutron activation of D_2O. Approximately 2 g or 6.8×10^{14} Bq of tritium is released in a year by a 900 MW(e) HWR. This amount is about 20 times greater than the amount of tritium released from an LWR, where tritium is mostly produced through the ternary fission of uranium and plutonium and mainly remains in the actual fuel until it is dissolved for treatment. During normal operation of a PWR, the tritium released is produced by a neutron capture reaction on ^{10}B, about 0.03–0.06 g (15–20 TBq) per year for a 1000 MW(e) reactor.

Most tritium produced in reactor fuel rods via ternary fission is retained within the fuel until it is reworked in a nuclear reprocessing facility. During reprocessing, the spent fuel is removed from its cladding material and then dissolved in concentrated nitric acid, releasing most of the tritium into the liquid waste stream. This stream may then be eventually released into the environment, after removal of radionuclides that can be chemically separated. Some tritium is also released in the dissolver off-gas, and the remainder is immobilized in the cladding as solid zirconium hydride or as tritium dissolved in the metal [6.23–6.26].

In northwestern Europe, HTO in the liquid effluent from the two SFRPs is released and dispersed into the English Channel and the Irish Sea. At the La Hague plant, approximately 30 g (1×10^{16} Bq) of tritium is released every year per 1600 t of nuclear material, whereas at Sellafield approximately 8 g/a (2.8×10^{15} Bq/a) had been released annually to the marine environment [6.14] before the end of operation in 2020 (see Fig. 5.7 for detailed variations from 1950 to the present). Traditionally, tritium recovery is considered too technically difficult and costly, and marine dispersal has remained the preferred option.

Apart from atmospheric and marine environments that receive major releases of human made tritium, freshwater systems (lakes and rivers) also receive some tritium releases from nuclear facilities. In the following, atmospheric and liquid tritium releases are compared for France and Canada to illustrate the impact of different nuclear technologies on freshwater systems.

6.2.3.1. France

In France, the total activities of tritium released from all nuclear industries are estimated to be 1.5×10^{16} Bq/a for liquid effluents and 3.3×10^{14} Bq/a for those delivered to the atmosphere [6.27]. This means that on average 96–99% of tritium releases enter aquatic systems. Although tritium is released from many sites as liquid effluents, several nuclear facilities also release significant amounts to the atmosphere (Fig. 6.3). These include the Creys-Malville NPPs (currently being dismantled); the CEA facilities at Saclay, Bruyères-le-Chatel and Grenoble; ANDRA CSFMA in Aube (almost 100%); Institut Laue-Langevin in Grenoble (98%); Centraco in Marcoule (91%); the CEA facility in Cadarache (90%); as well as the CEA facility in Valduc (75%) and the Marine Nationale facility in Brest (75%) — the percentage values quoted represent the atmospheric fraction.

Among French nuclear facilities, the La Hague SFRP and CEA Valduc are the sites releasing the greatest quantities of tritium to the environment (mean values of 1.2×10^{16} Bq/a and 2.6×10^{14} Bq/a, respectively, over 2011–2015). While the tritium releases from CEA Valduc are exclusively airborne, 99% of the releases from La Hague SFRP are liquid effluents delivered to the marine environment (Fig. 6.3).

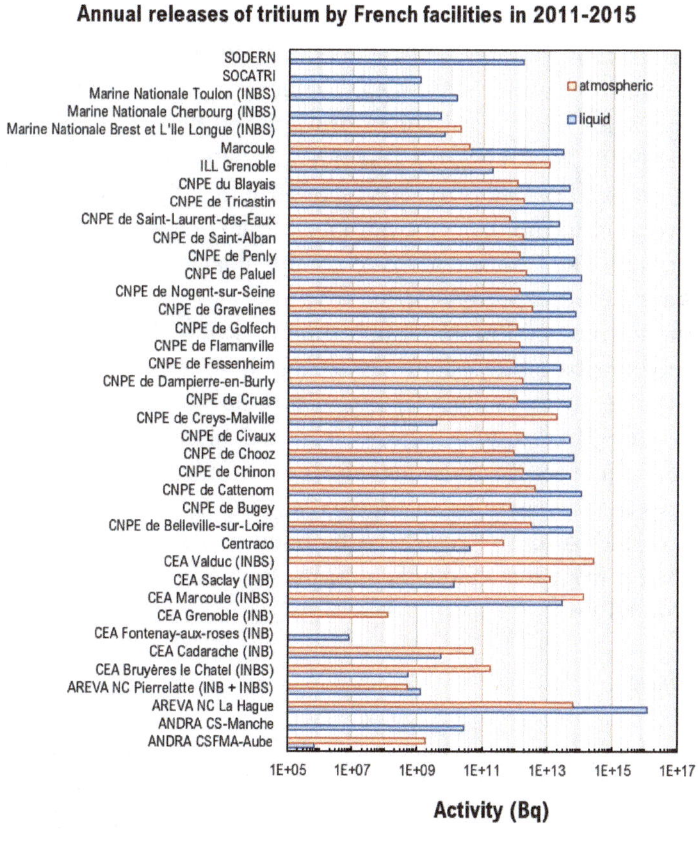

FIG. 6.3. Mean tritium atmospheric and liquid releases by French facilities in 2011–2015 [6.27].

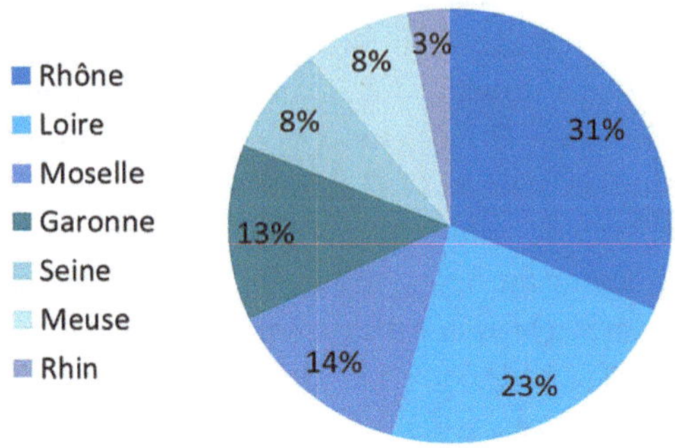

FIG. 6.4. Distribution of tritium liquid releases from French nuclear facilities to rivers in the period 2011–2015.

Apart from those releases, other operating PWRs account for significant additional tritium releases, with mean values in 2011–2015 ranging from 2.2×10^{13} Bq/a for the Saint-Laurent-des-Eaux NPP to 1.1×10^{14} Bq/a for the Cattenom NPP. In 2011–2015, the French NPPs were releasing into the environment on average 1.1×10^{15} Bq/s of tritium, of which 3.9×10^{14} Bq/a (35%) was released to the marine environment and 7.1×10^{14} Bq/a (65%) was delivered to rivers. The Rhône and Loire are the most affected rivers and receive around 31% and 23%, respectively (Fig. 6.4).

The Rhône river, 60 km upstream of the river mouth in the Mediterranean Sea, has received tritiated liquid effluents from the Centraco SFRP over several decades (Marcoule nuclear industrial site). Until the Complex started its planned dismantling in 1997, the liquid releases from the Marcoule site were the main source term of tritium for the river; namely, it contributed from 50% up to 80% of tritium releases at the beginning of the 1990s [6.26–6.28]. Over recent decades, only 5–25% of the tritium released by the nuclear industries located along the Rhône originated from the Marcoule site; most of the tritium in the river now originates from NPPs.

6.2.3.2. *Canada*

In Canada, tritium is produced as a by-product of the operation of the heavy water moderated CANDU reactors. A portion of the tritium produced by CANDU reactors is recovered and used by SRBT Inc., which produces tritium-powered, self-luminous light sources and luminescent paints. It is also used as a source of fuel for experimental fusion reactors such as JET at the Culham Centre for Fusion Energy (UK).

In Canada in 2006, approximately 3×10^{15} Bq of tritium was released by NPPs into the environment, mostly as DTO. Releases from NPPs were 7–8 times larger than releases at the CNL Chalk River site, which was responsible for 4×10^{14} Bq/a of additional tritium releases. Two-thirds of these releases were to the atmosphere [6.29]. For comparison, these quantities are around four times lower than those delivered by French NPPs (15×10^{15} Bq/a). Nevertheless, in contrast to French NPPs, where most releases are to aquatic environments (~96–99%), tritium releases from Canadian NPPs are almost equally distributed between liquid and gaseous effluents, leading to airborne tritium releases approximately five times higher than in France (15×10^{14} Bq/a and 3.3×10^{14} Bq/a, respectively). Furthermore, in France, approximately 3×10^{15} Bq/a is released to freshwater environments (large rivers), representing twice the amounts estimated for Canadian rivers (1.5×10^{15} Bq/a).

6.2.4. Other sources of tritium that may affect freshwater systems

The very small natural radioactivity levels of tritium impose the need for artificial production of this isotope for all industrial and nuclear purposes. Tritium can be produced in nuclear reactors by neutron irradiation of D_2O but is more efficiently produced by fast neutron capture reaction on ^6Li (see Section 4.4.5.5). Tritium can also be produced by irradiating ^3He with neutrons, although this approach has not been applied on a large scale. Tritium applications include use in industrial thickness gauges, luminous paints, tritium-powered self-luminous light sources, airport runway lights and watch dials. It is also used in fusion research, thermonuclear weapons, and as a tracer in chemical, biological and environmental studies. There is also some tritium-containing radioactive material with low solubility in aqueous media such as tritides of metals (e.g. Ti, Zr, Hf), tritiated luminous compounds, microfragments of glass, and carbon and beryllium particles contaminated with tritium. Worldwide, the releases of tritium associated with these sources are still low but can result in high concentrations in localized areas. The chemical form of tritium that might include nanoparticles can also cause significant levels of labelling in living species or sediment in localized areas.

Industrial and biomedical research and development processes often result in the release of high activity tritiated synthetic organic molecules in various matrices and has been termed ‘technogenic tritium’ [6.30]. For example, in the UK, such materials were released from approximately 1982 to 2000 into the Severn Estuary by the former radiopharmaceutical facility in Cardiff [6.31–6.33] (see also Chapter 5).

In Switzerland and France, luminous paint particulate residues containing tritium are derived from watchmaking industries (e.g. Refs [6.12, 6.34]) and are seen in the surface and buried sediments of rivers in the upper Rhône area [6.35–6.38]. These tritiated submicrometre particles are resistant, hardly biodegradable (i.e. refractory) and widely dispersed in the fluvial environment. The sources and their release histories are not well characterized and their impact in time and space on the components of the environment is not known. It is notable that these micrometre sized solid forms are often missed because the samples are generally distilled before scintillation analysis; only the water and tritiated water molecules are extracted during this analytical procedure. Additionally, the contribution of any tritiated submicrometre particles could be difficult to discern in solid matrices such as soils and sediments, as they are heavily diluted with coarser particles. No studies of the uptake of these particles by living organisms or their persistence and accumulation in the food chain have been conducted to date.

In Canada, luminous safety lighting factories, such as SRBT [6.29], are also sources of technogenic tritium.

A summary of the known sources of technogenic tritium is provided in Box 6.1 [6.30].

BOX 6.1. SOME KNOWN SOURCES OF TECHNOGENIC TRITIUM
(adapted from Ref. [6.30] with permission courtesy of Elsevier)

Tritiated molecules

Life sciences research and diagnostics

— Custom radiolabelled compounds for use in life science research and medical diagnostics were produced in a radiopharmaceutical facility in Cardiff. The range of isotopically labelled compounds included amino acids, lipids/fatty acids, nucleic acid precursors, carbohydrates, prostaglandins, vitamins, steroids, alkaloids and hydrocarbons [6.33].

— Many uncharacterized tritium labelled organic compounds (e.g. mixture of ^3H_2O, ^3H-labelled methanol, and methyl iodide) were produced as by-products during the chemical synthesis of the commercial products [6.31, 6.39].

— Isotopically labelled compounds were manufactured for use in life sciences diagnostics and research. Thousands of different organic chemicals were generated during the synthesis of these compounds, and some of these were released via authorized discharges as liquid waste into the Severn Estuary [6.32].

Tritiated solid compounds

Luminescent paints, panel lighting, self-illuminating aircraft, airport runway lights, light dials, gauges and watches, etc.

— Luminous compounds using radioactive substances have been used for several decades. Tritium in the form of tritiated polystyrene, having a specific activity up to 7.5 TBq/g, is mixed with a suitable fluorescent compound (phosphor), and this self-luminous mixture is applied to dials or signs. Gaseous tritium is also used in small glass capillaries, which are coated internally with a suitable fluorescent compound. Watches containing gaseous tritium are not permitted in many countries owing to the hazard of breakage and leakage [6.34].

— Tritium sources originating from the watchmaking industry include fine particles of tritiated compounds used in luminous paints. The form of these compounds is usually tiny tritiated polystyrene spherules of a few tens of micrometres in diameter [6.35, 6.36].

— During the manufacture of gaseous tritium light sources, borosilicate glass tubes of various lengths and diameters with an inner phosphor coating are filled with tritium gas.

Fusion reactors

— Tritium retention in plasma facing materials such as tungsten is a major concern for future fusion reactors. During operations of the ITER reactor, tritiated tungsten or boron dust-like particles may be generated, which need to be characterized in terms of amount of trapped tritium, tritium source and radiotoxicity [6.40].

6.3. TRANSPORT AND TRANSFER OF TRITIUM IN FRESHWATER ENVIRONMENTS

6.3.1. Transport, diffusion and dispersion

It is difficult to distinguish the transfer of tritium that occurs through precipitation from that occurring through direct water vapour exchange. However, a major difference exists between these two processes: while input via precipitation is irregular or sporadic (e.g. in some places, it may not rain for half a year), isotope exchange is a continuous process controlled by tritium gradients between the exchanging phases, temperature and relative humidity of the ground-level atmosphere.

The tritium concentrations in a body of surface water or in soil porewater respond to the recent history of tritium inputs and reflect a dynamic equilibrium between air and water. The relevant historical period depends on the residence time of water in the water body or in the uppermost soil layer. The residence time of water in soils is directly governed by infiltration rates and the thickness of the unsaturated zone, which determines the travel times of infiltrated water to reach the water table. The unsaturated zone is of major importance for tritium exchanges with the atmosphere, as it is where liquid and vapour phase water can exchange. Effective infiltration rates are climate dependent, and resaturation profiles are mainly related to the seasons and temperatures owing to the magnitude of evaporation processes. Infiltration rates at a site can be evaluated from the position of the tritium peaks at two different sampling times. For example, in a study conducted at semi-arid areas in China (see Section 6.3.4 for details), the displacement of the tritium peak over nine years was found to be approximately 5 m. Where there is a porosity of 40%, this would result in a water infiltration rate of ~220 mm/a. In groundwater, residence times differ according to the depth of the aquifer. Generally, surface aquifer waters, which are often hydraulically connected to rivers (shallow aquifer), can be renewed over timescales of a few days to several decades and are mainly recharged by precipitation. Deep aquifers, on the other hand, are renewed much more slowly, over hundreds or even thousands of years. Travel times also depend on the characteristics of the aquifer, with both the porosity and permeability of the ground playing important roles. Residence times of water in ponds and lakes range from weeks to hundreds of years, depending on water volumes and balance.

The amount of precipitation falling on a catchment and the duration of transfer of meteoric waters via surface and subsurface runoff control tritium content in surface waters. For this reason, tritium has been widely used as a tracer to determine the residence time of surface waters in catchments. Once tritium enters the given catchment as precipitation, its fate is controlled by the dynamics of water movement, mixing and dispersion processes acting both in the unsaturated zone and in the shallow aquifer systems involved. Additionally, transport, dispersion and mixing, supplemented by isotope exchange with the atmosphere, will occur in river courses. Such movements can be described by the classical laws of hydrodynamics and determine the field of volume activities under the effect of diffusion and dispersal transport mechanisms. In the context of isotope hydrology, there is a well established modelling framework for using tritium produced in nuclear tests as a water age indicator in groundwater systems [6.41]. The lumped parameter framework uses various combinations of transfer functions of tritium through investigated systems (e.g. piston flow, exponential and gamma functions applied to dispersion modelling). In practice, there are many methods of varying complexity to model the dispersion mechanisms of anthropogenic tritium in aquatic environments. Transfer functions for other inputs, mostly via the watershed, have been little studied and are poorly characterized.

Recent work carried out within the EMRAS programme has shown the relevance of several operational models used to assess transfers along a hydrological system [6.42]. One example of the application of such models is on the Loire river system [6.43], where slow flows are associated with high water levels (lakes, dams), and it is also necessary to take into account the thermal characteristics of the water column on which vertical exchanges depend. Such methods range from numerical models that describe the temporal evolution of volume activities at any point in the domain to dilution models that are based on assumptions of complete and instantaneous dilution of the flows of tritium within the volumes of water considered. The choice of method needs to take into account the dynamics of the tritiated flows (permanent or transient), characteristics of the hydrological system (e.g. river, lake) and the location of the area of interest (e.g. distance to discharge, good mixing zone).

From a modelling point of view, the discharge parameters of liquid effluents into an aquatic environment are input data that depend on space and time and are expressed in flow rates (Bq/s) [6.44]. However, the operation of nuclear facilities is rarely accompanied by constant steady releases; instead, they appear more often as frequent, non-uniform transient releases, informally termed ‘puffs’ or ‘spikes’. Such transient events create significant spatiotemporal variability of activity concentrations across the hydrological system. Such events bring into question the relevance of steady state models, particularly in the case of OBT, which is much less mobile than HTO. The higher mobility of HTO reaches a steady state quickly once tritium release rates and ambient concentrations stabilize (see Section 6.5).

In practice, contributions to a watershed are modelled by functions that reflect empirical transfers. These consider the characteristics of the watershed (e.g. slope, type and nature of the soil, plant cover), meteorology (rainfall) and

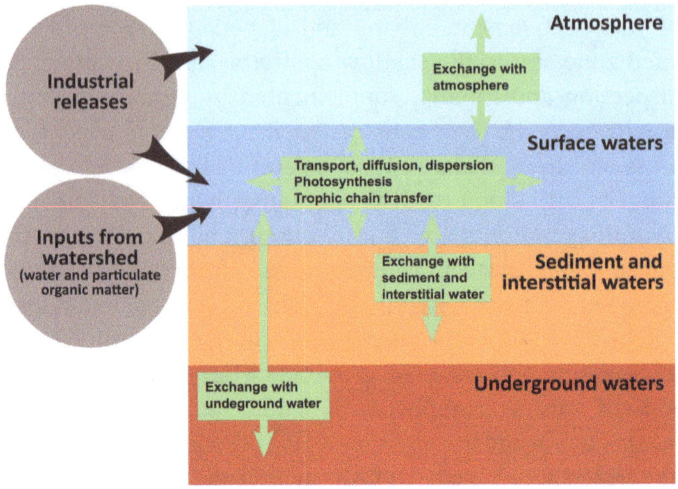

FIG. 6.5. Main mechanisms of transport and transfer of tritium in river systems. The interaction between groundwater and surface water describes the recharge rate for infiltration of surface (river) water to groundwater, controlled by local lithology, and the discharge rate describing the inflow of groundwater to the riverbed, determined by hydraulics and local lithology.

conditions of contamination of the basin slope (e.g. atmospheric deposition, partial or point contamination). For tritium, research [6.45] shows that there is little published work to characterize these transfer functions. From a deposit to a watershed, the few studies available indicate an average outflow transfer of 3.5 years [6.46, 6.47].

Exchange models have been proposed to describe exchange interactions with the atmosphere and interstitial water in sediment [6.48] (see Fig. 6.5). The transfers from air to soil water, or to small surface water bodies such as ponds, are best represented by specific activity models. These models apply when there is rapid isotopic exchange between environmental compartments, so that if one compartment is initially contaminated, the other compartment will tend to have the same tritium to hydrogen (T/H) ratio, except for any dilution effects. The T/H ratio is used here as an operational specific activity for modelling.

6.3.2. Exchange with the atmosphere

Tritium is exchanged between surface waters and the ambient atmosphere through natural processes. These relate to phase changes of water, such as evaporation, condensation and precipitation, as well as through isotope exchange across the water–air interface. Isotopic fractionation of tritium is not important for transfers between water and the atmosphere, although the vapour phase can be theoretically depleted in tritium compared with the liquid phase owing to isotopic fractionation. Where this occurs, such as in salt pans, it is accompanied by ^{18}O enrichment.

While fluxes and exchanges between these two compartments are of major importance at the scale of the catchment, direct exchanges between the river surface and the atmosphere are relatively more limited owing to the restricted surface of exchange involved. Investigations conducted for the Loire river to improve the understanding of HTO fluxes from surface water to air showed that during radioactive releases from NPPs, water vapour collected on the riverbank was enriched in tritium [6.49]. The average of these estimated exchange velocities was more than one order of magnitude higher than those calculated in the literature using indoor experimental data. The variability of these values was also larger, showing that outdoor studies cover a wide range of conditions influencing HTO flux, although no clear correlation has been identified between exchange velocities and meteorological conditions. Significant differences between day and night were also registered, with higher values observed during the day. Finally, a one-year simulation showed that in this case the exchange processes with the atmosphere are dosimetrically not very important compared with other pathways such as the ingestion of water and foodstuffs. The deposition of tritiated water vapour from the atmosphere to surface waters was studied by Noguchi et al. [6.50] in an area where elevated concentrations of tritiated water vapour existed in the atmosphere over several days. Exchange velocities of tritiated water vapour between the atmosphere and the surface water were estimated from field experiments. It was found that outdoor exchange velocities were about three times higher than those registered in, for example, a nuclear reactor control room, where the study was performed in support of a risk assessment. In the outdoor measurements, the wind speed was strongly correlated with the exchange velocity, and the temperature of the air had an enhancing effect on the velocity.

Tritium exchanges with the atmosphere can be modelled by considering two major mechanisms: (i) exchange between tritiated water in the river system and tritiated water vapour in the ambient atmosphere and (ii) input from precipitation. Input from precipitation is expressed by the product of the rain intensity and the tritium concentration in rainwater, as follows:

$$\Theta_{\mathrm{HTO}}^{\mathrm{rain}} = v_{\mathrm{rain}} [\mathrm{HTO}]_{\mathrm{rain}} \tag{6.1}$$

where

$\Theta_{\text{HTO}}^{\text{rain}}$	is the input of tritium from rain ($\text{Bq} \cdot \text{m}^{-2} \cdot \text{s}^{-1}$);
v_{rain}	is the rain intensity (m/s);

and $[\text{HTO}]_{\text{rain}}$ is the tritium concentration in rainwater (Bq/m^3).

Exchange between tritiated water in surface water systems and tritiated water vapour in the ambient atmosphere is essentially related to evaporation, isotope exchange at the water–air interface and condensation leading to local precipitation. Some studies consider the relevance of these mechanisms by identifying correlations between tritium radioactivity in the atmosphere close to river systems contaminated by tritium discharge. This exchange remains largely undocumented and is rarely treated in transfer models.

Exchange by diffusion always applies but is weighted according to the conditions for humidity saturation in the air layers in contact with the river system surface; namely, undersaturation, saturation or oversaturation.

For undersaturated conditions ($e_s > e_a$, where e_s is the saturation vapour pressure of water in the air and e_a is the water vapour pressure in the air), exchange is mainly due to evaporation from the water surface and occurs primarily from the surface water to the atmosphere. Modelling may then be based on the product of the evaporation flow from the surface water and the radioactivity of tritium in the water in the hydrological system, as follows:

$$\Theta_{\text{HTO}}^{\text{evap}} = \beta \, \Theta_{\text{water}}^{\text{evap}} [\text{HTO}]_{\text{hyd}} \tag{6.2}$$

where

$\Theta_{\text{HTO}}^{\text{evap}}$	is the flux density of tritium activity from the surface water to the adjacent atmosphere caused by evaporation process ($\text{L} \cdot \text{m}^{-2} \cdot \text{d}^{-1}$);
$\Theta_{\text{water}}^{\text{evap}}$	is the evaporation flux density ($\text{L} \cdot \text{m}^{-2} \cdot \text{d}^{-1}$);
$[\text{HTO}]_{\text{hyd}}$	is the tritium activity concentration in water (Bq/m^3);

and $\beta = 0.91$ is the inverse of the isotopic fractionation factor.

The evaporation flow from a water surface depends on temperature, atmospheric pressure, wind speed and the difference between vapour pressure and saturation vapour pressure. There are numerous expressions in the hydrological literature to calculate the evaporation flux [6.51, 6.52]. The Rohwer formula is frequently used:

$$\Theta_{\mathrm{HTO}}^{\mathrm{evap}} = 0.372\left(1-0.000374P_{\mathrm{a}}\right)\left(1+0.6\,u_{\mathrm{wind}}\right)\left(e_{\mathrm{s}}-e_{\mathrm{a}}\right) \tag{6.3}$$

where

P_{a} is the atmospheric pressure (hPa);
u_{wind} is the wind speed (m/s);
e_{s} is the saturation vapour pressure of water in the air (hPa);

and e_{a} is the water vapour pressure in the air (hPa).

At saturation ($e_{\mathrm{s}} = e_{\mathrm{a}}$), the water flow balance between the atmosphere and the river system is zero. Tritium exchange occurs primarily through isotope exchange controlled by the gradient in tritium concentrations in river water and in water vapour in the atmosphere.

$$\Theta_{\mathrm{HTO}}^{\mathrm{diff}} = v_{\mathrm{e}}\left(\left[\mathrm{HTO}\right]_{\mathrm{atm}} - \left[\mathrm{HTO}\right]_{\mathrm{hyd}}\right) \tag{6.4}$$

where

$\Theta_{\mathrm{HTO}}^{\mathrm{diff}}$ is the tritium flow from diffusion ($\mathrm{Bq/m^2}$);
v_{e} is the speed of tritium exchange between atmosphere and hydrosystem (m/d);
$[\mathrm{HTO}]_{\mathrm{atm}}$ is the tritium concentration in river system water (Bq/L);

and $[\mathrm{HTO}]_{\mathrm{hyd}}$ is the tritium concentration in water vapour in the atmosphere (Bq/L).

For oversaturated conditions ($e_{\mathrm{s}} < e_{\mathrm{a}}$), mechanisms for the transfer of tritium from the atmosphere to the river system can play a dominant role. Examples of mechanisms involved are the condensation of water vapour in the air and the interception of water droplets on the surface of the river system. Correspondingly, the modelling needs to consider the kinetics of water droplet formation and the speed of their deposition onto the surface.

6.3.3. Exchanges with interstitial water in sediments

HTO is transported by precipitation to the surface water system and a portion is retained by suspended particulate matter, which settles as sediment over time. Sediments generally consist of clays, organic matter, hydrated iron and manganese oxides, and represent soil and water mixtures transported by and deposited in aquatic environments. The relative amounts of these components and certain properties — such as particle size distribution, pH, organic content, oxidation–reduction conditions and solids content — will influence the interaction between sediment particles and contaminants such as tritium. In general, higher concentrations of tritium are reported in sediment samples containing elevated levels of organic matter. These characteristics change the attachment of the contaminants with the sediment and influence the potential for tritium movement. Tritium exchanges with the sediments are mainly governed by hydrological flows because HTO, unlike other particle reactive elements, does not strongly associate with solids, unless bonded within a compound (e.g. humic acids as OBT). Tritium attachment to inorganic solid particles is through various types of bonding of hydrous species (e.g. hydrogen bonding, waters of crystallisation, hydroxyl radical), which are characterized by various strengths depending on the nature of bonding and their closeness to solid surfaces. Many loosely bound hydrous phases are extracted by evaporation, distillation or freeze-drying. Association of tritium with inorganic parts of the sediment is mainly due to exchange with hydroxyl groups associated with clay minerals in the sediment and may then be accounted for together with the TFWT fraction associated with the sediment after evaporation, distillation or freeze-drying.

In certain cases, water flow through sediment can constitute the principal source of surface water contamination [6.53, 6.54]. This is interpreted using transfer models that are primarily concerned with the specific or surface activity of dry sediment. Sediment is, however, composed of water and solid particles, and the specific activity of this mix is not zero as soon as tritium is found in the interstitial water. Therefore, the following equation applies:

$$[\mathrm{HTO}]_{\mathrm{sed}} = \frac{[\mathrm{HTO}]_{\mathrm{its}}\, n}{\rho_{\mathrm{w}} n + \rho_{\mathrm{s}}(1-n)} \tag{6.5}$$

where

$[\mathrm{HTO}]_{\mathrm{sed}}$ is the activity concentration of tritium in sediments (Bq/kg wet weight);

$[\mathrm{HTO}]_{\mathrm{its}}$ is the activity concentration of tritium in interstitial water (Bq/m^3);

n is the sediment porosity (dimensionless);
ρ_w is the water density (kg/m^3);

and ρ_s is the density of solid particles (kg/m^3).

Variations of $[HTO]_{sed}$ depend on two mechanisms: interstitial diffusion at the interface of the water column and the sediment, and exchange with the alluvial aquifer.

$$\Theta_{HTO}^{its} = v_{its}\left([HTO]_{its} - [HTO]_{hds}\right) + \Theta_{HTO}^{all\ aquifer} \tag{6.6}$$

where

v_{its} is the exchange velocity with interstitial water (m/s);
$[HTO]_{hds}$ is the activity concentration of tritium in the water column (Bq/m^3);

and $\Theta_{HTO}^{all\ aquifer}$ is the flux density between the river system and the alluvial aquifer ($Bq \cdot m^{-2} \cdot s^{-1}$).

Subject to the nature of the soil (e.g. porous, fractured, impermeable) and flow conditions due to pressure gradients and the permeability of the aquifer, exchanges with the alluvial aquifer can be difficult to model. For this reason, they are generally considered empirically on the basis of tritium measurements.

For the diffusive exchange, v_{its} depends on the transport velocities of the river water (v_{hds}) and sediment (v_{sed}) at their interface, as follows:

$$v_{sed} = \frac{D_{sed}}{h_{sed}} \approx \frac{10^{-9} n}{h_{sed}\left[1 - \ln\left(n^2\right)\right]} \tag{6.7}$$

The exchange velocity in the river system, v_{hds}, can be defined differently. The following relationship serves as an example [6.55]:

$$v_{hds} = \frac{3\sqrt{3}}{2\pi} 0.1 \left(\frac{v_w}{D_m}\right)^{-2/3} u^* \approx 10^{-4} u \tag{6.8}$$

where

u^* is the flow friction velocity (m/s);
u is the average flow velocity (m/s);

ν_w is the kinetic viscosity of water ($\nu_w = 10^{-6}$ m^2/s);

and D_m is the molecular diffusion coefficient of tritium in water ($D_m \approx 10^{-9}$ m^2/s).

The exchange velocity for the sediment, v_{sed}, corresponds to the ratio of D_{sed}, the molecular diffusion coefficient of tritium in the superficial sediment layer, to h_{sed}, the thickness of the layer. The latter usually ranges from several millimetres to several centimetres. The diffusion coefficient D_{sed} (m^2/s) can be expressed via n, the porosity of this layer [6.56].

Following recent suggestions regarding the strong affinity of tritiated water for organic matter in suspended particulates and sediments, Jean-Baptiste and Fourré [6.56] performed equilibration experiments between sediment organic matter and tritiated water to look for potential tritium bioconcentration. The T/H ratios measured at the end of the experiments were found to be lower in the sediment organic matter than in the water, indicating that only a fraction of the hydrogen pool (between 14% and 20%) within the sediment equilibrated with the tritiated water. These results are consistent with the widely used concept of exchangeable and non-exchangeable tritium pools in organic matter and importantly show no sign of tritium bioaccumulation in the sediment relative to water.

It is concluded that hydrogen in the exchangeable hydrogen pool equilibrates isotopically with the ambient water, while the non-exchangeable pool retains the tritium content acquired by the organic matter when it was formed.

6.3.4. Exchanges with underground waters

Rivers are generally connected to underground waters (shallow aquifer) and act in almost all cases for their natural drainage. The detailed interaction in any such river–groundwater system depends on the hydraulic gradients (often variable in time owing to floods and drought periods) and the flow rates, which in aquifers depend further on the sediment/rock transmissivity. Commonly, in such systems, the underground waters show lower tritium values than the river water owing to the often long subsurface mean residence time of groundwater and the occurring radioactive decay. A hypothetical exception would be the case when a large fraction of the groundwater would have been recharged from precipitation in the 1960s, at the time of the tritium peak from atmospheric nuclear testing; see Fig. 6.2. More realistic scenarios involve enhanced tritium levels in groundwater caused by infiltration of tritium-rich river water (rivers downstream of NPPs) into a sedimentary basin aquifer, such as that existing at the Neckar river close to Heidelberg in southern Germany. In that location, the tritium plume of infiltrated river water in the recharged aquifer could be monitored over several kilometres of subsurface flow. At another site, the Semipalatinsk test site (STS) in Kazakhstan, local streams and a river are contaminated heavily by upwelling of tritium-contaminated underground water. For example, water in streams leaving one test site tunnel contained 100 kBq/L of tritium, while the tritium levels in the Chagan river, which is fed by waters from an underground test site, reached 240 kBq/L [6.57].

In normal situations, the residence time of underground waters can be roughly estimated by comparing the tritium contents between the surface and underground water. However, this estimation gets more complicated with time owing to the strongly time dependent tritium input function and the radioactive decay (Fig. 6.2). As mentioned, tritium and its stable progeny ^{3}He have been and continue to be widely used as hydrogeological tracers to estimate the origin, residence time and dynamics of mixing, storage volumes of groundwater and the zones of discharge in surface waters [6.58–6.66].

Depth profiles of tritium activity concentration in the unsaturated zone usually show a peak at a depth level corresponding to 1963. Studies in Japan [6.8] show a tritium peak in precipitation of approximately 100 Bq/L for 1963 and a corresponding peak in shallow groundwater (top of saturated zone) at approximately 60 Bq/L in 1964. The reduced concentration in groundwater in this zone was attributed mainly to dilution by discharge of deeper uncontaminated groundwater into the top layer of the aquifer. In this case study, the residence time in the top layer was estimated to be three years, which is insufficient for appreciable decay of tritium. A study in China [6.62] shows tritium in a 1988 core profile taken in an unsaturated zone peaking at 40 Bq/L at a depth of 6.5 m, representing the fallout signature from nuclear tests, and tritium in a 1997 core profile, at the same location, peaking at 27 Bq/L at a greater depth of 11 m (see Fig. 6.6). For ease of comparison, the data in this figure are decay corrected to 1997. The difference in concentration is partly due to radioactive decay and partly due to longitudinal dispersion during vertical migration. The infiltration rate was in this case study estimated at 0.25–0.30 m/a, characterized by an increase in the width of the peak at the mode height. Finally, a similar study in France [6.67] shows tritium in three 2002 core profiles peaking at a depth of 9–10 m, representing the fallout signature. The average infiltration rate over the 1963–2002 period was estimated at 0.20–0.21 m/a on the basis of the peak centre of tritium concentration and porosity of the cores.

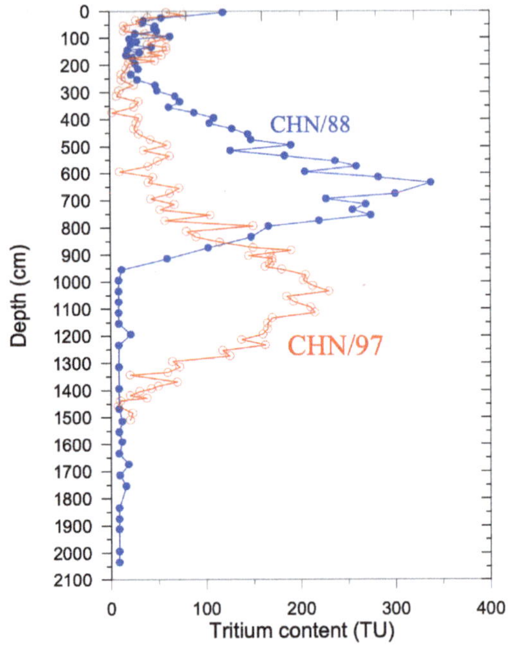

FIG. 6.6. Depth profile of tritium in the unsaturated zone measured in China in 1988 (blue) and in 1997 (red). Data obtained in 1988 were decay corrected to 1997. Note: 100 TU ≈ 12 Bq/L. Figure reproduced from Ref. [6.62].

6.3.5. Tritium behaviour in lake and river plumes

Gaussian type models for radionuclides released to large lakes and rivers have been applied [6.68] and have been widely used to estimate average annual concentrations of tritium in aquatic plumes arising from nuclear facilities. The large-lake model incorporates advection and dispersion processes, as well as radionuclide losses due to radioactive decay and sedimentation. Water travel times to receptor locations, such as drinking water intakes, are of the order of days, and therefore any losses from radioactive decay are considered negligible. Tritium losses from sedimentation are zero. The model as presented by Hart [6.69] includes a factor for the frequency of current direction towards the receptor location, so the calculated concentration at this location represents an annual average value. Other factors affecting the modelled concentration include water depth, current velocity and a lateral dispersion parameter.

There is abundant evidence in the hydrological literature on using artificial (injected) tritium as a tracer to derive flow rates of rivers. Numerical models describing river flow dispersion in the coastal and marine environment are based on tritium measurements [6.70–6.72]. Nie et al. [6.73] simulated the dispersion of tritium discharged from the Qinshan NPP in China and the river plume extending to Hangzhou Bay (China) under differing river flow conditions. They applied varying river flow rates and tidal scenarios to model normal and accidental tritium releases and exposure to determine potential environmental impact and to create a risk assessment.

6.4. ENVIRONMENTAL FACTORS AFFECTING TRITIUM CONTENTS IN FRESH WATERS

Tritiated water concentrations within the hydrosphere are expected to vary with environmental parameters including latitude, altitude, season, age of water bodies, hydrology, watershed size and precipitation rate [6.7, 6.74–6.77].

6.4.1. Latitudinal trend

Tritium contents in world river waters have been recently included in the Global Network of Isotopes in Rivers (GNIR) database[1] to complement the 45 year old GNIP.

Most of the datasets in the GNIR database relating to the Southern Hemisphere were acquired during the time of peak emissions from atmospheric nuclear testing in the 1960s. Tritium contents in rivers from both the Northern and Southern Hemispheres can be usefully compared only during this peak testing period, when signals were clear and strong. The well known latitudinal trend observed for tritium contents in precipitation is similarly registered in river waters (Fig. 6.7, Table 6.3). These observations were made within the latitudinal zone from 40° S to 65° N, since no datasets on fresh water or snow were available in the GNIR database for more extreme latitudes.

For the Southern Hemisphere, higher tritium contents were recorded in precipitation in the Antarctic because this zone is very sensitive to the intrusion of stratospheric air masses enriched in tritium [6.78]. This leads to the hypothesis that ice from Antarctica would sequester enriched tritium activities relative to the rest of the Southern Hemisphere.

[1] https://www.iaea.org/services/networks/gnir

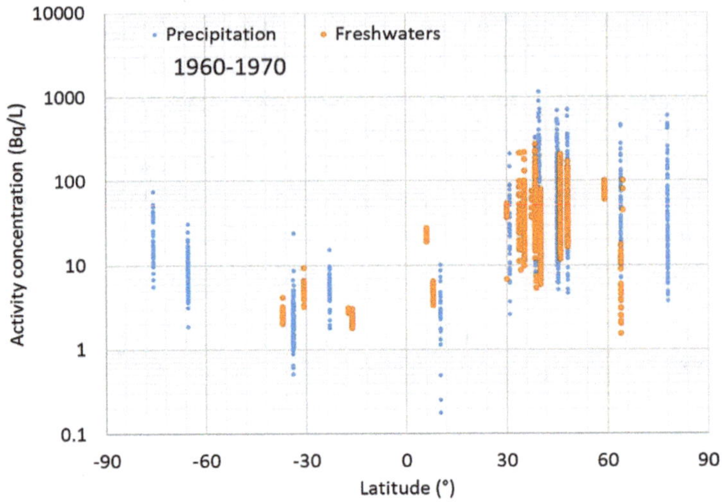

FIG. 6.7. Tritium concentrations measured in precipitation (blue symbols) and in rivers (orange symbols) for 1960–1970 at GNIP and GNIR stations (see Table 6.4), shown as a function of geographical latitude. Data from the GNIP and GNIR databases.

TABLE 6.3. GNIP AND GNIR STATIONS PROVIDING THE DATA SHOWN IN FIG. 6.7

Station name	Country or region	WMO code	Latitude (°)	Longitude (°)	Altitude (m)
		Precipitation			
Ottawa (Ontario)	Canada	7162800	45.32	−75.67	114
Vienna (Hohe Warte)	Austria	1103500	48.25	16.36	198
Isfjord radio	Norway	100500	78.07	13.63	6
Reykjavik	Iceland	403000	64.13	−21.93	14
Kano	Nigeria	6504600	12.05	8.53	476
Maracay	Venezuela	8041300	10.25	−67.65	442
Cape Town airport	South Africa	6881600	−33.97	18.60	44
Windhoek	Namibia	6811000	−22.95	17.15	1685

TABLE 6.3. GNIP AND GNIR STATIONS PROVIDING THE DATA SHOWN IN FIG. 6.7 (cont.)

Station name	Country or region	WMO code	Latitude (°)	Longitude (°)	Altitude (m)
Alexandria	Egypt	6231800	31.18	29.95	7
St. Louis (MO)	USA	7243400	38.75	−90.38	163
Ankara (central)	Türkiye	1713000	39.95	32.88	902
Halley bay	Antarctica	8902200	−75.58	−20.57	30
Denver (CO)	USA	7246900	39.77	−104.88	1610
Portland (OR)	USA	7269800	45.60	−122.60	6
Juan Fernandez Island (Pacific O.)	Chile	8558500	−33.62	−78.83	6
Vernadsky (Argentine Is.)	Antarctica	8906300	−65.08	−63.98	20
		Rivers			
Arkansas — Van Buren (AR)	USA	R723440	35.35	−94.30	113
Bio Bio — Concepcion	Chile	R856820	−36.87	−73.05	7
Colorado — Cisco (UT)	USA	R724773	38.80	−109.29	1247
Columbia — Pasco (WA)	USA	R727843	46.22	−119.10	120
Danube — Vienna	Austria	R110350	48.25	1638	156.48
Glomma — Sarpsborg	Norway	R013841	59.28	11.12	20
Grande — San Acacia (NM)	USA	R723620	34.26	−106.89	1420
Grande — Socorro (NM)	USA	R722711	33.93	−106.85	1406

TABLE 6.3. GNIP AND GNIR STATIONS PROVIDING THE DATA SHOWN IN FIG. 6.7 (cont.)

Station name	Country or region	WMO code	Latitude (°)	Longitude (°)	Altitude (m)
Mississippi — St. Louis (MO)	USA	R722351	38.86	−90.14	129
Niger — Onitsha	Nigeria	R652571	6.14	6.76	39
Nile — Cairo	Egypt	R623660	30.10	31.38	102
Orange — Gariep Dam	South Africa	R682900	−30.62	25.51	1253
Orinoco — Ciudad Bolivar	Venezuela	R804440	8.15	−63.55	34
Ottawa — Deep River	Canada		46.11	−77.48	113
Sao Francisco — Pirapora (MG)	Brazil	R834371	−17.36	−44.95	492
Sog — Thingvallavatn	Iceland	R404000	64.20	−21.17	109
Susquehanna — Harrisburg (PA)	USA	R725112	40.26	−76.89	88
Tigris — Rezuk	Türkiye	R172101	37.66	41.83	429
Zambezi — Tete	Mozambique	R672610	−16.16	33.59	123
Yukon — Ruby (AK)	USA	R749222	64.74	−155.49	53

6.4.2. Altitude effect

The natural tritium activity concentrations in precipitation are expected to increase with altitude [6.79, 6.80], and the tritium contents in rivers were initially hypothesized to also depend on the altitude of the catchment. The main mechanism controlling the increases in tritium activity in precipitation with altitude is related to temperature, which induces variable condensation. Additionally, the high troposphere could contain more tritium-rich water vapour owing to the intensity of the exchanges with the stratosphere. In addition, soil surface water vapour is

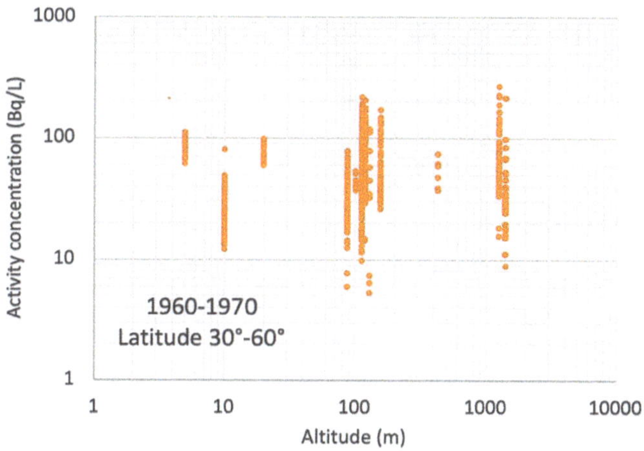

FIG. 6.8. Tritium activity concentration measured in river water for 1960–1970 at GNIR stations in the latitudinal zone from 30° to 60°, shown as a function of altitude. Data from the GNIR database.

expected to have significantly different concentration of tritium from streams or precipitation, depending on soil type and water fluxes. There are no reported studies on the influence of altitude on tritium concentration in rivers.

Data from several sources were investigated to evaluate the altitude effect. These included river water data from the GNIR database acquired over 1960–1970, peak emission data from the global atmospheric nuclear fallout, and river data distributed over the latitudinal area ranging from 30° to 60° (i.e. the zone experiencing the greatest atmospheric fallout). Results show that for fresh waters, the altitude of the catchment does not play a significant role in tritium contents (see Fig. 6.8 and Table 6.4).

TABLE 6.4. NAMES AND LOCATIONS OF GNIP STATIONS PROVIDING THE DATA SHOWN IN FIG. 6.8

Station name	Country or region	WMO code	Latitude (°)	Longitude (°)	Altitude (m)
Grande — San Acacia (NM)	USA	R723620	34.26	−106.89	1420
Mississippi — St. Louis (MO)	USA	R722351	38.86	−90.14	129

TABLE 6.4. NAMES AND LOCATIONS OF GNIP STATIONS PROVIDING THE DATA SHOWN IN FIG. 6.8 (cont.)

Station name	Country or region	WMO code	Latitude (°)	Longitude (°)	Altitude (m)
Susquehanna — Harrisburg (PA)	USA	R725112	40.26	−76.89	88
Grande — Socorro (NM)	USA	R722711	33.93	−106.85	1406
Ottawa — Deep River	Canada		46.11	−77.48	113
Colorado — Cisco (UT)	USA	R724773	38.80	−109.29	1247
Arkansas — Van Buren (US)	USA	R723440	35.35	−94.30	113
Columbia — Pasco (WA)	USA	R727843	46.22	−119.10	120
Danube — Vienna	Austria	R110350	48.25	16.38	156.48
Glomma — Sarpsborg	Norway	R013841	59.28	11.12	20
Nile — Cairo	Egypt	R623660	30.10	31.38	102
St. Lawrence — Champlain	Canada	R717141	46.44	−72.33	5
Weser — Langwedel	Germany	R1022401	52.96	9.15	10

6.4.3. Seasonality and residence time of water masses

River waters have tritium concentrations that are generally significantly lower than those in precipitation. The main mechanisms controlling tritium concentration are the transit time of the water masses between the atmosphere and the terrestrial hydrosphere, any radioactive decay within the systems, and any dilution with underground waters that are generally depleted in tritium. Two aqueous components contribute to water flow in rivers: (i) surface runoff, with typical transit times in the catchment of the order of days to weeks, and (ii) groundwater, with typical transit times of the order of years to tens of years. Whereas surface waters carry the tritium signature of local precipitation, the tritium content in groundwater can vary widely, from much lower tritium contents compared with modern precipitation to tritium levels higher than modern precipitation. This depends on the dynamics of groundwater flow and mean transit time of this component. Both components may feed the river along its course with varying intensity, depending on the morphology of the catchment and the lithology of the underlying geological strata.

From 2014 to 2016, river water samples were collected along 17 small rivers in the south of France. The sampling locations were positioned away from the influence of direct releases from nuclear plants, and HTO concentrations ranged from 0.12 ± 0.11 to 0.86 ± 0.15 Bq/L. This compared with HTO in rain, which ranged from 0.59 Bq/L (first two months of 2014) to 1.55 Bq/L (July 2014) at Thonon-les-Bains, which was the nearest sampling station with a similar altitude (385 m) and latitude (46.3708° N) [6.30]. In that study, the results of monthly averaged tritium concentrations in rivers showed no correlation with tritium in precipitation and had values approximately half of tritium concentrations in precipitation (see Fig. 6.9). The application of multiple linear regression to the data showed that the HTO concentration in rain, the catchment area and the watershed altitude were the main parameters that influenced the variability of HTO levels in the studied rivers [6.30]. The difference between HTO concentrations in rain and rivers reflected both the residence time of water masses at the scale of the catchment and the mixing of river waters with groundwater most generally depleted in tritium compared with superficial waters.

A negative correlation was expected between tritium content and the distance from the river source, since increased distance allows for more input from groundwater and from lower elevation runoff. This runoff leads to dilution of the higher tritium concentrations from the upper part of the catchment with shorter flow paths [6.81]. It was also anticipated that tritium concentrations would be lower below dammed reservoirs, with a dependency based on reservoir size, turnover times and possible groundwater input. Surface water bodies are strongly coupled to the atmospheric water vapour reservoir via isotope exchange across the water–air interface. Correspondingly, waters released from reservoirs are expected to have lower tritium concentration than waters flowing into reservoirs because of radioactive decay of tritium in the reservoir and because of input of low tritium groundwater into the reservoir.

The strong seasonality observed in rain is not reflected in the river series studied by Ducros et al. [6.82] (see Fig. 6.9), in contrast to inferences from several other studies [6.74, 6.83]. Different processes (e.g. mixing, storage) operate to smoothen the amplitude of the seasonality in rivers, similar to those occurring in lakes, reservoirs and groundwater [6.84]. A seasonality effect for tritium content in precipitation is carried to rivers by surface runoff. In groundwater, the seasonal variations of tritium in precipitation are largely smoothed out if the mean transit time of water is longer than about four years [6.85]. The influxes of younger

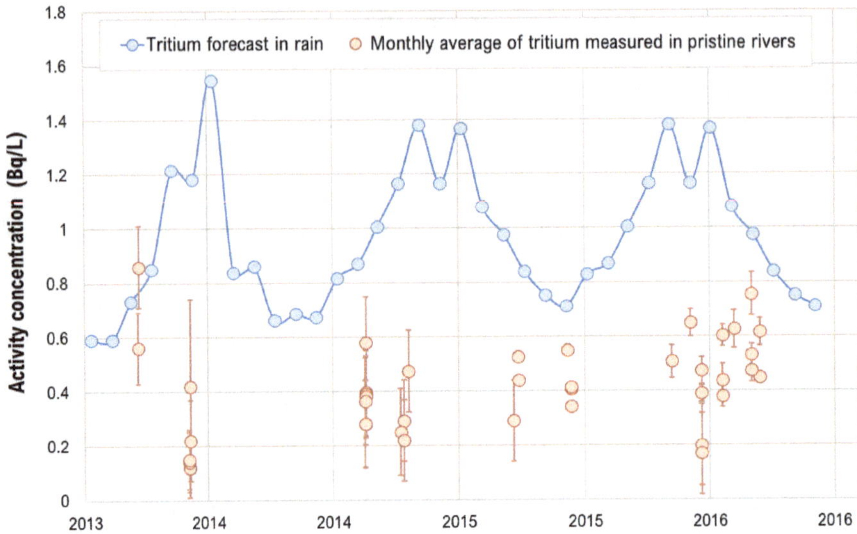

FIG. 6.9. Comparison between monthly tritium concentrations forecasted in rain at Thonon-les-Bains (blue) and monthly averages of tritium measured in the studied rivers (red) [6.82].

water or recent rain with low tritium concentration compared with base flow for most Northern Hemisphere rivers also combine to decrease the seasonality effect [6.83, 6.86]. Harms et al. [6.87] did not find a clear relationship between seasonal variation of tritium in rainfall and tritium levels in river water. Overall, the data indicate that there is no systematic seasonality of tritium contents in the river systems.

6.4.4. Drainage basin area

Studies on the relationship between tritium concentrations and drainage catchments show variations. Cartwright and Morgenstern [6.88] reported no positive correlation between an Australian catchment area and tritium concentrations in the draining river. However, a positive relationship was described by Radwan et al. [6.76], which underlined lower annual average tritium contents in rivers characterized by smaller draining surfaces. Momoshima et al. [6.7] indicated that the larger surface area of some islands in Japan would cause a longer residence time of groundwater owing to the larger size of the water reservoir. This could explain, together with the hydrogeological background, the differences observed in tritium concentrations in the various rivers of different islands. Indeed, the base flow of many rivers is generally composed of water that was retained in the basin over decadal timescales and labelled by global fallout. It may also reflect the higher tritium concentrations found during the earlier part of the tritium transient [6.83].

6.4.5. Hydrology and mixing with underground waters

The tritium concentration in the surface waters throughout California were characterized by Harms et al. [6.87]. The authors showed that tritium concentrations in most surface waters decreased between summer–autumn 2013 and winter–spring 2014 owing to an increased groundwater signal because of drought conditions in 2014. The same authors reported that surface waters in the Sierra Nevada and Central Valley areas showed large to moderate differences in tritium between seasons. The most significant seasonal difference was observed in the South Fork of the American River, above Folsom Lake. Tritium concentrations decreased from summer to autumn to winter to spring in most of the Sierra Nevada and Central Valley locations. This decrease was attributed to a greater contribution of low tritium groundwater to surface waters in the 2013–2014 winter–spring samples collected under drought conditions. Furthermore, a relationship between tritium and electrical conductivity in surface water was found to be indicative of water provenance and anthropogenic influences such as agricultural runoff.

Despite low initial concentrations in precipitation, tritium continues to be a valuable hydrological tracer in the nuclear era owing to the labelling of waters by atmospheric global fallout from nuclear tests.

6.4.6. Distance to the marine coast

Little information exists in the literature on the effect of coastal distance. Some reported work indicates that a trend of changing tritium concentrations with distance from coasts would likely be due to a decreased effect of dilution by marine derived moisture and the lower mixing between water surface and groundwater at higher altitudes. Harms et al. [6.87] showed that surface water samples collected in California presented a trend of increasing tritium contents with increasing distance from the marine coast. Superimposed on that trend, elevated tritium concentrations are found in the San Francisco Bay area compared with other coastal areas, resulting from municipal water imported from inland mountain sources and local anthropogenic sources.

6.4.7. Potential impact of climate change

Recent studies indicate that climate change, through its impact on natural processes, will remobilize tritium that was trapped in freshwater reservoirs at the time of global atmospheric nuclear fallout. The thawing of vast areas of Arctic and Himalayan permafrost by global warming will remobilize and transfer initially trapped HTO to lakes and rivers. Similarly, increased glacier melting is expected to enhance buried tritium concentrations in surface and underground waters owing to their remobilization from such reservoirs.

Radioactive decay of tritium released in atmospheric nuclear tests in 1963 can reduce its activity 30 times by 2023, resulting in activity concentrations ranging from 10 to 30 Bq/L. However, additional environmental transfer processes, such as dilution, exchange and transport, result in significantly lower values of tritium activity concentrations. Typically, less than 1 Bq/L is observed currently in rain and river waters (cf. Fig. 6.2).

6.4.7.1. Permafrost thaw

Permafrost melting is particularly sensitive to climate change, with recent studies providing evidence that permafrost is warming and retreating, leading to a deepening of the active top layer of soil that thaws during the summer and freezes again during the autumn [6.89–6.93]. Recent studies have identified a general increasing trend in the thickness of the active layer in many regions, including the Russian Federation [6.94], Tibet [6.95], Northern Europe [6.90], Alaska [6.96] and the Canadian Arctic [6.91, 6.93].

Widespread future warming in permafrost regions will result in substantial changes in terrestrial hydrology [6.97], vegetation composition [6.98], landscape topography [6.99], methane fluxes [6.89], ecosystem function [6.100] and the fate and transport of contaminants [6.101–6.104], including radionuclides [6.105]. Some research on permafrost degradation induced by global warming has focused on tritium concentrations in surrounding freshwater systems. Gibson et al. [6.106] were the first to measure tritium in various lakes, wetlands and groundwaters in the oil sands region of northeastern Alberta (Canada). They compared the tritium contents in recent precipitations and those sampled during the period of peak emission from nuclear tests in the 1960s. Their study included water samples from several thaw lakes that had higher runoff attributed to melting of permafrost in plateaux peats within their watersheds. It was shown that while tritium concentrations in all lakes were intermediate between recent and 1960s tritium concentrations in precipitation, the thaw lakes were significantly enriched compared with other lakes. This provided evidence that melting of modern permafrost, formed since the 1950s, is occurring in these systems. In addition, Bond and Carr [6.107] showed that stratigraphic layers of permafrost contained trapped tritium from the 1950s and 1960s. This isolated and/or trapped tritium has undergone only radioactive decay and has substantially lengthened its environmental half-life compared with areas with little or no permafrost. In the latter areas, the weapon test era precipitation has been subjected to both decay and hydrodynamic dilution and dispersion.

Recent data show that the Arctic is warming three times faster than other parts of the world, with northern regions suffering some of the most pronounced effects of climate change, resulting in permafrost degradation. Various water bodies across the Canadian subarctic region, spanning the continuous, discontinuous and isolated patches of permafrost zones in northern Manitoba, Northwest Territories and Labrador, were studied. These showed excess tritium contents in several water samples from the sporadic discontinuous and extensive discontinuous permafrost areas [6.107]. The greatest tritium enrichment (up to 128 TU) was observed in lakes near Jean Marie River in the Mackenzie River valley, a region known to be experiencing extensive permafrost degradation. These results demonstrate a significant permafrost degradation in the central Mackenzie River basin and that tritium is becoming increasingly mobile in the subarctic environment at concentrations higher than expected because of the warming climate.

6.4.7.2. Glacier melting

Permafrost soils and mountain glaciers have acted as repositories of sequestered tritium deposited during the nuclear weapon testing era. Supporting evidence comes from the wide use of tritium to date ice cores and study ice accretion and degradation mechanisms [6.78, 6.108–6.113].

A decade after peak tritium from atmospheric nuclear tests, Ambach et al. [6.108] showed seasonal fluctuations of tritium concentration in glacier melt water discharges, which are known to contain practically no tritium. They observed low tritium contents in the runoff during the period of maximum ice ablation. They also showed that the mean residence time of the base flow in the groundwater system of a catchment area was estimated to be a few years on account of the tritium content in the winter runoff and in the precipitations of past years. Later studies by Strasser et al. [6.110] on samples of debris-rich basal ice from Matanuska Glacier (Alaska) collected in the early 1990s showed tritium concentrations ranging from 5 to 40 TU, which indicated ice containing an early 1950s tritium nuclear test component. Michel and Naftz [6.114] also reported low tritium concentrations in some runoff samples from an alpine glacier in Wyoming (USA) collected over the same period, which indicates that some water was likely derived from melting of pre-1960 glacial ice layers.

Inputs from glacier melting are of interest because the residence time of waters in such reservoirs would be close to decadal. This will potentially lead to the remobilization of free waters initially trapped as ice over the period of nuclear weapon test fallout. In the European Alpine area, where the mean residence time of ice in glaciers is estimated to be approximately 60 years, such removal is expected to occur by the 2020s, although this has so far not been clearly demonstrated [6.115]. Not only tritium, but also other radionuclide components of radioactive fallout from the nuclear test period, will be mobilized.

6.5. ORIGIN AND FATE OF ORGANICALLY BOUND TRITIUM IN FRESHWATER SYSTEMS

It is established that OBT tritium is likely to be of greater dosimetric importance than HTO but, unlike HTO, its behaviour is not well understood under dynamic tritium exposure conditions both in aquatic and in terrestrial ecosystems [6.116, 6.117]. Despite limited knowledge, a few dynamic models have been developed to evaluate OBT concentrations in aquatic organisms [6.116, 6.118, 6.119]. For example, the model presented by Melintescu and Galeriu [6.118] has been successful in predicting OBT concentrations in small fish and in larger trout [6.11, 6.120–6.122]. Data relating to larger fish are currently limited, and well designed experiments are still required to obtain those data, as established by the Tritium Working Group of EMRAS II [6.42]. Such studies are beneficial, as they provide insights regarding the dynamics of HTO relative to the much less mobile OBT in freshwater systems (see Section 6.3.1). For example, in the case of sudden changes in ambient tritium concentrations in fresh surface waters, TFWT concentrations in aquatic biota will reach steady state within hours, whereas OBT in biota tissues can take years to reach a steady state [6.43, 6.116, 6.123, 6.124] (see Table 6.5).

A few tritium models assume that tritium is evenly distributed within the body. This is essentially true with respect to a single acute intake of HTO [6.125]. However, a small fraction of the HTO intake is metabolized and becomes incorporated into cellular and tissue components. Under conditions of continuous intake, the amount of tritium incorporated as OBT will gradually accumulate under constant intake conditions until an equilibrium position is reached. It is also known that different tissues accumulate or metabolize OBT differently. Animal models (farmed and wild fauna) clearly demonstrate that tritium is not uniformly distributed within the body [6.121, 6.126–6.129].

For aquatic ecosystems, especially fish, the models developed up to now assume that tritium is uniformly distributed within the body, because those dedicated models do not divide the fish body into organs or tissues, but there is

TABLE 6.5. ESTIMATED TIME NEEDED FOR TRANSPLANTED FRESHWATER BARNES MUSSELS (*elliptio complanata*) TO REACH STEADY STATE FOLLOWING ABRUPT CHANGES IN AMBIENT TRITIUM CONCENTRATIONS IN A FRESHWATER LAKE

Exposure conditions	Parameter	Total time needed for mussel to reach steady state (d)	References
Abrupt increase in ambient HTO	HTO concentration	0.08	[6.116, 6.123]
	OBT concentration	767 [a]	
Abrupt decrease in ambient HTO	HTO concentration	0.08	[6.42]
	OBT concentration	613 [b]	

[a] Projected value, as OBT concentrations had not reached a steady state by the end of the 88th day mussel transplantation study.

[b] Projected value, as OBT concentrations had not reached a steady state by the end of the 117th day mussel transplantation study.

some experimental evidence indicating that this is a significant oversimplification [6.130, 6.131].

6.5.1. Tritium bound to naturally occurring organic compounds

In river systems, tritium is present as OBT in both the fresh and detrital organic matter reservoir. POM in rivers displays two main origins.

Autochthonous POM is produced directly within rivers mainly through photosynthesis of aquatic plants. This will be variably transferred along food chains to fish and shellfish that directly and/or indirectly feed on plants [6.116, 6.132]. Such autochthonous organic material is expected to reflect the HTO contents of the surrounding waters.

Allochthonous detrital POM is transported by rivers and originates from the erosion of the soils within the watershed. Those detrital allochthonous materials also carrying OBT would display OBT contents that are different from those of autochthonous POM, since this material is produced in the terrestrial environment of the catchment.

Tritium contents within the hydrosphere are not at equilibrium and depend on water fluxes and water mass recycling between various environmental compartments. Similarly, the OBT content in river POM is not expected to be at equilibrium with the HTO content of the river waters. In this context, disequilibrium in OBT/HTO ratios is expected to mostly depend both on the proportion of allochthonous/autochthonous POM and on the age of the contained organic material. In practical terms, this will apply to organic matter formed during the last 60 years, which will have a component of weapon fallout tritium. OBT concentrations in the organic matter formed before 1958–1963 are essentially not accessible to precise measurements owing to radioactive decay of tritium.

6.5.2. Tritium bound to human made organic compounds

In addition to tritium biointegration in 'natural' organic molecules, which conventionally characterize OBT, some specialized industrial processes have released technogenic organic compounds, termed here as T-TOM or technogenic tritiated organic particles. Examples of such industrial sources of technogenic tritium compounds are the following:

(a) The former radiopharmaceutical site near Cardiff that released T-TOM into the Severn Estuary [6.31–6.33].
(b) The workshops used to support the watchmaking industries found in the upstream waters of the Rhône river, releasing tritium loaded paint residues that became admixed in riverine sediment [6.35, 6.36, 6.38, 6.48, 6.133].
(c) Residues from plants manufacturing tritium powered luminescent items (e.g. SRBT factory in Pembroke, Canada[2]).

6.5.2.1. Luminescent paints used in watchmaking workshops: Rhône river

Radioluminescent paint has been used in the watchmaking industry since 1918, when radium paints were used. Owing to the radiotoxicity of radium, the practice was abandoned by the 1960s and tritium was introduced as a safer, alternative luminizing agent. Tritiated luminescent materials were integrated into the paints applied to watches and clocks, leading to technogenic tritium production and wastes. The practice ended in 1992 (France) and 2008 (Switzerland), but high levels of T-TOM as particulates remain in the sediments of the Rhône river, downstream of Lake Geneva. The French and Swiss watchmakers produced almost 350 million watches and used more than 28 000 TBq of tritium over the

[2] https://nuclearsafety.gc.ca/eng/resources/maps-of-nuclear-facilities/iemp/srb-tech.cfm

periods of production [6.35, 6.37]. More than 95% of this would have disappeared owing to the radioactive decay of tritium.

The watchmaking and clockmaking industries were mainly located in the Rhône watershed and particularly in the part farthest upstream, the Doubs (tributary of the Saône). The workshops were mostly located in three cities: Besançon and Morteau in France, and La Chaux-de-Fond in Switzerland. Tritium analyses carried out on the sediments of the Rhône and its tributaries indicate significant concentrations along the river, as well as in several northern tributaries, in particular the Ognon and the Tille rivers (tributaries of the Saone) and the Doubs, Loue and Arve rivers [6.82]. The recorded levels in the sediments of these rivers (10 to over 20 000 Bq/L at the end of the 2010s) are inferred to be linked to the synthetic tritiated particles used by the workshops. The significant contamination levels decrease from north to south in the Rhône watershed and further decrease over time owing to radioactive decay.

The tritiated radioactive particles are poorly characterized and their temporal and spatial impacts in the riverine environment are relatively understudied. To date, the assimilation of these particles by living organisms, their persistence, and their accumulation along the trophic chain have not been studied significantly. The particles are predicted to be submicrometre sized polystyrene fragments, chemically resistant and widely dispersed in the accumulated river sediment, like many other non-radioactive microplastics. They tend to escape identification and analysis and they are generally difficult to discern within solid matrices because of their high dilution with coarser particles.

Investigators noted that the tritium contents within the various aquatic compartments of the Rhône (fish, plant, sediment) showed significant enrichments both upstream and downstream of the Creys-Malville NPP [6.36, 6.134, 6.135]. The elevated tritium in the suspended matter and sediments of the Rhône were not seen in other French rivers hosting NPPs. Jean-Baptiste et al. [6.36] showed that while TFWT concentrations are in equilibrium with the river water, the non-exchangeable fraction of OBT in plants and fishes showed values that exceeded the river water background by one to two orders of magnitude, and up to four to five orders of magnitude in the sediments. Since these anomalous tritium levels were present 100 km upstream of the NPP (at the Swiss border), this clearly indicated that another, non-NPP-related, source would be responsible.

Other studies involved radiometric dating of sediment cores and found tritium maxima in sediments corresponding to 1980, which was the time of intensive use of tritium by watchmakers [6.12]. These T-TOM spikes reach 688 ± 26 Bq/kg for the upstream core and 87.7 ± 6.1 Bq/kg for the downstream core in the lower layers and are 10^3 to 10^4 times higher than current natural background levels in the riverine sediments (<0.1 Bq/kg in 2023). The lifetime of tritium in the environment was determined using trajectory analysis to estimate the return time to insignificant levels. These sediment core studies show that the Rhône returns to undetectable and/or insignificant levels in 14–28 years for the downstream part and 14–70 years for the upstream part of the river.

The technogenic tritium particulate contamination from the watchmaking industry is now known to have reached the coastal marine environment at the mouth of the Rhône [6.133]. The tritium content in the suspended matter in sea water ranges between 85 and 1540 TU (10–184 Bq/L).

6.5.2.2. *Tritium labelled organic compounds used in life sciences research: Radiopharmaceutical centre in Cardiff*

The levels of tritium observed in fish and other marine life in Cardiff Bay (Severn Estuary) were much higher than predicted by steady state models [6.32, 6.136]. Concentration factors of the order of 10 000 between seawater and marine species were observed, while standard models assume a concentration factor of 1. A later study on muddy sediments by Croudace et al. [6.33] postulated that some of the classes of technogenic compounds with T-TOM released from the Cardiff radiopharmaceutical facility became attached to finely divided and relatively abundant coal dust. These acted as carbonaceous geosorbents to trap the T-TOM. The coal dust particles were introduced into the sedimentary system from the significant South Wales coal mining operations of the last two centuries. They were continually released, dispersed and mixed into Severn Estuary sediment over the period 1800–1980. The industrial facility produced a wide range of isotopically labelled compounds for use in life sciences diagnostics and research [6.137]. In the late 1990s, the recognition that elevated levels of tritium were accumulating in marine species led to a programme of investigative work by specialist organizations [6.13, 6.27, 6.33, 6.37, 6.48, 6.138, 6.139].

So far, no similar releases of T-TOM into the environment at these magnitudes have been reported and the Cardiff case appears to have been unique. The combination of T-TOM released into macerated sewage with significant interaction and uptake by biota and estuarine muddy sediment, with carbonaceous geosorbent properties, provide an important environmental impact case study [6.33]. In principle, however, other research centres using and releasing such tritiated compounds could generate similar bioaccumulation effects of tritium in freshwater systems (see also Section 5.5.6).

6.5.2.3. *Tungsten particles from thermonuclear facilities: ITER international project*

The development of fusion technology for power generation is expected to be associated with environmental tritium releases [6.140]. The ITER liquid and gaseous releases, after 50 years of operation with regular maintenance schedules, are predicted to generate a total effective dose of around 2.2 μSv/a for an adult in Saint-Paul-lès-Durance (less than 4 km from ITER). This dose is negligible compared with the ICRP regulatory limit set at 1 mSv/a for the public. Tritium in the form of the tritiated water molecule (in gaseous effluent) provides the major contribution to this dose (96%). Other contributions are ^{14}C (about 3%) and ^{41}Ar (<1%), with all other beta-gamma emitters contributing <0.1%.

Assessments are modelled on the basis of increased usage of tungsten–hydrogen (W–H system) in components of ITER. Until now, the walls of the Tokamak prototypes consisted mainly of carbon material in which significant trapping of tritium occurred, inducing the release of tritiated carbon particles with activities sometimes higher than 1 GBq/g [6.141]. During ITER operating mode, the facility could generate tritiated tungsten and other dust-like particles, which need to be characterized in terms of the amount of trapped tritium, tritium desorption and radiotoxicity. Some research has shown that much more tritium is incorporated in tungsten particles than in massive tungsten components, indicating important surface effects on tritium absorption, desorption and trapping in tungsten [6.40]. Tritium desorption from particles is expected to occur at different temperatures that will be related to the different interactions on the particle surface and in the bulk metal. It is expected that the tritium behaviour in massive samples will also depend on the metal microstructure. Following a theoretical loss of vacuum accident or other accident scenarios, tritiated particles could escape through the different protection barriers and filters.

6.5.3. OBT/HTO disequilibria in sediments and aquatic biota

The persistence of tritium derived from atmospheric nuclear weapon testing has led to significant imbalances in the OBT/HTO ratios in riverine sediments. These ratios were about 45 in the 1980s and decreased gradually over time because of (a) the radioactive decay of tritium and (b) the degradation and turnover of organic matter [6.132]. Currently, the OBT/HTO ratios can still reach values of about 15 in the sediments of rivers draining large watersheds, where the fallout from atmospheric nuclear weapon testing was the most significant [6.82]. As previously stated (Section 6.5.1), riverine sediments generally contain various types of organic particulate matter. These are categorized as allochthonous POM, derived from soil erosion in the catchment, and autochthonous POM, which is more labile and more rapidly degraded than allochthonous POM. In the organic horizons of undisturbed soils, the OBT/HTO ratios increase with soil depth (i.e. in relation to the age of soil organic matter and its higher OBT content) [6.11, 6.142, 6.143]. This helps to explain the origin of excess OBT contents in riverine sediments far from nuclear industry releases.

In aquatic systems exposed to tritium releases from nuclear industries, all impacted environmental compartments are expected to display OBT/HTO ratios close to equilibrium. Nevertheless, kinetic differences between water mass transfer and metabolic processes, which lead to OBT formation, may generate significant disequilibria in OBT/HTO. This was observed by Fiévet et al. [6.144] in seaweeds collected close to the La Hague SFRP owing to discontinuous (spiky) releases of liquid tritium effluents from the site. A similar disequilibrium was also reported by Kim et al. [6.145] for Perch Lake, a small shallow shield lake located on the Canadian CRL site that contains elevated levels of tritium due to long lived inputs from a nearby nuclear waste management area. Although tritium levels in Perch Lake have been gradually decreasing since 2000, lake water, sediments, aquatic plants, clams and fish show that OBT concentrations in biota did not always correlate well with the lake water HTO levels. OBT/HTO was found to be less than 1 for aquatic plants, around 1 for clams and fish, and above 1 for birds reared on the shore of the lake. The sediment OBT content did not correlate with the sediment HTO content, presumably owing to inputs from both aquatic and terrestrial sources (e.g. the sediments are partly made up of fallen tree leaves).

In most cases, living organisms reflect HTO contents in rivers, although the presence of enriched OBT contents due to technogenic tritium releases can lead to enhanced concentrations such as those observed in the Rhône (Fig. 6.10).

In river systems, while the mobile forms of tritium (HTO and E-OBT) rapidly dilute and exchange with water molecules in the surrounding environment, the bound forms produced during metabolic processes persist longer, depending on the recycling of organic components involved. Recycling kinetics explain most of the cases of OBT/HTO disequilibria in natural and impacted environments.

The behaviour of tritium in aquatic organisms is reasonably well established, and it is expected that tritium transfer in terrestrial organisms, which share many physiological processes, will be similar. Indeed, in its HTO form, tritium is very easily transferred to plants, as well as to plant-eating animals. The exchangeable fraction of organic tritium equilibrates in a few hours with the environment. Its non-exchangeable fraction exhibits a different behaviour, depending on the nature of the labelled organic molecules, with slower kinetics, which can lead to a notably long retention of tritium in the various environmental compartments. Most information is derived from studies on freshwater microalgae [6.146–6.148]. The incorporation of tritium into the organic matter of these plants results from the use of tritiated water during photosynthesis (simplified reaction: $CO_2 + HTO \rightarrow CHTO + O_2$). These plants are also able to use certain tritiated organic compounds, the fate of which depends closely on their nature [6.57, 6.149, 6.150]. Rare data on tritium transfer in higher plants revealed an imbalance between the organism and water for semi-aquatic plants, probably

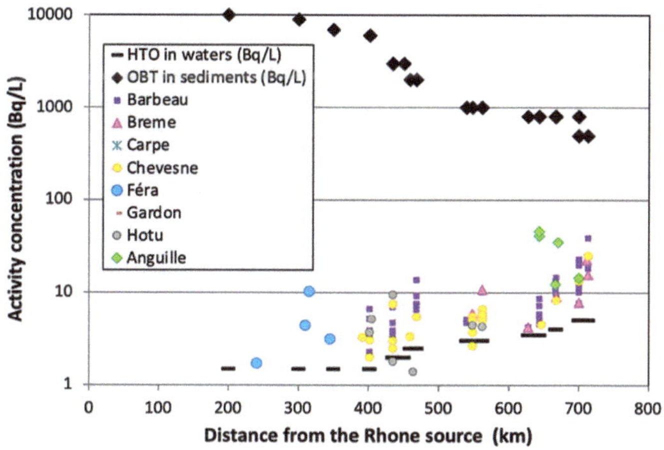

(a)

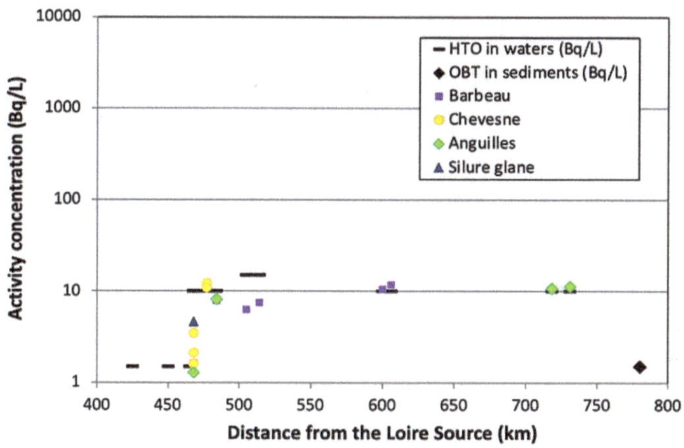

(b)

FIG. 6.10. HTO and OBT content in waters, sediments and fish flesh along the rivers Rhône (top) and Loire (bottom). Barbeau: Barbus; Breme: Abramis Brama; Carpe: Cyprinus carpio; Chevesne: Squaius cephalus; Féra: Coregonus Lavaretus; Gardon: Rutilus; Hotu: Chondrostoma nasus; Anguilles: Anguila Anguilla; Silure glane: Silurus glanis.

related to exchanges between emerged foliage and atmospheric tritium [6.151]. OBT enrichment associated with the tritium remaining from atmospheric tests was less visible in plants. In such cases, the organic matter of allochthonous origin was diluted by that produced locally by plants [6.132]. The same is true for aquatic animals. A phytoplankton bloom, after incorporating tritium into organic forms in the vicinity of an HTO release site, could then be transported by currents to pristine sites with negligible tritium concentrations [6.152].

Hydrogen in exchangeable hydrogen pools is expected to equilibrate isotopically with the ambient water, whereas in a non-exchangeable pool the tritium content acquired by the organic matter remains unchanged from the time when it was formed. This does not mean that tritium enrichments relative to water do not exist, as such enrichments are known in some marine biota. In these cases, this enrichment is not due to an unknown isotopic effect, but to the uptake of tritiated compounds already present in the aquatic medium. These compounds may be either of technogenic origin discharged into the environment [6.32] or organic matter formed in a tritiated water plume close to nuclear discharges and subsequently dispersed by currents. Thus, concentration factors can be misleading when applied to tritium, since they imply that tritium concentrated in aquatic organisms is derived from the tritiated water molecules themselves. It is thus important to remember that whereas there is evidence for bioaccumulation of tritiated compounds or tritiated suspended matter in the aquatic environment, there is no evidence to demonstrate a significant tritium bioaccumulation mechanism involving conversion of HTO to OBT from surrounding water.

6.6. TRITIUM CONTENT IN FRESH WATERS OVER THE NUCLEAR ERA

6.6.1. Past and present tritium background contents in river systems

6.6.1.1. HTO contents in river waters

Tritium concentrations in rivers are poorly documented in the literature, except for specific hydrological studies. Small variations exist in the background levels of tritium in lakes and rivers today, depending to some extent on latitude and proximity to the sea. Radwan et al. [6.76] reported background levels for inland rivers and tributaries in Poland at 1.5–1.9 Bq/L over the 1994–1999 period. Lakes ranged from 1.4 to 1.8 Bq/L, while rivers near the Baltic Sea ranged from 0.7 to 1.3 Bq/L. The concentrations in inland waters were similar to those in precipitation, which ranged from 1.3 to 2.0 Bq/L over the same period. Miyamoto et al. [6.8] reported tritium levels in Japanese rivers at 2.0 Bq/L in 1985, which were also similar to the average levels in precipitation.

In France, tritium background levels have been reported for 17 small rivers in the south of France [6.81] that were outside the influence of direct releases from nuclear facilities (see Section 6.4.3). The tritiated water (HTO) activity concentrations ranged from 0.12 ± 0.11 to 0.86 ± 0.15 Bq/L over the period 2013–2016, showing significant variability and being less than half that in rainwater over the same period. Multiple linear regression showed that for HTO concentration in rainfall, watershed areas and altitude were responsible for the variability of HTO concentrations in the studied rivers. Estimates of HTO fluxes into the Mediterranean Sea indicate that French coastal rivers not influenced by nuclear industry releases deliver only 1% of HTO to the sea, while the remaining 99% is brought by the Rhône river, where several nuclear facilities are located.

In Canada, Kim et al. [6.146] recently reported a likely reference value of 1.9 ± 0.4 Bq/L for tritium contents in freshwater systems by sampling in Lake Nipissing, an area considered to be a tritium background location.

The datasets from the GNIR database show that the maximum tritium contents in river waters were mostly observed during the peak tritium emissions from thermonuclear weapon testing in the atmosphere in the early 1960s (see Figs 6.11 and 6.12, with the list of stations in Table 6.6). This is generally observed in samples from the Northern Hemisphere, where most nuclear tests occurred. Interestingly, various rivers display peak tritium contents in waters delayed to the summer of 1964 (e.g. the Danube in Vienna, the Colorado River in Cisco, the Ottawa River in Deep River) or the summer of 1965 in some cases (e.g. the Mississippi River in St Louis). This time lag is explained by the transit time of the water masses through the catchment and the spring leak effect (see Section 4.3.1).

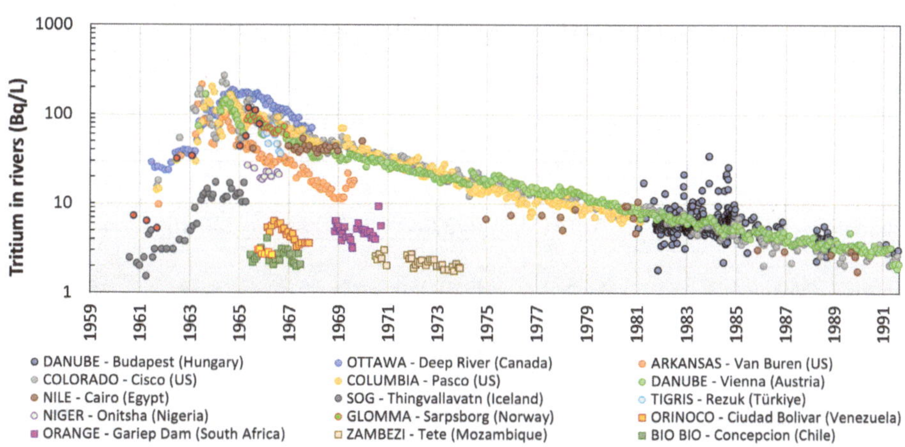

FIG. 6.11. HTO variations in river waters over time at the displayed locations in the Northern Hemisphere (circles) and the Southern Hemisphere (squares) in the period 1959–1991. Data from the GNIR database.

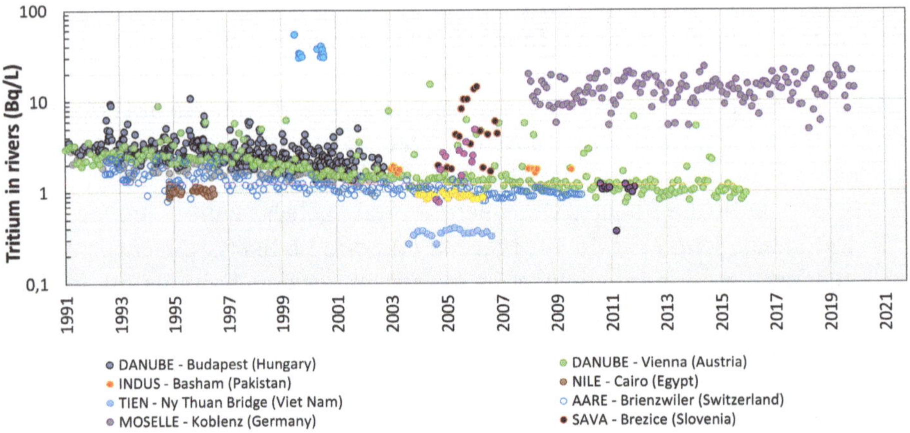

FIG. 6.12. HTO variations in river waters over time at the displayed locations in the Northern Hemisphere in the period 1991–2020. Data from the GNIR database.

TABLE 6.6. LOCATIONS OF SAMPLING STATIONS FOR THE DATA SHOWN IN Figs 6.11 AND 6.12

Station name	Country or region	WMO code	Latitude (°)	Longitude (°)	Altitude (m)
Arkansas — Van Buren (AR)	USA	R723440	35.35	−94.30	113
Bio Bio — Concepcion	Chile	R856820	−36.87	−73.05	7
Colorado — Cisco (UT)	USA	R724773	38.80	−109.29	1247
Columbia — Pasco (WA)	USA	R727843	46.22	−119.10	120
Danube — Vienna	Austria	R110350	48.25	16.38	156.48
Danube — Budapest	Hungary	R128440	47.39	19.01	99
Glomma — Sarpsborg	Norway	R013841	59.28	11.12	20
Grande — San Acacia (NM)	USA	R723620	34.26	−106.89	1420

TABLE 6.6. LOCATIONS OF SAMPLING STATIONS FOR THE DATA SHOWN IN Figs 6.11 AND 6.12 (cont.)

Station name	Country or region	WMO code	Latitude (°)	Longitude (°)	Altitude (m)
Grande — Socorro (NM)	USA	R722711	33.93	−106.85	1406
Mississippi — St. Louis (MO)	USA	R722351	38.86	−90.14	129
Niger — Onitsha	Nigeria	R652571	6.14	6.76	39
Nile — Cairo	Egypt	R623660	30.10	31.38	102
Orange — Gariep Dam	South Africa	R682900	−30.62	25.51	1253
Orinoco — Ciudad Bolivar	Venezuela	R804440	8.15	−63.55	34
Ottawa — Deep River	Canada		46.11	−77.48	113
Sao Francisco — Pirapora (MG)	Brazil	R834371	−17.36	−44.95	492
Sog — Thingvallavatn	Iceland	R404000	64.20	−21.17	109
Susquehanna — Harrisburg (PA)	USA	R725112	40.26	−76.89	88
Tigris — Rezuk	Türkiye	R172101	37.66	41.83	429
Zambezi — Tete	Mozambique	R672610	−16.16	33.59	123
Yukon — Ruby (AK)	USA	R749222	64.74	−155.49	53

Generally, tritium contents in most river waters returned to background values (<1 Bq/L) over the decades following the peak values after the cessation of atmospheric nuclear tests. Examples include the Aare river (Switzerland), the Tien River (Vietnam) and the Euphrates (Syria). By contrast, there are several rivers displaying higher tritium contents than expected background values, and these reflect the local and/or regional impact of nuclear industries on the Danube in Vienna (Austria) and the Moselle in Koblenz (Germany). In both cases, several NNPs are located upstream. Excess tritium contents are also observed in the Terelj River in Mongolia at the beginning of the 2000s, when a survey was performed there. This was linked to the proximity to the Chinese nuclear test site Lop Nur, where atmospheric nuclear weapon tests took place from 1964 to 1980.

The first measurements available for the Southern Hemisphere were performed in the Bio Bio River (Chile) in June 1965 (Fig. 6.11). The data acquired show that the rivers in the Southern Hemisphere displayed lower tritium contents than those in the Northern Hemisphere by one to two orders of magnitude over the 1965–1975 period.

Overall, the tritium contents in river waters and precipitation (see Fig. 6.13) were generally similar at the global scale (approaching background values). The exception to this was river systems that had been impacted for decades by nuclear industry releases, where tritium contents may have reached more than 10 times background values.

6.6.1.2. OBT contents in river sediments

Terrestrial biomass pools, contaminated by global atmospheric fallout from nuclear testing, have constituted a significant delayed source of OBT for aquatic systems, resulting in an apparent enrichment of OBT compared with HTO [6.132]. This finding helps to explain CRs greater than one, i.e. tritium concentration in biota/water observed in areas that are not directly affected by industrial radioactive wastes. This also sheds light on controversies regarding tritium bioaccumulation. Such apparent enrichment of OBT in riverine systems is expected to be more pronounced in the Northern Hemisphere, where fallout was greatest, depending on the nature and biodegradability of terrestrial biomass at the regional scale. OBT transfers from the continent to the oceans have been sufficient to affect tritium concentrations in coastal marine biota (i.e. near river mouths). The persistence of terrestrial organic tritium explains imbalances between OBT and HTO in most river systems, particularly those not impacted by releases from nuclear facilities. These findings were recently confirmed by Eyrolle et al. [6.20] on the basis of data acquired by measuring both free and bound forms of tritium in subsamples taken from a continuous sediment core collected in the Loire river basin. The core archive collected represented approximately 70 years of sediment accumulation. The authors demonstrated that tritium from global atmospheric fallout was largely immobilized for decades as organically bound forms in highly resistant organic matter (POM) in accumulating sediment.

By contrast, the same study showed that tritium released by nuclear industries as HTO into rivers would not become trapped for the long term in organic matter of riverine sediments. This is explained by the relatively high

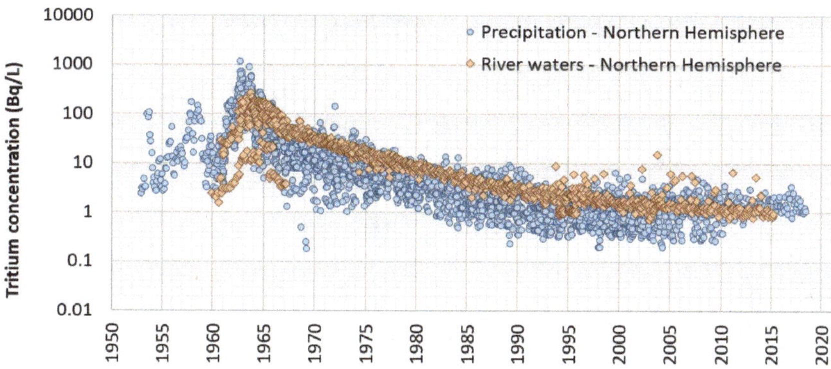

(a)

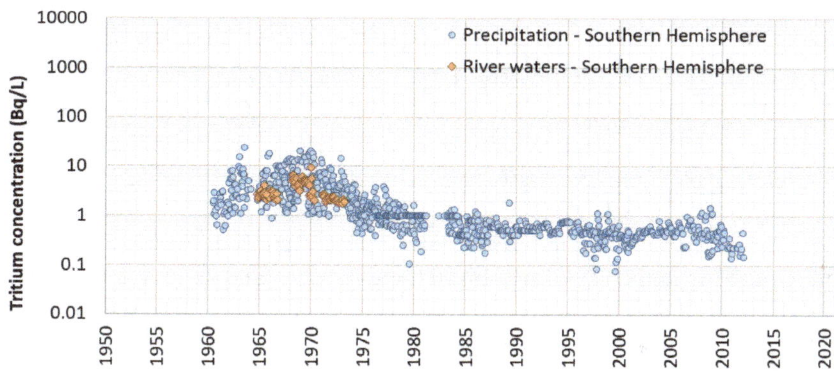

(b)

FIG. 6.13. HTO contents in precipitation (blue circles) and river waters (orange diamonds) over time in (a) the Northern Hemisphere and (b) the Southern Hemisphere. Data from the GNIR database.

degradation rates for aquatic organic compounds. Additionally, it was found that OBT concentrations in riverine sediments reach a maximum several years after peaking in the atmosphere owing to the time lag between the production of organic matter on the given catchment and its erosion and transport to the river systems. Such tritium storage, as well as the decadal time lag of tritium contamination levels between atmosphere and river systems, are assumed to behave in this way globally. These sedimentary reservoirs act as both tritium sinks and potential delayed sources of mobile and bioavailable tritium for freshwaters and living organisms, even decades after the atmospheric contamination had ceased.

6.6.1.3. OBT contents in living organisms in fresh waters

Tritium background reference values for living organisms in aquatic systems are poorly documented in the literature. However, a survey of eastern Canadian biota determined background OBT in fish samples. In fish samples from Lake Nipissing with averaged TFWT of 1.6 ± 0.1 Bq/L, the average tissue OBT value was reported as 4.0 ± 0.4 Bq/L [6.146].

The reference activities of the main artificial radionuclides, including tritium, in foodstuffs produced in France are given in Fig. 6.14. These activities are mainly linked to the presence of fallout from nuclear tests and the Chernobyl accident (for ^{137}Cs). The $^{239+240}$Pu activities are lowest for three main reasons: (1) during the nuclear fallout, $^{239+240}$Pu deposits were of the order of 100 times lower than those of either ^{137}Cs or ^{90}Sr; (2) ^{137}Cs from the fallout from the Chernobyl accident was added to that deposited during nuclear tests; and (3) the root transfer of plutonium is much lower than that of caesium or strontium. Although the soils that constitute the reservoir of these radionuclides clearly have higher ^{137}Cs than ^{90}Sr content, the activities of these two radionuclides in foodstuffs are of the same order owing to the greater root transfer of strontium. Wild foods (mushrooms, berries and game) are a special case, with the persistence of the fallout from the Chernobyl accident being significantly greater in forest environments. The activities of tritium in the terrestrial and aquatic environments are very similar, whatever the type of food, and are in equilibrium with the lower atmosphere. Similarly, ^{14}C has a natural component, but most of it was derived from nuclear weapon tests.

The dataset highlights that reference OBT contents in freshwater fish are close to 1 Bq/kg (fresh weight), namely, close to 1 Bq/L of combustion water, by considering a mean stable hydrogen content in fish of approximately 10%. These contents are similar to those observed for HTO in most reference fresh waters.

6.6.2. Impact of nuclear tests on the local environment: Semipalatinsk test site

The STS in the northeastern part of Kazakhstan was among the world's largest proving grounds for testing nuclear weapons. The test site covers 18 300 km^2, where 456 tests using 616 nuclear devices were conducted by the USSR from 1949 to 1989 [6.148, 6.149]. Up to 1963, most nuclear tests were conducted on the ground or in the atmosphere. Following the adoption of the Partial Nuclear Test-Ban Treaty (August 1963), signed by the USA, the USSR and the UK, the tests were conducted in underground locations until 1989. Overall, from 1949–1989, the combined STS nuclear explosions had a yield of 17.4 Mt [6.148].

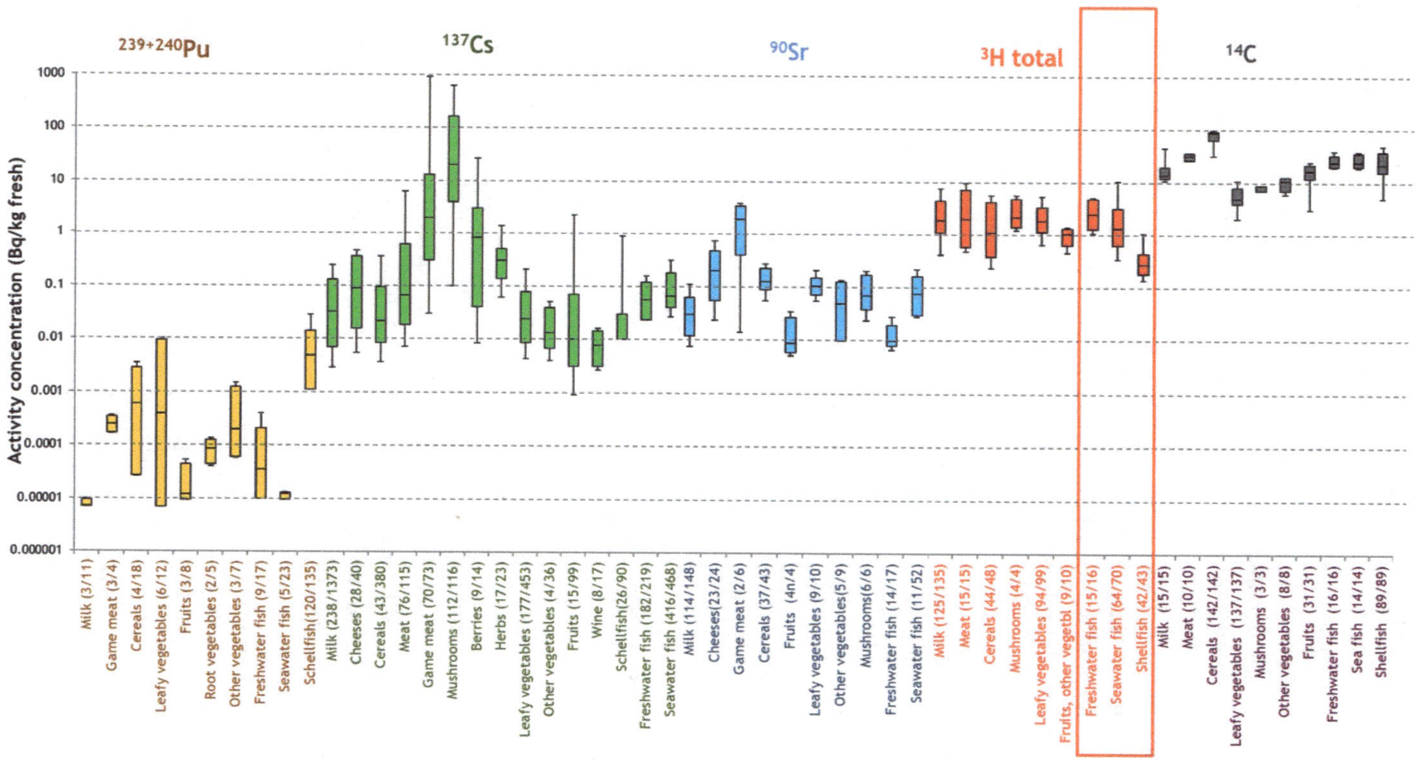

FIG. 6.14. Activities of the main artificial radionuclides in foodstuffs in France (2015–2017). For each category, the numbers in brackets show the number of samples analysed (denominator) and the number of samples with activity concentration values higher than the detection limit (numerator) (data from Ref. [6.147]).

The tests conducted at the STS varied in terms of device placement (e.g. ground, air, underground in tunnels, underground in wells, excavational), type (nuclear, thermonuclear) and the scale of energy released (ultralow, low, medium and high power). These factors, in combination with the different landscape and geological conditions of the test sites, led to highly diverse patterns of radioactive contamination. Air and ground tests were conducted at the site 'Experimental Field' ('Opytnoe Pole' in Russian), underground tests were conducted in horizontal tunnels at the Degelen mountain range and in vertical wells at the Balapan and Sary-Uzen sites [6.57, 6.148, 6.149].

Along with the production of fission and activation products, anthropogenic tritium was created by the following mechanisms:

— Ternary fission;
— High energy photonuclear reactions, produced by de-excitation of fission fragments with nuclei of light elements in the environment, for example, via the reaction $^{7}Li(\gamma, \alpha)^{3}H$;
— Neutron activation of light nuclei in the environment, for example, via the reactions $^{6}Li(n, \alpha)^{3}H$, $^{10}B(n, 2\alpha)^{3}H$ and $^{14}N(n, ^{3}H)^{12}C$;
— Dispersion of unreacted tritium present in the thermonuclear device;
— Tritiated charge components not fully reacted during a test explosion.

Given the abundance of naturally occurring isotopes in the environment, their neutron interaction cross-sections, the energy spectrum and the fluence of neutrons, it was estimated that the main source of tritium in soils was the reaction $^{6}Li(n, \alpha)^{3}H$ [6.150].

The Experimental Field was the first STS and was used for surface and airborne nuclear tests from 1949 to 1962. The site is a steppe plain about 20 km in diameter, surrounded on three sides by low mountains. There, on 12 August 1953, the first neutron-enhanced nuclear device (a prototype of a thermonuclear bomb) was tested, followed by the first Soviet thermonuclear weapon explosion on 22 November 1955. In total, 116 nuclear tests were carried out at the Experimental Field site, of which 86 were airborne and 30 were carried out on the ground surface [6.149].

An extensive radiological characterization study of the Experimental Field test site was completed in 2012–2016 with a spatial resolution of 10–20 m. [6.149]. The characterization included several radionuclides produced in the tests as fission and neutron activation products and non-reacted fissile or fusion components, such as ^{137}Cs, ^{90}Sr, $^{239+240}Pu$, ^{241}Am, ^{152}Eu, ^{60}Co and ^{3}H.

In the samples collected at the hypocentres of the test explosions, the activity concentrations in soil A (kBq/kg) of ^{3}H and ^{152}Eu were found to be correlated as follows:

$$A(^{3}\text{H}) \propto 1.63 A\left(^{152}\text{Eu}\right) \tag{6.9}$$

The average of the observed activity concentrations ratios ^{3}H/^{152}Eu in soil at the Experimental Field site was found to vary from 1.6 to 2.5. The correlation suggests that neutron activation was the main production mechanism for tritium at such locations.

Apart from neutron activation, the elevated levels of tritium in soils were also found to originate from the non-reacted tritium component originally present in the thermonuclear devices, captured by the resulting condensation particles and aerosols. The highest concentrations of ^{3}H in the soil, recorded in samples taken at test sites P-1 and P-5, close to their hypocentres, were 50 and 20 kBq/kg, respectively. At the hypocentres of sites P-2 and P-7, markedly lower levels of tritium in soil were found, which did not exceed 11 kBq/kg. Soil tritium activities of up to 10 Bq/kg were found in the P-3 hypocentre. The total area of plots in which soil contained significant amounts of tritium did not exceed 2 km^{2} [6.149, 6.151]. Tritium in the soil also led to its presence in the air; for example, at the P-1 hypocentre, the activity concentration of HTO in air was found to be up to 0.8 Bq/m^{3}. The activity concentration of HTO in plants in some areas reached 120 Bq/kg [6.152].

The Degelen mountain range test site was used to conduct low to medium power nuclear tests. The site (300 km^{2}) has a dome shape with a diameter of 17–18 km and an elevation of about 1000 m. Underground nuclear tests were carried out between 1961 and 1989, resulting in 209 nuclear tests in 181 tunnels [6.149]. The combined impact of the underground nuclear explosions in tunnels generated significant deformation of the rock, with the formation of numerous crushing zones, failed tunnels, faults and fissures. Tritium activity concentrations in water flowing from some tunnels reached 100 kBq/L. Tritium activities in water at the tunnel exit from the Degelen mountain range reached 300 kBq/L, and at a distance of 10 km from the mountains, it decreased to 30 kBq/L. At a distance of 15–20 km from the mountains, the concentration of tritium in surface waters did not exceed 1 kBq/L. In numerous springs within 3 km of the territory of Degelen, the tritium activity concentration in water ranged between 1 and 100 kBq/L [6.149, 6.153–6.154].

Across the Degelen site, elevated activity concentrations of tritium were found both in vegetation and in the air. The activity concentration of HTO in the free water of plants (TFWT) was related to the activity in water and on average was at the level of 10–100 kBq/kg [6.155, 6.156]. In the air of the Degelen site, tritium activities reached 1 kBq/m^3. At the same time, it was found that the tritium concentration in the atmospheric air was proportional to the concentration of tritium in stream water and groundwater, soil air and plants, and could also depend on the productivity of the vegetation cover, with significant concentrations of tritium in plants. The maximum tritium concentrations in the air were found near open water surfaces, and its concentrations dropped exponentially with distance from the source, reaching background concentrations at distances of 100–200 m [6.149].

The test sites at Balapan and Sary-Uzen were used for underground nuclear explosions in wells. At Balapan, 105 tests were carried out in 106 wells, and at Sary-Uzen, 24 tests were carried out in 25 wells from 1965 and continued until the 1980s. The average depths at which nuclear devices were placed were 650 m on the Balapan site and about 250 m on the Sary-Uzen site, with explosion yields varying up to 150 kt. Most the underground tests at these sites were not accompanied by a release of radioactive materials to the surface [6.149]. However, a specific feature of these sites was emanation of tritium from the ground surface of many sites near their mouth (openings), in the form of both HTO vapours and other gaseous compounds (methane, volatile organic compounds, hydrogen). At the Balapan site, the concentration of gaseous tritium compounds in air was comparable to the concentration of HTO. For example, the maximum recorded activity concentration of gaseous tritium compounds in wells was 190 Bq/m^3 and in the form of HTO it was 45 Bq/m^3 [6.152, 6.153, 6.157].

The so-called ‘Atomic Lake’ was created as an experimental excavation nuclear explosion. At the STS from 1965 to 1968, four experimental peaceful nuclear explosions were carried out with the purpose to develop and justify methods to create artificial reservoirs and channels. The first and most powerful explosion, with a yield of 140 kt, was carried out in 1965, producing the Atomic Lake — an artificial lake with a depth of about 100 m and a diameter of 500 m. The explosion that created the Atomic Lake was 93% thermonuclear, with tritium being present as an original nuclear component, and a high neutron flux, which also ensured a high production rate of tritium by activation reactions with primordial elements [6.158–6.160]. As a result, the materials that form the lake contain significant amounts of tritium, with activity concentrations ranging from 1 to 10 kBq/kg throughout its depth and over the entire area. The activity concentration of tritium in soils (kBq/kg) was found to be correlated with the activity concentration of ^{152}Eu (kBq/kg) as follows:

$$A(^{3}\mathrm{H}) \propto 5A\left(^{152}\mathrm{Eu}\right) \qquad (6.10)$$

This indicates that neutron activation of natural elements in soil might play a significant role in the production of these anthropogenic radionuclides.

The tritium concentration in the surface water layers is lower and amounts to only tens of becquerels per litre. However, at a depth of 10–20 m, the tritium activity concentration sharply increases and reaches a maximum in the bottom layers equal to about 20 kBq/L. There are also elevated activity concentrations of tritium in the air (10–90 Bq/m^3), which cannot be explained only by evaporation from the surface of the lake. The mechanism of tritium release into the air in this case can be attributed to emanation of tritium directly from the material forming the lake bottom and its shores [6.152].

Additional information is provided in Refs [6.150, 6.161] in relation to the drainage of Degelen test tunnels to local streams and the Balapan test area to the Chagan River. The latter is particularly interesting because of groundwater stratification, which depends on which strata are draining into the river. For this reason, tritium levels at the same collection site can vary from tens of becquerels per litre to 240 kBq/L.

6.6.3. Impact of recorded tritium releases from nuclear industries on rivers

6.6.3.1. Canada

In Canada, tritium is closely monitored because it represents one of the major radionuclides released to aquatic environments by the nuclear industry. A survey of eastern Canada rivers and lakes was conducted by Kim et al. [6.146] to determine the concentrations of OBT in aquatic biota. Fish samples were collected from Lake Ontario and the St Lawrence River in areas continuously receiving inputs of tritiated water (HTO) from operating NPPs, and from Lake Nipissing, which is categorized as a background area. Average fish TFWT values were found to be 1.6 ± 0.1 Bq/L in Lake Nipissing and 3.1 ± 0.3 Bq/L in Lake Ontario, while TFWT ranged from 11.1 to 80.8 Bq/L in the St Lawrence River near the Gentilly-2 CANDU site. Fish tissue OBT levels were 4.0 ± 0.4 Bq/L and 5.3 ± 0.2 Bq/L for Lake Nipissing and Lake Ontario, respectively, and between 18.1 and 134.2 Bq/L for CANDU sites. The activity concentrations of TFWT and OBT were determined in algae, freshwater mussels and various fish samples collected near Gentilly-2 and the Pickering and Darlington Nuclear Generating Stations.

TFWT in aquatic biota was found to be correlated with the tritium activity concentrations measured in waters at the time of sampling (TFWT/HTO of ambient water was 0.3–4.3). OBT concentration factors (OBT/HTO of ambient water) were found to be higher in freshwater mussels (17–47), which are filter feeders, than in algae and fish (0.3–10). These results point to a heterogeneous distribution of OBT in biota within these aquatic ecosystems. It was also noted that all fish and algae samples were found to be within the range of tritium activity concentrations that have been measured in the past in the same waters. Values in freshwater mussels were above this range.

6.6.3.2. France

Continuous sampling devices are distributed among the French metropolitan cities and include atmospheric samples (aerosols, volatile radionuclides such as tritium and ^{14}C) and surface water samples (taken using hydrocollectors [6.147]). Most of the tritium present in the fluvial environment downstream of the NPPs comes from liquid effluent discharges from nuclear facilities [6.147, 6.161]. Activities measured from samples taken by IRSN hydrocollectors are linked to the liquid discharges from upstream sites and the flow rate of the river, which governs dispersal patterns, as well as sites located downstream. Several other NPPs contribute to a cumulative effect.

Over the 2015–2017 period, the lowest annual average tritium activities were measured in the waters of the Grand Canal of Alsace and in the Rhône river, despite releases from several NPPs. Reported data are 3 Bq/L in the Grand Canal of Alsace downstream from the Fessenheim NPP, 6 Bq/L downstream from the St Alban NPP and up to 11 Bq/L downstream from the Bugey NPP. The Bugey NPP lies ahead of several major tributaries of the Rhône and has a flow rate of only 450–600 m^3/s. Tritium activities in the Loire are relatively uniform, between 14 Bq/L downstream from the Belleville NPP and 21 Bq/L downstream from the Dampierre NPP. The streams with lower flow reveal the highest average tritium activities, ranging from 22 Bq/L on the Vienne River downstream from the Civaux NPP to 35 Bq/L on the Seine river downstream from the Nogent NPP. These activities are below the regulatory limits and are of the order of 300–3000 times lower than the guideline value recommended by WHO for tritium in drinking water (10 000 Bq/L).

Temporal variations for the average annual tritium activity measured by IRSN since the 1980s in the waters of the Loire demonstrate the uniformity of liquid releases of tritium from NPPs since the 2000s (Fig. 6.15). Higher tritium activities downstream of the Chinon NPP until 2000 can be noted. After 2000, the average annual activities fluctuate by less than one order of magnitude year after year. This may result in part from a coordinated management of discharges by the different NPPs. The Saint-Laurent-des-Eaux NPP coordinates the discharges of all the NPPs discharging into the Loire river and, unless there is an exemption, two discharges cannot be carried out simultaneously. The drought in May–June 2011, with a significant rainfall deficit, led to an increase in the activity of tritium in

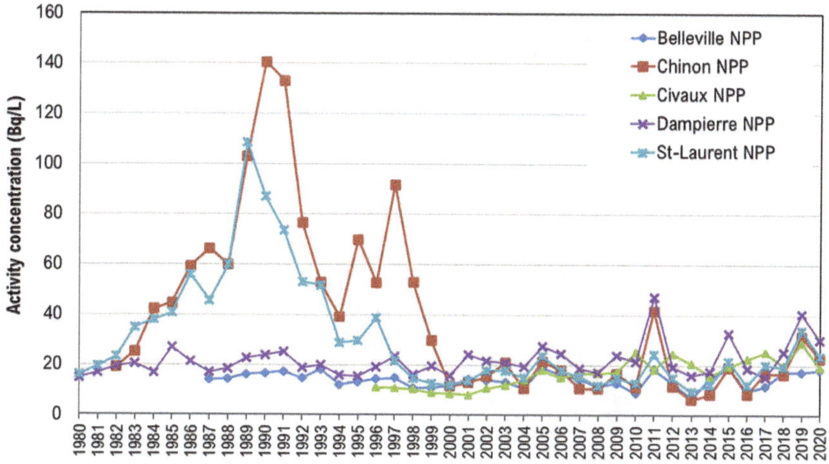

FIG. 6.15. Average annual activities of tritium in river water from the Loire river downstream of the Belleville, Dampierre, Chinon, Saint-Laurent-des-Eaux and Civaux NPPs (data from Ref. [6.147]).

the Loire that had not been observed since 1996. The activity of tritium in the water of the Loire downstream from the Dampierre NPP site between July and December 2015 was slightly above the mean level for 2000–2020.

The final shutdown of the Chooz-A NPP in 1991 led to a sharp drop of tritium activity in the Meuse river in the early 1990s, from 40 Bq/L to 10 Bq/L as an annual average (Fig. 6.16). Since the mid-1990s, the annual average tritium levels in the Meuse have increased slightly, reaching about 30 Bq/L around 2020; this increase was most probably due to a combination of the slight increase of tritium releases from the Chooz-B NPP and the decrease of almost 13% of the mean annual flow rate of the Meuse over this time. The average annual activity in the Moselle river downstream of the Cattenom NPP has increased from 15 Bq/L in 1985 to around 30–40 Bq/L at present. The average annual activity of tritium downstream of the Fessenheim NPP has been declining steadily since 1976. The average annual activity measured since 2014 in the Seine river is between 27 and 48 Bq/L.

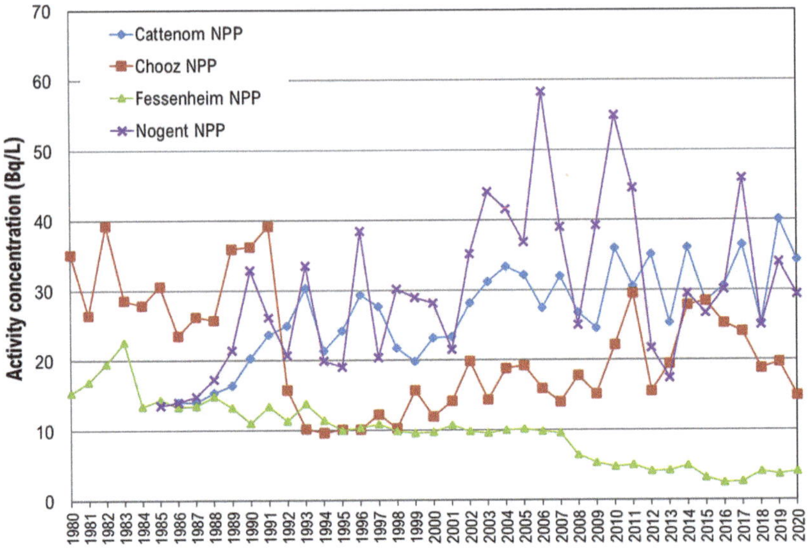

FIG. 6.16. Average annual activities of tritium in water of the Meuse and Moselle rivers, the Grand Canal of Alsace and the Seine river, downstream of the Chooz, Cattenom, Fessenheim and Nogent NPPs, respectively (data from Ref. [6.147]).

6.6.4. Impact of accidental releases

6.6.4.1. Impact of Chernobyl accident on tritium contents in freshwater systems

After the Chernobyl accident, tritium was released into the environment and measurements of terrestrial waters showed increases of 2–5 times compared with the levels observed before the accident. However, within a few months after the accident, tritium concentrations had returned close to background levels [6.162]. According to various estimates, the total supply of tritium into the Black Sea from the accident was about 5.2 PBq. This was equivalent to 1% of the inventory of tritium in the Black Sea and therefore had only a limited effect on tritium concentrations in the sea water [6.163]. No tritium was found in rainwater sampled in Poland, Austria and Japan after the accident [6.164], so the dispersal of Chernobyl tritium was reported only within the former USSR.

6.6.4.2. *Accident at the Fukushima Daiichi nuclear power plant and tritium content in freshwater systems*

There are few research articles reporting on the behaviour in freshwater systems of tritium emitted to the terrestrial environment from the Fukushima Daiichi NPP accident. Tritium concentrations in river water collected in the Fukushima Prefecture from 2011 to 2014 have been reported by Ueda et al. [6.165]. Concentrations in river water, collected at short time intervals at two rivers under base flow conditions and flood events, were higher in 2011 compared with the background levels. The concentrations decreased with time and returned close to background levels by 2013. During the flood event in July 2011, tritium concentrations tracked flow rate variations. However, the maximum tritium concentration appeared in September 2011, after the peak of the flood event. The differences in mean annual tritium concentrations during flood events and base flow conditions were indistinguishable after 2012. Ueda et al. [6.165] interpreted this to be a result of water with high tritium concentration infiltrating to deeper levels and being diluted by soil water or groundwater. In addition, river water collected annually (2012 to 2014) from 16 rivers and one dam under base flow conditions also showed elevated tritium concentrations compared with background levels. Their mean annual concentration decreased gradually from 2012 to 2014. Yabusaki et al. [6.166] reported tritium concentrations for seven groundwater and spring water samples collected from the northern coastal region in the Fukushima Prefecture and pointed out the possibility that the tritium concentration was elevated by recharge from precipitation with high tritium concentrations after the Fukushima Daiichi NPP accident. Spatial variations in tritium concentrations in groundwater were identified in the southern part of the coastal region in the Fukushima Prefecture by Kashiwaya et al. [6.167] and higher tritium concentrations were measured in wells near the Fukushima Daiichi NPP.

6.6.5. Impact of releases from non-nuclear facilities

6.6.5.1. *Technogenic tritium in the Rhône river*

As mentioned in Section 6.5.2.1, substantial quantities of radioluminescent tritium compounds were used in eastern France and Switzerland between 1960 and 1990 by the watchmaking industry [6.35]. Several production units manufacturing fluorescent paints based on tritiated polystyrene supplied numerous watchmaking workshops. The use of this technogenic tritiated organic material was linked to the discovery of significant amounts of tritium in sediments and suspended matter of the Rhône river and its estuary in the form of relatively non-labile, micrometre sized particulates [6.19, 6.30, 6.38, 6.133, 6.168]. It was unclear whether these particulates originated from past direct liquid releases into the Rhône or indirect supply from its catchment. The tritium levels are highest in the upper Rhône, upstream of Creys Malville, the northernmost nuclear power station on the Rhône, with current values in sediments between 105 and 106 TU [6.36], but are similarly observed all along the river [6.12, 6.19]. It is notable that this situation is found only in the Rhône, and no such tritium contamination of the sediments is observed in other major French rivers where nuclear power stations are also in operation [6.38].

6.6.5.2. Peaceful industrial low yield underground thermonuclear explosions in the Russian Federation

Current tritium contamination of surface waters and groundwaters at some sites in the Russian Federation are associated with the consequences of peaceful underground nuclear explosions carried out in the territory of the former USSR in the second half of the 20th century [6.169, 6.170]. The site of the peaceful underground nuclear explosions, Dnepr, located at the Khibiny mountain range, Kola Peninsula in the Murmansk region, can serve as a representative example of such activities. Low yield thermonuclear explosions (1.7–2.1 kt) were carried out inside Mount Kuel'por in 1972 and 1984. The aim of the explosions was to crush the ore body (apatite minerals) before its extraction and transfer to the Earth's surface [6.169, 6.170].

The main radiological problem generated by these explosions was the persisting discharge of tritium-contaminated groundwater to the surface. In 2002, the activity concentration of tritium in the mine water reached a level of 32 000 Bq/L [6.170], which was higher by a factor of four than the permissible tritium level in drinking water established in the Russian Federation at that time (7700 Bq/L). By 2008, the tritium activity concentration in the mine water declined to the level of about 7500 Bq/L, comparable with the intervention level of 7600 Bq/L, adopted for drinking water in the Russian Federation at present [6.171]. The monitoring surveys conducted in 2013 and 2019 revealed further significant decreases of tritium activity concentration in the mine water to the level of about 1200 Bq/kg, recorded in 2019. The effective environmental half-life of tritium activity concentration in the mine water in 2008–2019 was estimated as 4.2 years [6.171]. Some impact of the Dnepr peaceful underground nuclear explosions on the environment can be observed in the Kuniyok river, which is flowing at the western base of Mount Kuel'por. The tritium activity concentration in the river water samples collected in 2019 was found to range from 4.7 ± 0.8 to 6.9 ± 1.1 Bq/L. These values were only slightly higher than the regional background tritium level in the surface waters (about 2 Bq/L). Ramzaev et al. [6.171] conservatively estimated an annual effective dose of 0.17 μSv/a due to intake of tritium from drinking water from the Kuniyok river in 2019. This value is insignificant when compared with the average annual effective dose of around 3 mSv received by the population of the Murmansk region from natural exposure sources [6.172].

6.6.6. Impact of releases from facilities producing tritium

6.6.6.1. Chalk River Laboratories

The CNL CRL complex in Canada has existed since 1942. The control of tritium dispersal within and around the research and power stations of the Canadian nuclear programme has always been important because of the high production of tritium in heavy water moderated reactors. At CRL, two major research reactors have operated for more than 30 years. Over the years, airborne tritium emissions have ranged from 300 to 700 TBq/a (8–19 kCi/a) and liquid tritium effluents to local water systems from 100 to 200 TBq/a (3–5 kCi/a)[6.173].

6.6.6.2. Mayak Production Association

The Mayak PA industrial complex in the Russian Federation was originally established in 1948 in Chelyabinsk-65 (now known as Ozersk). The site is the source of tritium releases to air and water associated with reactor operations, tritium production and the reprocessing of the spent nuclear fuel (see Chapter 3 for the dynamics of radionuclide emissions and discharges from this complex since 2002). As of 2017, the level of tritium emissions into the atmosphere has been approximately 1.2 PBq/a.

Disposal of liquid radioactive waste in the early years of the Mayak PA operation, when priority was given to production, resulted in significant contamination of nearby ponds/lakes and the Techa River. To stop the discharge of liquid radioactive waste into the open hydrological network through the Techa River, Mayak PA built several dams on the river. They also created two artificial reservoirs, R-10 and R-11, and bypass channels to allow clean water to flow from Lake Irtyash to the Techa River [6.174] in the 1960s. Current water use of lakes and artificial reservoirs for water supply and waste disposal is shown in Fig. 6.17 [6.175]. The industrial reservoir R-9 (Lake Karachay) was used to store highly radioactive waste and by 2015 was completely infilled by covering with soil and paving with concrete on top. The industrial reservoir R-17 (Old Swamp), containing highly radioactive waste, is planned to be closed by 2025 [6.176]. Lake Kyzyltash (reservoir R-2) serves as a cooling pond and low active liquid waste storage and is used as a technological water supply and for disposal.

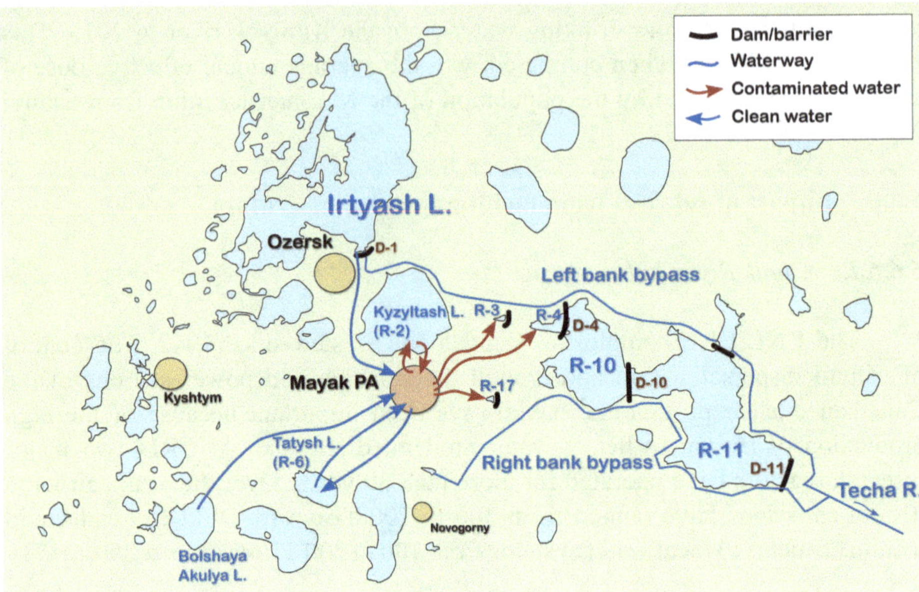

FIG. 6.17. Water supply and liquid waste disposal at the Mayak PA site (based on the ecological safety report of Mayak PA [6.175]).

The open reservoirs used as water supplies to the population and personnel of Mayak PA and for industrial purposes currently have the status of liquid radioactive waste storage facilities and were studied for tritium content in 1971–2015. The summary of tritium activity concentration in the water of open reservoirs and lakes in the vicinity of Mayak PA are shown in Table 6.7.

Specifically, tritium activity concentrations in eight open reservoirs used for economic purposes and located 7–70 km from the main site of Mayak PA were studied from 1978 to 1983 [6.177]. The study demonstrated that average levels of tritium activities in the water were in the range from 60 Bq/L (Lake Uvildy) to 630 Bq/L (Lake Ulagach) and decreased with distance from Mayak PA. In the summer, tritium activities in the open reservoirs were found to be approximately twice as high as in the winter period. At the same time, tritium activities measured in water samples from four artesian wells did not show seasonal variations. Therefore, it was concluded that tritium activities in surface waters were affected by airborne tritium releases from Mayak PA.

The studies conducted in 2001–2003 [6.178] in 34 water reservoirs used for economic purposes and located 6–90 km from Mayak PA showed tritium activity concentrations in the range from 8 Bq/L (Lake Kabanskoe) to 113 Bq/L (Lake Ulagach). The levels of tritium activity concentrations in the reservoirs decreased by 3–16 times in comparison with the levels observed in 1978–1983.

TABLE 6.7. TRITIUM ACTIVITY CONCENTRATION IN WATER BODIES AROUND MAYAK PA

Period	Number of studied water bodies	Distance from Mayak PA (km)	Tritium activity concentration (Bq/L)	Reference
1978–1983	8	7–70	60–630	[6.177]
2001–2003	34	6–90	8–113	[6.178]
2009–2012	33	7.5–77	10–62	[6.179]
2014–2015	23	6–78	<8–61	[6.180]

The relationship between tritium activity concentration in open water reservoirs and the distance from Mayak PA was approximated by the following power law model:

$$A = A_0 R^b \tag{6.11}$$

where

A is the tritium activity concentration in water (Bq/L);
R is the distance from Mayak PA (km);

and A_0 (Bq/L) and b (unitless) are the fitting parameters.

From 2009 to 2012, tritium activity concentration in water from 33 open reservoirs, used for economic purposes and located at distances from 7.5 to 77 km from Mayak PA, were measured. They were in the range 10–62 Bq/L. Kazachyonok et al. [6.179] determined the dependence of tritium activity concentration in water as a function of the distance and direction from Mayak PA and found a statistically significant negative correlation [6.181].

From 2014 to 2015, a new survey of tritium concentrations in 23 water reservoirs used for economic purposes was made. These reservoirs were located 6–78 km from the Mayak PA site. The survey revealed tritium activity concentrations in water ranging from below the detection limit (8 Bq/L) to a maximum value of 61 Bq/L, which was found in a lake located in proximity to the settlement at Novogorny (see Fig. 6.18). These data can be approximated by the power law model of Eq. (6.11) with the following fitting parameters: $A_0 = 245 \pm 66$ Bq/L and $b = 0.83 \pm 0.11$ (see Fig. 6.17). In contrast to the previous study, no dependence of tritium activity concentration in water on the distance from Mayak PA was found in 2009–2012. Also, there was no statistically significant difference between tritium activity concentrations in the same reservoirs obtained in the 2014–2015 and 2009–2012 surveys, while the values of tritium content in these reservoirs sampled during the period 2001–2003 were 1.4 times higher than those in the 2014–15 study [6.181].

Among the industrial water reservoirs of Mayak PA, the one most studied for tritium content was R-17 (see Fig. 6.17). Kuzmenkova et al. [6.182] reported an average tritium activity concentration of 300 ± 200 kBq/L in the reservoir during 1971–1979, on the basis of 18 samples. In the period 1980–2005, increased tritium activity concentrations, by approximately an order of magnitude, were reported in R-17 [6.183] with an average value of 2.9 ± 1.2 kBq/L (48 samples). A more recent study for 2012–2013 showed that the tritium activities in reservoir R-17 had decreased by approximately three times, to 900 kBq/L.

The levels of tritium activity concentration in the Techensky cascade of reservoirs (R-2, R-3, R-4, R-10 and R-11; see Fig. 6.17) for 2007–2009 were

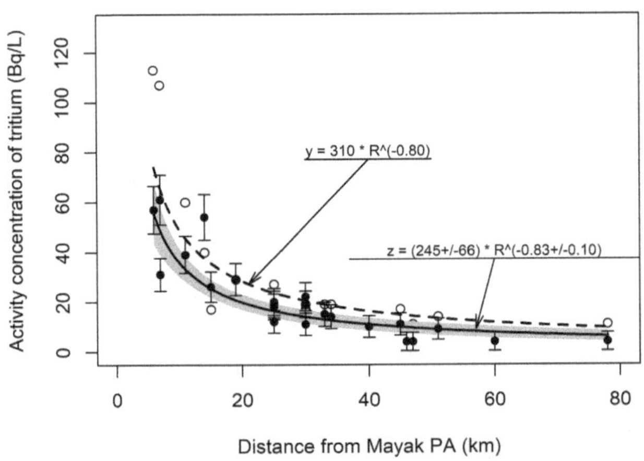

FIG. 6.18. Activity concentration of tritium measured in open water reservoirs of economic significance at various distances from Mayak PA. Data obtained in 2001–2003 are indicated by empty circles (data) and a dashed line (fit). The data obtained in 2014–2015 are shown as solid circles with error bars (data) and a solid line (fit). The shaded area represents the 95% confidence interval of the latter fit. Adapted from Ref. [6.180] with permission courtesy of IOP Publishing.

TABLE 6.8. TRITIUM ACTIVITY CONCENTRATION IN WATER OF INDUSTRIAL RESERVOIRS OF MAYAK PA

Period	Activity concentration in reservoir water (kBq/L)						Ref.
	R-17	R-2	R-3	R-4	R-10	R-11	
1971–1979	300±200	—[a]	—	—	—	—	[6.183]
1980–2005	2900±1200	—	—	—	—	—	
2007–2009	—	6.7±0.8	5.0±1.0	4.0±0.7	2.9±0.3	0.9±0.7	[6.184]
2012–2013	900	—	—	16.0	—	—	[6.182]
2015	—	11.6	—	—	—	—	[6.180]

[a] —: no data

Note: Numerical values shown represent the mean or the mean ± standard deviation.

found in the range from 0.9 kBq/L (R-11) to 6.7 kBq/L (R-2). A significant exponential decrease of tritium activities in the reservoirs of the Techensky cascade were found to vary as a function of the distance from the source of the Techa River [6.184]. The level of tritium activity in Lake Kyzyltash (R-2) increased by 1.7 times in 2015 to a value of 11.6 kBq/L [6.180].

The main results of studies of tritium activity concentration in the water of open reservoirs used for industrial purposes by Mayak PA are given in Table 6.8.

6.7. CONCLUSIONS

For human beings and other terrestrial life, fresh water is a critically important resource on the Earth but represents only a small fraction (2.5%) of the total water reservoir on the planet. Fresh water is sequestered in ice sheets, glaciers and permafrost (up to 99%), with the remaining mobile portion forming surface waters, lakes, rivers, soils, swamps and the water bound in the biosphere. Owing to the high mobility of water molecules, the free form of tritium, HTO, is its predominant species on Earth. It exchanges quickly with other water components and, at the global scale, water exchange with the atmosphere constitutes the predominant mechanism controlling tritium concentrations in surface mobile waters.

Atmospheric nuclear testing carried out from 1945 to 1980 led to significantly increased tritium contents in precipitation and rivers, particularly in the Northern Hemisphere, where most of the testing occurred. Rivers in the Northern Hemisphere displayed concentrations over 100 times more elevated than those in the Southern Hemisphere during peak emissions at the beginning of the 1960s and over the following decades. At present, rivers in the Northern Hemisphere still contain ten times more tritium than those in the Southern Hemisphere.

Tritium contents are affected to varying degrees by latitude, altitude, season, distance to the coast, surface area of the draining basin and mixing with underground waters. Climate change could increase tritium contents in lakes and rivers owing to the remobilization of sequestered waters (e.g. via melting of permafrost) containing tritium from the nuclear testing era, as is currently being observed in northern Canada and presumably in other permafrost areas. The clearly defined temporal input of tritium makes it a powerful tool for hydrological dating, as well as for quantifying mixing of different water masses and tracing the circulation of water in the terrestrial environment.

Over recent decades, improvements in methodologies and evaluation of environmental tritium species (HTO, E-OBT, NE-OBT) have made significant advances. This improved understanding of OBT speciation has provided insights into mechanisms of integration of tritium into living organisms through photosynthesis and other metabolic processes and transfers into the food chain. The OBT form of tritium generally remains locked into organic compounds until biodegradation occurs. Thus, the behaviour and fate of free and bound forms of tritium in the various environmental components are not governed by a single process and may operate under different kinetics. Kinetic gaps between these two forms of tritium commonly generate apparent OBT–HTO disequilibria in various environmental components. In riverine sediments, OBT concentrations higher than those in surrounding waters originate in most cases from biogenic processes operating on the tritium originating from nuclear tests. These concentrations have now reduced, within analytical uncertainty, owing to both radioactive disintegration and biodegradation of the remaining organic compounds over time. In river systems affected by industrial releases, OBT concentrations in living organisms or detritus higher than background values in surrounding waters can be observed owing to biological integration of tritium at the time of the peak nuclear testing releases.

Technogenic tritium from non-nuclear industries relates to specific sources releasing highly tritiated manufactured compounds that have been released and dispersed into the environment (e.g. site in Cardiff, site in Rhône valley). In the case of the Rhône river (France), substantial quantities of micrometre sized tritiated paint fragments (as a substitute for radium luminescent paint) were used between 1960 and 1990 by the watchmaking industry. These synthetic tritiated compounds led to the contamination of Rhône sediments all along the course of the river and to OBT contents in fish more than ten times higher than the HTO concentration in waters.

Apart from tritium dispersal at the global scale originating from the nuclear testing era, much of the technogenic tritium in rivers originates from nuclear industry releases over the past seven decades. Impacted rivers around the world often display tritium concentrations more than one order of magnitude higher than the background tritium contents in pristine fresh waters. The effect of tritium associated with the Chernobyl accident was confirmed only in the former USSR. In the case of the Fukushima Daiichi NPP accident, tritium concentrations in rivers decreased rapidly over time and returned to regional background values relatively quickly and within three years after the accident. The impact of tritium releases from nuclear and industrial facilities on freshwater systems is well documented. The tritium levels in the aquatic environment are observed to rapidly decline with the distance from the source.

The environmental behaviour of tritium in freshwater domains on the Earth is highly complex, depending on the water mass origin, circulation, mixing and exchanges with the atmosphere. The bound forms of tritium in organic matter pools are often not in equilibrium with its mobile form, in many cases reflecting kinetic discrepancies arising from water circulation and biomass production and biodegradation.

6.8. ACKNOWLEDGEMENTS

Contributions to this chapter made by S. Lukashenko (Russian Federation) and V. Ramzaev (Russian Federation) on the environmental effects of nuclear weapon tests at the STS and industrial underground thermonuclear explosions on Kola Peninsula, respectively, are gratefully acknowledged.

REFERENCES TO CHAPTER 6

[6.1] SHIKLOMANOV, I., "World fresh water resources", Water in Crisis: A Guide to the World's Fresh Water Resources (GLEICK, P.H., Ed.), Oxford University Press, Oxford (1993).

[6.2] BROWN, R.M., "Tritium in precipitation at Canadian sites (1953–1963)", Tritium and Other Environmental Isotopes in the Hydrological Cycle, Technical Reports Series No. 73, IAEA, Vienna (1967) 71–75.

[6.3] ERIKSSON, E., An account of the major pulses of tritium and their effects in the atmosphere, Tellus A **17** 1 (1965) 118–130,
https://a.tellusjournals.se/articles/10.3402/tellusa.v17i1.8998

[6.4] ISRAEL, G., ROETHER, W., SCHUMANN, G., Seasonal variations of bomb-produced tritium in rain, J. Geophys. Res. **68** 13 (1963) 3771–3773,
https://doi.org/10.1029/JZ068i013p03771

[6.5] SCHELL, W., SAUZAY, G., PAYNE, B., "World distribution of environmental tritium", Symposium on Physical Behavior of Radioactive Contaminants in the Atmosphere, Vienna, 1973, IAEA, Vienna (1974) 375–400.

[6.6] WEISS, W., BULLACHER, J., ROETHER, W., "Evidence of pulsed discharges of tritium from nuclear energy installations in Central European precipitation", Behaviour of Tritium in the Environment (San Francisco, 16–20 Oct. 1978), Proceedings Series, IAEA, Vienna (1979).

[6.7] MOMOSHIMA, N., OKAI, T., KAJI, T., TAKASHIMA, Y., Distribution and transformation of various chemical forms of tritium in the environment, Radiochim. Acta **54** 3 (1991) 129–132,
https://doi.org/10.1524/ract.1991.54.3.129

[6.8] MIYAMOTO, K., KIMURA, K., HONGO, S., A transfer model of tritium in a local hydrosphere, Fusion Technol. **28** 3P1 (1995) 910–917,
https://doi.org/10.13182/FST95-A30521

[6.9] BARRY, P.J., ROBERTSON, E., WILDSMITH, D.P., Water and Tritium Budgets for Perch Lake, 1970–1977, Report No. AECL-6404, Atomic Energy of Canada Limited, Chalk River (1979).

[6.10] ROZANSKI, K., FROEHLICH K., MOOK, W.G., Environmental isotopes in the hydrological cycle: Principles and applications, Vol. III (MOOK, W.G., Ed.), UNESCO, Paris (2001),
https://www.hydrology.nl/images/docs/ihp/Mook_III.pdf

[6.11] KIM, S.B., BAGLAN, N., DAVIS, P.A., Current understanding of organically bound tritium (OBT) in the environment, J. Environ. Radioact. **126** (2013) 83–91,
https://doi.org/10.1016/j.jenvrad.2013.07.011

[6.12] MOREREAU, A., et al., Trajectories of technogenic tritium in the Rhône River (France), J. Environ. Radioact. **223–224** (2020) 106370,
https://doi.org/10.1016/j.jenvrad.2020.106370

[6.13] INSTITUT DE RADIOPROTECTION ET DE SÛRETÉ NUCLÉAIRE, Tritium and the Environment, IRSN, Fontenay-aux-Roses (2010).

[6.14] LEBARON-JACOBS, L., "État des connaissances de l'impact du tritium sur la santé", Tritium Conference Days, Paris, 2009, Société Française de Radioprotection, Paris (2009).

[6.15] UNITED NATIONS, Exposure to the public from man-made sources of radiation, 2000 Report of the United Nations Scientific Committee on the Effects of Atomic Radiation (UNSCEAR) to the General Assembly, with Scientific Annexes, Annex C, UN, New York (2000).

[6.16] GUÉTAT, P., DOUCHE, C., HUBINOIS, J.C., Le tritium et l'environnement: Sources, mesures et transferts, Radioprotection **43** 4 (2008) 547–569, https://doi.org/10.1051/radiopro:2008027

[6.17] EISENBUD, M., GESELL, T.F., Environmental Radioactivity from Natural, Industrial and Military Sources: From Natural, Industrial and Military Sources, 4th edn, Academic Press, Cambridge, MA (1997), https://doi.org/10.1016/B978-0-12-235154-9.X5000-2

[6.18] MORONEY, J.R., Radioactive Fallout in the Southern Hemisphere from Nuclear Weapons Tests, ARL/TR013, Australian Radiation Laboratory, Melbourne (1979).

[6.19] INTERNATIONAL ATOMIC ENERGY AGENCY, Global Network of Isotopes in Precipitation (GNIP), https://www.iaea.org/services/networks/gnip

[6.20] EYROLLE, F., et al., Evidence for tritium persistence as organically bound forms in river sediments since the past nuclear weapon tests, Sci. Rep. **9** 1 (2019) 11487, https://doi.org/10.1038/s41598-019-47821-1

[6.21] UNITED NATIONS, Sources and Effects of Ionizing Radiation, Vol. I: Sources, 2008 Report of the United Nations Scientific Committee on the Effects of Atomic Radiation (UNSCEAR) to the General Assembly, UN, New York (2008).

[6.22] UNITED NATIONS, Sources and Effects of Ionizing Radiation, 2016 Report of the United Nations Scientific Committee on the Effects of Atomic Radiation (UNSCEAR) to the General Assembly, UN, New York (2017).

[6.23] ALBENESIUS, E., ONDREJOIN, R., Nuclear fission produces tritium, Nucleonics (US) Ceased Publ. **18** (1960).

[6.24] BLOMEKE, J.O., Management of Fission Product Tritium in Fuel Processing Wastes, ORNL-TM-851, Oak Ridge National Laboratory, Oak Ridge, TN (1964).

[6.25] HOLMQUIST, C., "Hazards of tritium oxides induced in large heavy water reactor stations", AED-CONF-65-116-7, American Industrial Hygiene Conf., Houston, 1965, Vattenfall, Solna (1965).

[6.26] WORLD HEALTH ORGANIZATION, Selected Radionuclides — Environmental Health Criteria 25, WHO, Geneva (1983).

[6.27] AUTORITÉ DE SÛRETÉ NUCLÉAIRE, Livre Blanc du Tritium, ASN, Montrouge (2024), https://www.asn.fr/sites/tritium/

[6.28] DINGWALL, S., MILLS, C., PHAN, N., TAYLOR, K., BOREHAM, D., Human health and the biological effects of tritium in drinking water: Prudent policy through science–addressing the ODWAC new recommendation, Dose-Response **9** 1 (2011), https://doi.org/10.2203/dose-response.10-048.Boreham

[6.29] CANADIAN NUCLEAR SAFETY COMMISSION, Tritium Studies Project Synthesis Report, CNSC, Ottawa (2011).

[6.30] EYROLLE, F., et al., An updated review on tritium in the environment, J. Environ. Radioact. **181** (2018) 128–137, https://doi.org/10.1016/j.jenvrad.2017.11.001

[6.31] McCUBBIN, D., LEONARD, K.S., BAILEY, T.A., WILLIAMS, J., TOSSELL, P., Incorporation of organic tritium (^{3}H) by marine organisms and sediment in the Severn Estuary/Bristol Channel (UK), Mar. Pollut. Bull. **42** 10 (2001) 852–863, https://doi.org/10.1016/S0025-326X(01)00039-X

[6.32] WILLIAMS, J., RUSS, R., McCUBBIN, D., KNOWLES, J., An overview of tritium behaviour in the Severn Estuary (UK), J. Radiol. Prot. **21** 4 (2001) 337, https://doi.org/10.1088/0952-4746/21/4/301

[6.33] CROUDACE, I.W., WARWICK, P.E., MORRIS, J.E., Evidence for the preservation of technogenic tritiated organic compounds in an estuarine sedimentary environment, Environ. Sci. Technol. **46** 11 (2012) 5704–5712, https://doi.org/10.1021/es204247f

[6.34] SCHÖNHOFER, F., POCK, K., Incorporation of tritium from wrist watches, J. Radioanal. Nucl. Chem. **197** 1 (1995) 195–202, https://doi.org/10.1007/BF02040231

[6.35] KREJCI, K., ZELLER, J., "Tritium Pollution in the Swiss Luminous Compound Industry", Proc. Behaviour of Tritium in the Environment, San Francisco, 1978, IAEA, Vienna (1979).

[6.36] JEAN-BAPTISTE, P., BAUMIER, D., FOURRÉ, E., DAPOIGNY, A., CLAVEL, B., The distribution of tritium in the terrestrial and aquatic environments of the Creys-Malville nuclear power plant (2002–2005), J. Environ. Radioact. **94** 2 (2007) 107–118, https://doi.org/10.1016/j.jenvrad.2007.01.010

[6.37] INSTITUT DE RADIOPROTECTION ET DE SÛRETÉ NUCLÉAIRE, Actualisation des Connaissances Acquises sur le Tritium dans l'Environnement, Report PRP-ENV/SERIS/2017-004, IRSN, Fontenay-aux-Roses (2017).

[6.38] GONTIER, G., SICLET, F., Le tritium organique dans les écosystèmes d'eau douce: évolution à long terme dans l'environnement des centres nucléaires de production d'électricité français, Radioprotection **46** 4 (2011) 457–491, https://doi.org/10.1051/radiopro/2011139

[6.39] GALERIU, D., TAKEDA, H., MELINTESCU, A., TRIVEDI, A., Energy metabolism and human dosimetry of tritium, Fusion Sci. Technol. **48** 1 (2005) 795–798, https://doi.org/10.13182/FST05-A1040

[6.40] EL-KHARBACHI, A., et al., Tritium absorption/desorption in ITER-like tungsten particles, Int. J. Hydrog. Energy **39** 20 (2014) 10525–10536, https://doi.org/10.1016/j.ijhydene.2014.05.023

[6.41] INTERNATIONAL ATOMIC ENERGY AGENCY, Manual on Mathematical Models in Isotope Hydrogeology, IAEA-TECDOC-910, IAEA, Vienna (1996).

[6.42] INTERNATIONAL ATOMIC ENERGY AGENCY, Modelling the Environmental Transfer of Tritium and Carbon-14 to Biota and Man, Report of the Tritium and Carbon-14 Working Group of EMRAS Theme 1, IAEA, Vienna (2012).

[6.43] GOUTAL, N., et al., Assessment, validation and intercomparison of operational models for predicting tritium migration from routine discharges of nuclear power plants: The case of Loire River, J. Environ. Radioact. **99** 2 (2008) 367–382, https://doi.org/10.1016/j.jenvrad.2007.10.016

[6.44] INTERNATIONAL ATOMIC ENERGY AGENCY, Generic Models for Use in Assessing the Impact of Discharges of Radioactive Substances to the Environment, Safety Reports Series No. 19, IAEA, Vienna (2001).

[6.45] GARCIA-SANCHEZ, L., Watershed wash-off of atmospherically deposited radionuclides: Review of the fluxes and their evolution with time, J. Environ. Radioact. **99** 4 (2008) 563–573, https://doi.org/10.1016/j.jenvrad.2007.11.013

[6.46] DOMINIK, J., BURRUS, D., VERNET, J.-P., Transport of the environmental radionuclides in an alpine watershed, Earth Planet. Sci. Lett. **84** 2–3 (1987) 165–180, https://doi.org/10.1016/0012-821X(87)90083-5

[6.47] JOSHI, S.R., McCREA, R.C., Sources and behavior of anthropogenic radionuclides in the Ottawa river waters, Water. Air. Soil Pollut. **62** 1–2 (1992) 167–184, https://doi.org/10.1007/BF00478459

[6.48] INSTITUT DE RADIOPROTECTION ET DE SÛRETÉ NUCLÉAIRE, Le Tritium Dans l'Environnement: Synthèse Critique des Connaissances et Perspectives de Recherches, IRSN/DEI/2009-05, IRSN, Fontenay-aux-Roses (2009).

[6.49] MARANG, L., et al., Modelling tritium flux from water to atmosphere: Application to the Loire River, J. Environ. Radioact. **102** 3 (2011) 244–251, https://doi.org/10.1016/j.jenvrad.2010.11.015

[6.50] NOGUCHI, H., FUKUTANI, S., YOKOYAMA, S., KINOUCHI, N., Deposition of tritiated water vapour to a water surface in an outdoor field, Radiat. Prot. Dosimetry **93** 2 (2001) 167–172, https://doi.org/10.1093/oxfordjournals.rpd.a006425

[6.51] MORTON, F.I., Operational estimates of lake evaporation, J. Hydrol. **66** 1–4 (1983) 77–100, https://doi.org/10.1016/0022-1694(83)90178-6

[6.52] MORTON, F.I., Practical estimates of lake evaporation, J. Appl. Meteorol. Climatol. **25** 3 (1986) 371–387, https://doi.org/10.1175/1520-0450(1986)025<0371:PEOLE>2.0.CO;2

[6.53] BOLSUNOVSKY, A.Ya., BONDAREVA, L.G., Tritium in surface waters of the Yenisei River basin, J. Environ. Radioact. **66** 3 (2003) 285–294, https://doi.org/10.1016/S0265-931X(02)00132-7

[6.54] HIGASHINO, M., GANTZER, C.J., STEFAN, H.G., Unsteady diffusional mass transfer at the sediment/water interface: Theory and significance for SOD measurement, Water Res. **38** 1 (2004) 1–12, https://doi.org/10.1016/j.watres.2003.08.030

[6.55] SCHULZ, H.D., ZABEL, M., Marine Geochemistry, Springer, Berlin Heidelberg New York (2000), https://doi.org/10.1007/978-3-662-04242-7

[6.56] JEAN-BAPTISTE, P., FOURRÉ, E., The distribution of tritium between water and suspended matter in a laboratory experiment exposing sediment to tritiated water, J. Environ. Radioact. **116** (2013) 193–196, https://doi.org/10.1016/j.jenvrad.2012.11.004

[6.57] PRIEST, N., BURKITBAYEV, M., ARTEMYEV, O., LUKASHENKO, S., MITCHELL, P., "The radiological situation in the Sarzhal region of the Semipalatinsk Nuclear Test Site – The NATO SEMIRAD project", Proc. Int. Conf. on Radioactivity in the Environment, Monaco, 2002, Norwegian Radiation Protection Authority, Østerås (2002) 591–594.

[6.58] GUSYEV, M.A., et al., Application of tritium in precipitation and baseflow in Japan: A case study of groundwater transit times and storage in Hokkaido watersheds, Hydrol. Earth Syst. Sci. **20** 7 (2016) 3043–3058, https://doi.org/10.5194/hess-20-3043-2016

[6.59] MAYER, A., et al., A multi-tracer study of groundwater origin and transit-time in the aquifers of the Venice region (Italy), Appl. Geochem. **50** (2014) 177–198, https://doi.org/10.1016/j.apgeochem.2013.10.009

[6.60] MICHEL, R.L., Residence times in river basins as determined by analysis of long-term tritium records, J. Hydrol. **130** 1–4 (1992) 367–378, https://doi.org/10.1016/0022-1694(92)90117-E

[6.61] LIN, C.C., "Water and impurity activation products", Radiochemistry in Nuclear Power Reactors, National Academy Press, Washington, DC (1996) 113–124; "Radiation chemistry in reactor coolant", ibid., 125–142.

[6.62] LIN, R., WEI, K., "Environmental isotope profiles of the soil water in loess unsaturated zone in semi-arid areas of China", Isotope Based Assessment of Groundwater Renewal in Water Scarce Regions, IAEA-TECDOC-1246, IAEA, Vienna (2001) 101–118.

[6.63] MICHEL, R.L., Tritium hydrology of the Mississippi River basin, Hydrol. Process. **18** 7 (2004) 1255–1269, https://doi.org/10.1002/hyp.1403

[6.64] MUNDSCHENK, H., KRAUSE, W., Behaviour and radiological significance of tritium from nuclear power plants and other sources in the Rhine river basin, J. Environ. Radioact. **14** 4 (1991) 341–360, https://doi.org/10.1016/0265-931X(91)90024-A

[6.65] RAVIKUMAR, P., SOMASHEKAR, R., Environmental Tritium (^{3}H) and hydrochemical investigations to evaluate groundwater in Varahi and Markandeya river basins, Karnataka, India, J. Environ. Radioact. **102** 2 (2011) 153–162, https://doi.org/10.1016/j.jenvrad.2010.11.006

[6.66] SUBBOTIN, S., AIDARKHANOV, A., DUBASOV, Y.V., Migration of tritium with underground waters on the former Semipalatinsk test site, Radiochemistry **55** 5 (2013) 557–565, https://doi.org/10.1134/S1066362213050184

[6.67] BARAN, N., RICHERT, J., MOUVET, C., Field data and modelling of water and nitrate movement through deep unsaturated loess, J. Hydrol. **345** 1–2 (2007) 27–37, https://doi.org/10.1016/j.jhydrol.2007.07.006

[6.68] NATIONAL COUNCIL ON RADIATION PROTECTION AND MEASUREMENTS, Screening Models for Releases of Radionuclides to Atmosphere, Surface Water, and Ground – Vol. I and Vol. II, Rep. 123, NCRP, Bethesda, MD (1996).

[6.69] HART, D., Derived Release Limits Guidance, COG-06-3090-R2-1, CANDU Owners Group, Toronto (2008).

[6.70] BAILLY DU BOIS, P., GERMAIN, P., ROZET, M., SOLIER, L., "Water masses circulation and residence time in the Celtic sea and English channel approaches, characterisation based on radionuclides labelling from industrial releases", Int. Conf. on Radioactivity in Environment 2002, Monaco (2002) 395–399.

[6.71] BAILLY DU BOIS, P., DUMAS, F., Fast hydrodynamic model for medium- and long-term dispersion in seawater in the English Channel and southern North Sea, qualitative and quantitative validation by radionuclide tracers, Ocean Model. **9** 2 (2005) 169–210, https://doi.org/10.1016/j.ocemod.2004.07.004

[6.72] BAILLY DU BOIS, P., DUMAS, F., SOLIER, L., VOISEUX, C., Controlled tritium liquid releases from Areva-NC reprocessing plant [dataset], PANGAEA (2011), https://doi.org/10.1594/PANGAEA.762428

[6.73] NIE, B., et al., Numerical study on tritium dispersion in coastal waters: The case of Hangzhou Bay, China, J. Hydrol. **590** (2020) 125532, https://doi.org/10.1016/j.jhydrol.2020.125532

[6.74] HADŽIŠEHOVIĆ, M., et al., Determination of tritium in natural waters, J. Radioanal. Chem. **44** 2 (1978) 301–306, https://doi.org/10.1007/BF02519622

[6.75] HADŽIŠEHOVIĆ, M., et al., Characteristics of the tritium distribution in the Danube basin region in Yugoslavia, J. Radioanal. Chem. **74** 1 (1982) 239–248, https://doi.org/10.1007/BF02520373

[6.76] RADWAN, I., PIETRZAK-FLIS, Z., WARDASZKO, T., Tritium in surface waters, tap water and in precipitation in Poland during the 1994–1999 period, J. Radioanal. Nucl. Chem. **247** 1 (2001) 71–77, https://doi.org/10.1023/A:1006775300812

[6.77] PALOMO, M., PEÑALVER, A., AGUILAR, C., BORRULL, F., Tritium activity levels in environmental water samples from different origins, Appl. Radiat. Isot. **65** 9 (2007) 1048–1056, https://doi.org/10.1016/j.apradiso.2007.03.013

[6.78] FOURRÉ, E., et al., Past and recent tritium levels in Arctic and Antarctic polar caps, Earth Planet. Sci. Lett. **245** 1 (2006) 56–64, https://doi.org/10.1016/j.epsl.2006.03.003

[6.79] GAT, J., MOOK, W., MEIJER, H., "Tritium in the atmosphere", Environmental Isotopes in the Hydrological Cycle, Principles and Applications, Vol. II, UNESCO/IAEA, Vienna (2001) 63–74.

[6.80] TADROS, C.V., HUGHES, C.E., CRAWFORD, J., HOLLINS, S.E., CHISARI, R., Tritium in Australian precipitation: A 50 year record, J. Hydrol. **513** (2014) 262–273, https://doi.org/10.1016/j.jhydrol.2014.03.031

[6.81] MORGENSTERN, U., STEWART, M.K., STENGER, R., Dating of streamwater using tritium in a post nuclear bomb pulse world: Continuous variation of mean transit time with streamflow, Hydrol. Earth Syst. Sci. **14** 11 (2010) 2289–2301, https://doi.org/10.5194/hess-14-2289-2010

[6.82] DUCROS, L., et al., Tritium in river waters from French Mediterranean catchments: Background levels and variability, Sci. Total Environ. **612** (2018) 672–682, https://doi.org/10.1016/j.scitotenv.2017.08.026

[6.83] MICHEL, R., AGGARWAL, P., ARAGUAS ARAGUAS, L., NEWMAN, B., KURTTAS, T., "Use of isotopic data to determine influence of seasonal effects in rivers", Proc. Isotopes in Hydrology, Marine Ecosystems and Climate Change Studies, Monaco, 2011, IAEA, Vienna (2013) 203–210.

[6.84] HALDER, J., TERZER, S., WASSENAAR, L., ARAGUÁS-ARAGUÁS, L., AGGARWAL, P., The Global Network of Isotopes in Rivers (GNIR): Integration of water isotopes in watershed observation and riverine research, Hydrol. Earth Syst. Sci. **19** 8 (2015) 3419–3431, https://doi.org/10.5194/hess-19-3419-2015

[6.85] MAŁOSZEWSKI, P., RAUERT, W., STICHLER, W., HERRMANN, A., Application of flow models in an alpine catchment area using tritium and deuterium data, J. Hydrol. **66** 1–4 (1983) 319–330, https://doi.org/10.1016/0022-1694(83)90193-2

[6.86] RANK, D., et al., "Hydrological parameters and climatic signals derived from long term tritium and stable isotope time series of the river Danube", Proc. Isotope Techniques in the Study of Environmental Change, Vienna, 1997, IAEA, Vienna (1998).

[6.87] HARMS, P.A., VISSER, A., MORAN, J.E., ESSER, B.K., Distribution of tritium in precipitation and surface water in California, J. Hydrol. **534** (2016) 63–72, https://doi.org/10.1016/j.jhydrol.2015.12.046

[6.88] CARTWRIGHT, I., MORGENSTERN, U., Transit times from rainfall to baseflow in headwater catchments estimated using tritium: the Ovens River, Australia, Hydrol. Earth Syst. Sci. **19** 9 (2015) 3771–3785, https://doi.org/10.5194/hess-19-3771-2015

[6.89] CHRISTENSEN, T.R., et al., Thawing sub-arctic permafrost: Effects on vegetation and methane emissions, Geophys. Res. Lett. **31** (2004) L04501, https://doi.org/10.1029/2003GL018680

[6.90] ÅKERMAN, H.J., JOHANSSON, M., Thawing permafrost and thicker active layers in sub-arctic Sweden, Permafr. Periglac. Process. **19** 3 (2008) 279–292, https://doi.org/10.1002/ppp.626

[6.91] SMITH, S.L., WOLFE, S.A., RISEBOROUGH, D.W., NIXON, F.M., Active-layer characteristics and summer climatic indices, Mackenzie Valley, Northwest Territories, Canada, Permafr. Periglac. Process. **20** 2 (2009) 201–220, https://doi.org/10.1002/ppp.651

[6.92] JORGENSON, M.T., SHUR, Y.L., PULLMAN, E.R., Abrupt increase in permafrost degradation in Arctic Alaska, Geophys. Res. Lett. **33** 2 (2006) L02503, https://doi.org/10.1029/2005GL024960

[6.93] QUINTON, W., BALTZER, J., The active-layer hydrology of a peat plateau with thawing permafrost (Scotty Creek, Canada), Hydrogeol. J. **21** 1 (2013) 201–220, https://doi.org/10.1007/s10040-012-0935-2

[6.94] ANISIMOV, O., RENEVA, S., Permafrost and changing climate: The Russian perspective, AMBIO J. Hum. Environ. **35** 4 (2006) 169–175, https://doi.org/10.1579/0044-7447(2006)35[169:PACCTR]2.0.CO;2

[6.95] WU, Q., ZHANG, T., LIU, Y., Permafrost temperatures and thickness on the Qinghai-Tibet Plateau, Glob. Planet. Change **72** 1–2 (2010) 32–38, https://doi.org/10.1016/j.gloplacha.2010.03.001

[6.96] JAFAROV, E.E., ROMANOVSKY, V.E., GENET, H., McGUIRE, A.D., MARCHENKO, S.S., The effects of fire on the thermal stability of permafrost in lowland and upland black spruce forests of interior Alaska in a changing climate, Environ. Res. Lett. **8** 3 (2013) 035030, https://doi.org/10.1088/1748-9326/8/3/035030

[6.97] WOO, M., KANE, D.L., CAREY, S.K., YANG, D., Progress in permafrost hydrology in the new millennium, Permafr. Periglac. Process. **19** 2 (2008) 237–254, https://doi.org/10.1002/ppp.613

[6.98] DANBY, R.K., HIK, D.S., Variability, contingency and rapid change in recent subarctic alpine tree line dynamics, J. Ecol. **95** 2 (2007) 352–363, https://doi.org/10.1111/j.1365-2745.2006.01200.x

[6.99] ROWLAND, J., et al., Arctic landscapes in transition: Responses to thawing permafrost, Eos Trans. Am. Geophys. Union **91** 26 (2010) 229–230, https://doi.org/10.1029/2010EO260001

[6.100] GROSSE, G., GOETZ, S., McGUIRE, A.D., ROMANOVSKY, V.E., SCHUUR, E.A., Changing permafrost in a warming world and feedbacks to the Earth system, Environ. Res. Lett. **11** 4 (2016) 040201, https://doi.org/10.1088/1748-9326/11/4/040201

[6.101] KLAMINDER, J., YOO, K., RYDBERG, J., GIESLER, R., An explorative study of mercury export from a thawing palsa mire, J. Geophys. Res. Biogeosciences **113** (2008) G04034, https://doi.org/10.1029/2008JG000776

[6.102] KLAMINDER, J., HAMMARLUND, D., KOKFELT, U., VONK, J.E., BIGLER, C., Lead contamination of subarctic lakes and its response to reduced atmospheric fallout: Can the recovery process be counteracted by the ongoing climate change? Environ. Sci. Technol. **44** 7 (2010) 2335–2340, https://doi.org/10.1021/es903025z

[6.103] RYDBERG, J., KLAMINDER, J., ROSÉN, P., BINDLER, R., Climate driven release of carbon and mercury from permafrost mires increases mercury loading to sub-arctic lakes, Sci. Total Environ. **408** 20 (2010) 4778–4783, https://doi.org/10.1016/j.scitotenv.2010.06.056

[6.104] SCHUSTER, P.F., et al., Permafrost stores a globally significant amount of mercury, Geophys. Res. Lett. **45** 3 (2018) 1463–1471, https://doi.org/10.1002/2017GL075571

[6.105] KALLENBORN, R., et al., Combined Effects of Selected Pollutants and Climate Change in the Arctic Environment, AMAP Technical Report No. 5 (2011), Arctic Monitoring and Assessment Programme, Oslo (2011).

[6.106] GIBSON, J., BIRKS, S., YI, Y., Higher tritium concentrations measured in permafrost thaw lakes in northern Alberta, Hydrol. Process. **30** 2 (2016) 245–249, https://doi.org/10.1002/hyp.10599

[6.107] BOND, M.J., CARR, J., Permafrost thaw and implications for the fate and transport of tritium in the Canadian north, J. Environ. Radioact. **192** (2018) 295–311, https://doi.org/10.1016/j.jenvrad.2018.07.006

[6.108] AMBACH, W., et al., Deuterium, tritium and gross-beta-activity investigations on Alpine glaciers (Ötztal Alps), J. Glaciol. **17** 77 (1976) 383–400, https://doi.org/10.3189/S0022143000013691

[6.109] KAMIYAMA, K., KITAOKA, K., WATANABE, O., Characteristics of the Yala Glacier from the viewpoint of tritium content, J. Geophys. Res. Atmospheres **91** D11 (1986) 11841–11844, https://doi.org/10.1029/JD091iD11p11841

[6.110] STRASSER, J.C., LAWSON, D.E., LARSON, G.J., EVENSON, E.B., ALLEY, R.B., Preliminary results of tritium analyses in basal ice, Matanuska Glacier, Alaska, USA: Evidence for subglacial ice accretion, Ann. Glaciol. **22** (1996) 126–133, https://doi.org/10.3189/1996AoG22-1-126-133

[6.111] KERN, Z., MOLNÁR, M., SVINGOR, É., PERŞOIU, A., NAGY, B., High-resolution, well-preserved tritium record in the ice of Bortig Ice Cave, Bihor Mountains, Romania, The Holocene **19** 5 (2009) 729–736, https://doi.org/10.1177/0959683609105296

[6.112] KANG, S., et al., Dramatic loss of glacier accumulation area on the Tibetan Plateau revealed by ice core tritium and mercury records, The Cryosphere **9** 3 (2015) 1213–1222, https://doi.org/10.5194/tc-9-1213-2015

[6.113] KEHRWALD, N.M., et al., Mass loss on Himalayan glacier endangers water resources, Geophys. Res. Lett. **35** 22 (2008) L22503, https://doi.org/10.1029/2008GL035556

[6.114] MICHEL, R.L., NAFTZ, D.L., Use of sulfur-35 and tritium to study runoff from an alpine glacier, Wind River Range, Wyoming, IAHS Publ. **228** (1995) 441–444.

[6.115] ZEKOLLARI, H., HUSS, M., FARINOTTI, D., Modelling the future evolution of glaciers in the European Alps under the EURO-CORDEX RCM ensemble, The Cryosphere **13** 4 (2019) 1125–1146, https://doi.org/10.5194/tc-13-1125-2019

[6.116] YANKOVICH, T., et al., Measured and modelled tritium concentrations in freshwater Barnes mussels (Elliptio complanata) exposed to an abrupt increase in ambient tritium levels, J. Environ. Radioact. **102** 1 (2011) 26–34, https://doi.org/10.1016/j.jenvrad.2010.06.014

[6.117] AUDETTE-STUART, M., et al., Adaptive response in frogs chronically exposed to low doses of ionizing radiation in the environment, J. Environ. Radioact. **102** 6 (2011) 566–573, https://doi.org/10.1016/j.jenvrad.2011.03.001

[6.118] MELINTESCU, A., GALERIU, D., Dynamic model for tritium transfer in an aquatic food chain, Radiat. Environ. Biophys. **50** 3 (2011) 459–473, https://doi.org/10.1007/s00411-011-0362-0

[6.119] BAUMGAERTNER, F., YANKOVICH, T.L., KIM, S.B., Logistic analysis of OBT dynamics, Health Phys. **97** 6 (2009) 572–580, https://doi.org/10.1097/01.HP.0000363837.32428.01

[6.120] RODGERS, D.W., Tritium dynamics in juvenile rainbow trout, Salmo gairdneri, Health Phys. **50** 1 (1986) 89–98 https://doi.org/10.1097/00004032-198601000-00008

[6.121] GALERIU, D., MELINTESCU, A., Retention of tritium in reference persons: A metabolic model. Derivation of parameters and application of the model to the general public and to workers, J. Radiol. Prot. **30** 3 (2010) 445–468, https://doi.org/10.1088/0952-4746/30/3/003

[6.122] MELINTESCU, A., GALERIU, D., KIM, S.B., Tritium dynamics in large fish – a model test, Radioprotection **46** 6 (2011) S431–S436, https://doi.org/10.1051/radiopro/20116547s

[6.123] YANKOVICH, T., KOARASHI, J., KIM, S., DAVIS, P., International study on the validation of models for the environmental transfer of tritium and carbon-14, Appl. Radiat. Isot. **66** 11 (2008) 1726–1729, https://doi.org/10.1016/j.apradiso.2007.12.021

[6.124] INTERNATIONAL ATOMIC ENERGY AGENCY, Environmental Modelling for Radiation Safety (EMRAS), IAEA-TECDOC-1678, IAEA, Vienna (2012).

[6.125] PRIEST, N.D., et al., Tritium (^{3}H) retention in mice: Administered as HTO, DTO or as ^{3}H-labeled amino-acids, Health Phys. **112** 5 (2017) 439–444, https://doi.org/10.1097/HP.0000000000000637

[6.126] TAKEDA, H., et al., Comparative biokinetics of tritium in rats during continuous ingestion of tritiated water and tritium-labeled food, Int. J. Radiat. Biol. **77** 3 (2001) 375–381, https://doi.org/10.1080/09553000010017117

[6.127] TAKEDA, H., et al., Comparative biokinetics of radiocarbon ingested as compounds or foods in rats, Health Phys. **99** 5 (2010) 668–673, https://doi.org/10.1097/HP.0b013e3181de6f71

[6.128] GALERIU, D., MELINTESCU, A., BERESFORD, N.A., TAKEDA, H., CROUT, N.M.J., The dynamic transfer of ^{3}H and ^{14}C in mammals: A proposed generic model, Radiat. Environ. Biophys. **48** 1 (2009) 29–45, https://doi.org/10.1007/s00411-008-0193-9

[6.129] MELINTESCU, A., GALERIU, D., Energy metabolism used as a tool to model the transfer of ^{14}C and ^{3}H in animals, Radiat. Environ. Biophys. **49** 4 (2010) 657–672, https://doi.org/10.1007/s00411-010-0302-4

[6.130] KOMATSU, K., HIGUCHI, M., SAKKA, M., Accumulation of tritium in aquatic organisms through a food chain with three trophic levels, J. Radiat. Res. **22** 2 (1981) 226–241, https://doi.org/10.1269/jrr.22.226

[6.131] KIM, S., SHULTZ, C., STUART, M., FESTARINI, A., Tritium uptake in rainbow trout (*Oncorhynchus mykiss*): HTO and OBT-spiked feed exposures simultaneously, Appl. Radiat. Isot. **98** (2015) 96–102, https://doi.org/10.1016/j.apradiso.2015.01.020

[6.132] EYROLLE-BOYER, F., ANTONELLI, C., RENAUD, P., TOURNIEUX, D., Origins and trend of radionuclides within the lower Rhône River over the last decades, Radioprotection **50** 1 (2015) 27–34, https://doi.org/10.1051/radiopro/2014022

[6.133] JEAN-BAPTISTE, P., et al., Tritium and radiocarbon levels in the Rhône river delta and along the French Mediterranean coastline, J. Environ. Radioact. **187** (2018) 53–64, https://doi.org/10.1016/j.jenvrad.2018.01.031

[6.134] LUMIA, G., JOURD'HEUIL, L., PALLY, M., Etude Radioécologique du Site de Creys-Malville en 1993, CEA/IPSN/DPEI, Rapport SERE/94/051 (P), Cadarache (1994).

[6.135] BEAUGELIN, K., THIRION, J., Bilan Radioécologique Décennal du CNPE de Creys-Malville-1996, CEA/IPSN/DPRE, Rapport SERE/97/029 (P), Cadarache (1997).

[6.136] LAMBERT, B., Welsh tritium, J. Radiol. Prot. **21** 4 (2001) 333–336, https://doi.org/10.1088/0952-4746/21/4/002

[6.137] KITSON, S.L., Amersham radiochemistry to GE Healthcare, J. Label. Compd. Radiopharm. **50** 9–10 (2007) 737–745, https://doi.org/10.1002/jlcr.1427

[6.138] ENVIRONMENT AGENCY, FOOD STANDARDS AGENCY, FOOD STANDARDS SCOTLAND, NATURAL RESOURCES WALES, NORTHERN IRELAND ENVIRONMENT AGENCY, SCOTTISH ENVIRONMENT PROTECTION AGENCY, Radioactivity in Food and the Environment, 2023: RIFE-29, CEFA (2024), https://www.gov.uk/government/publications/radioactivity-in-food-and-the-environment-rife-reports/rife-29-radioactivity-in-food-and-the-environment-2023

[6.139] HEALTH PROTECTION AGENCY, Review of Risks from Tritium: Report of the Independent Advisory Group on Ionising Radiation, Vol. 4, Health Protection Agency, Didcot (2007).

[6.140] NIE, B., NI, M., JIANG, J., WU, Y., A dynamic modeling ^{3}H transfer to the environment under accidental release from the fusion reactor, J. Fusion Energy **34** 4 (2015) 739–745,
https://doi.org/10.1007/s10894-015-9878-7

[6.141] Di PACE, L., LETELLIER, E., MAUBERT, H., PATEL, B., RASKOB, W., Biological hazard issues from potential releases of tritiated dust from ITER, Fusion Eng. Des. **83** 10–12 (2008) 1729–1732,
https://doi.org/10.1016/j.fusengdes.2008.06.011

[6.142] WIERCZINSKI, B., AIGN, J., BAUMGÄRTNER, F., MÜLLEN, G., TÜRLER, A., Proton exchange and tritium enrichment in inorganic systems: Inorganic metal salts, J. Radioanal. Nucl. Chem. **266** 1 (2005) 39–42,
https://doi.org/10.1007/s10967-005-0865-8

[6.143] KIM, S., BREDLAW, M., KOROLEVYCH, V., HTO and OBT activity concentrations in soil at the historical atmospheric HT release site (Chalk River Laboratories), J. Environ. Radioact. **103** 1 (2012) 34–40,
https://doi.org/10.1016/j.jenvrad.2011.08.013

[6.144] FIÉVET, B., et al., Transfer of tritium released into the marine environment by French nuclear facilities bordering the English Channel, Environ. Sci. Technol. **47** 12 (2013) 6696–6703,
https://doi.org/10.1021/es400896t

[6.145] KIM, S., FARROW, F., BREDLAW, M., STUART, M., Changes in HTO and OBT activity concentrations in the Perch Lake aquatic ecosystem, J. Environ. Radioact. **165** (2016) 280–285,
https://doi.org/10.1016/j.jenvrad.2016.10.015

[6.146] KIM, S.B., et al., Organically bound tritium (OBT) activity concentrations in surface soil at the Chalk River Laboratories, Canada, J. Environ. Radioact. **208–209** (2019) 105999,
https://doi.org/10.1016/j.jenvrad.2019.105999

[6.147] INSTITUT DE RADIOPROTECTION ET DE SÛRETÉ NUCLÉAIRE, Bilan de l'État Radiologique de l'Environnement Français de 2018 à 2020, IRSN, Fontenay-aux-Roses (2021).

[6.148] INTERNATIONAL ATOMIC ENERGY AGENCY, Radiological Conditions at the Semipalatinsk Test Site, Kazakhstan: Preliminary Assessment and Recommendations for Further Study, Radiological Assessment Reports Series, IAEA, Vienna (1998).

[6.149] NATIONAL NUCLEAR CENTER OF THE REPUBLIC OF KAZAKHSTAN, Topical Issues of Radioecology of Kazakhstan, Issue 5: Optimization of Studies of the Territories of the Semipalatinsk Test Site to Allow Their Future Economic Use (Lukashenko, S.N., Ed.), KNNC, Pavlodar (2015).

[6.150] LYAKHOVA, O., LUKASHENKO, S., LARIONOVA, N., SUBBOTIN, S., Tritium as an indicator of nuclear test sites, NNC RK Bulletin **3** (2011) 125–128 (in Russian).

[6.151] TIMONOVA, L., LYAKHOVA, O., LUKASHENKO, S., AIDARKHANOV, A., Study on tritium content in soil at sites of nuclear explosions on the territory of Semipalatinsk test site, Radiat. Biol. Radioecol. **55** 6 (2015) 667–672 (in Russian).

[6.152] LYAKHOVA, O., LUKASHENKO, S., TIMONOVA, L., LARIONOVA, N., TURCHENKO, D., Evaluation of the concentration of gaseous compound of tritium in places of nuclear tests at the Semipalatinsk test site, Radiat. Biol. Radioecol. **60** 6 (2020) 635–646 (in Russian).

[6.153] MITCHELL, P.I., et al., Tritium in well waters, streams and atomic lakes in the East Kazakhstan Oblast of the Semipalatinsk Nuclear Test Site, J. Radiol. Prot. **25** 2 (2005) 141–148,
https://doi.org/10.1088/0952-4746/25/2/002

[6.154] LYAKHOVA, O., LUKASHENKO, S., AYDARKHANOV, A., Identification of the main sources of tritium input to the atmospheric air and assessment of the level of tritium pollution of the air in the Degelen mountain range, Problemy Biogeokhimiyi i Geokhimicheskoy Ekologii **3** 14 (2010) 73 (in Russian).

[6.155] POLIVKINA, Y.N., LARIONOVA, N.V., LYAKHOVA, O.N., Assessment of tritium uptake by Helianthus Annuus culture continuously exposed to HTO at the Semipalatinsk test site, Radiat. Risk **29** 1 (2020) 79–89,
https://doi.org/10.21870/0131-3878-2020-29-1-79-89

[6.156] LARIONOVA, N., et al., Transfer of radionuclides to plants of natural ecosystems at the Semipalatinsk test site, J. Environ. Radioact. **186** (2018) 63–70,
https://doi.org/10.1016/j.jenvrad.2017.09.006

[6.157] POURCELOT, L., et al., Hydrological behaviour of tritium on the former Semipalatinsk nuclear test site (Kazakhstan) determined using stable isotope measurements, Eurasian Chem.-Technol. J. **15** 4 (2013) 293–299,
https://doi.org/10.18321/ectj234

[6.158] AIDARKHANOV, A., LUKASHENKO, S., UMAROV, M., YAKOVENKO, Y.Y., Results of radiological survey of the territory adjacent to the “Atomic” lake, Eurasian J. Phys. Funct. Mater. **1** 2 (2017) 5–18,
https://doi.org/10.29317/ejpfm.2017010201

[6.159] AKTAEV, M., et al., Research of tritium content in soil in the places of nuclear testing on the territory of Semipalatinsk test site, Radiat. Biol. Radioecol. **59** 3 (2019) 311–320 (in Russian),
https://doi.org/10.1134/S0869803119030044

[6.160] MITCHELL, P., et al., “Tritium in streams, well waters and atomic lakes at the Semi-Palatinsk Nuclear Test Site: Present status and future perspectives”, Proc. Nuclear Risks in Central Asia, Springer, Dordrecht (2008) 41–60,
https://doi.org/10.1007/978-1-4020-8317-4_5

[6.161] PALLY, M., BARRÉ, A., FOULQUIER, L., Tritium associé à la matière organique de sédiments, végétaux et poissons des principaux cours d’eau français, Int. Ver. Für Theor. Angew. Limnol. Verhandlungen **25** 1 (1993) 285–289,
https://doi.org/10.1080/03680770.1992.11900116

[6.162] KRYSHEV, I., Radioactive contamination of aquatic ecosystems following the Chernobyl accident, J. Environ. Radioact. **27** 3 (1995) 207–219, https://doi.org/10.1016/0265-931X(94)00042-U

[6.163] KATRICH, I.Y., Tritium in natural water after the Chernobyl accident, Sov. Meteorol. Hydrol. **5** (1990) 80–84.

[6.164] KOGA, T., MORISHIMA, H., NIWA, T., KAWAI, H., Tritium precipitation in European cities and in Osaka, Japan owing to the Chernobyl nuclear accident, J. Radiat. Res. **32** 3 (1991) 267–276, https://doi.org/10.1269/jrr.32.267

[6.165] UEDA, S., et al., Nuclear accident-derived ^{3}H in river water of Fukushima Prefecture during 2011–2014, J. Environ. Radioact. **146** (2015) 102–109, https://doi.org/10.1016/j.jenvrad.2015.04.014

[6.166] YABUSAKI, S., SHIBASAKI, N., TAKAGAI, Y., Water quality and tritium concentrtion of groundwater and spring water at vicinity of coast area in Minamisoma City, Fukushima Prefecture, Report of the observations in 2013, J. Cent. Reg. Aff. Fukushima Univ. **27** (2015) 37.

[6.167] KASHIWAYA, K., et al., Spatial variations of tritium concentrations in groundwater collected in the southern coastal region of Fukushima, Japan, after the nuclear accident, Sci. Rep. **7** (2017) 12578, https://doi.org/10.1038/s41598-017-12840-3

[6.168] JEAN-BAPTISTE, P., et al., Organically bound tritium (OBT) and carbon-14 accumulation in the sediments off the mouth of the Rhône river, Environ. Earth Sci. **78** (2019) 78, https://doi.org/10.1007/s12665-019-8081-y

[6.169] NORDYKE, M.D., The Soviet program for peaceful uses of nuclear explosions, Sci. Glob. Secur. **7** 1 (1998) 1–117, https://doi.org/10.1080/08929889808426448

[6.170] LOGACHEV, V.A., et al., Modern Radioecological Situation in the Areas of Peaceful Nuclear Explosions on the Territory of the Russian Federation: Facts and Witnesses, IzdAT, Moscow (2005) (in Russian).

[6.171] RAMZAEV, V.P., BIBLIN, A.M., REPIN, V.S., KHRAMTSOV, E.V., VARFOLOMEEVA, K.V., Tritium contamination of surface and ground waters at the “Dnepr” peaceful underground nuclear explosions site, Radiat. Hyg. **15** 1 (2022) 6–26 (in Russian), https://doi.org/10.21514/1998-426X-2022-15-1-6-26

[6.172] BARKOVSKY, A., et al., Radiation doses to the population of the Russian Federation in 2020, Radiat. Hyg. **14** 4 (2021) 103–113 (in Russian), https://doi.org/10.21514/1998-426X-2021-14-4-103-113

[6.173] BROWN, R.M., WORKMAN, W.J., KOTZER, T.G., Tritium program at Chalk River Laboratories, Trans. Am. Nucl. Soc. **69** (1993) 20.

[6.174] MOKROV, Yu.G., ALEKSAKHIN, A.I., PA Mayak activity to ensure the safe use of the Techensky cascade of water reservoirs, J. Rad. Saf. Issues **4**(92) (2018) 13 (in Russian).

[6.175] MAYAK PA, 2020 Report on Ecological Safety, Federal State Unitary Enterprise "Mayak Production Association", Ozersk (2021) (in Russian).

[6.176] MAYAK PA, 2021 Report on Ecological Safety, Federal State Unitary Enterprise "Mayak Production Association", Ozersk (2022) (in Russian), https://xn-btb4bfrm9d.xn-p1ai/upload/iblock/b2d/b2dc09e6683e4e3ef8d14ba659ead824.pdf

[6.177] DEMIN, S.N., TELUSHKINA, E.L., Radiation-hygienic assessment of the environmental contamination by tritium and radiation doses to public in proximity of a radiochemistry plant, Bull. Radiat. Med. **1** (1987) 23 (in Russian).

[6.178] CHEBOTINA, M., NIKOLIN, O., Tritium water bodies in the area of PA Mayak, J. Rad. Saf. Issues **1** (2005) 79 (in Russian).

[6.179] KAZACHYONOK, N., et al., Regularities of distribution ^{3}H in open water systems and sources of drinking water in the zone of influence of the Mayak PA, ANRI Magazine **3** (74) (2013) 43 (in Russian).

[6.180] VOSTROTIN, V.V., YANOV, A.Y., FINASHOV, L.V., Tritium in environmental objects in the area affected by FSUE Mayak Production Association in 2014–15, J. Radiol. Prot. **41** 2 (2021) S56–S66, https://iopscience.iop.org/article/10.1088/1361-6498/abe8c8

[6.181] KAZACHYONOK, N., et al. Contemporary radioactive contamination of the EURT and other areas influenced by the Mayak PA, J. Rad. Saf. Issues **1** (2014) 34 (in Russian).

[6.182] KUZMENKOVA, N., et al., Radionuclides distribution between biotic and abiotic components of radioactive contaminated water reservoirs R-17 and R-4, J. Rad. Saf. Issues **1** (2017) 54 (in Russian).

[6.183] STUKALOV, P., Radioactive contamination of the "Mayak" PA industrial reservoir Old swamp. Review of results of investigations during period from 1949 to 2006, Suppl. J. Rad. Saf. Issues (2007) (in Russian).

[6.184] CHEBOTINA, M.Y., NIKOLIN, O.A., SMAGIN, A.I., MURASHOVA, E.L., Tritium in the ponds of industrial and non-industrial usage around the "Mayak" enterprise, Water Sect. Russ. **4** (2011) 75–84 (in Russian), https://doi.org/10.35567/1999-4508-2011-4-7

Chapter 7

TRITIUM IN THE TERRESTRIAL ENVIRONMENT

N. BAGLAN, S.B. KIM

7.1. INTRODUCTION

Tritiated water vapour from the atmosphere is easily incorporated as TFWT into living organisms and can lead to labelling of organic matter as OBT following biointegration processes. OBT is currently classified into two categories: E-OBT, which is mainly formed through isotope exchange between TFWT and labile hydrogen isotopes in living organisms, and NE-OBT, which is formed when the tritium is integrated into biomolecules through metabolic processes. In the case of plants, this integration happens mostly via photosynthesis. Fast equilibrium is reached between TFWT and tritiated water in the environment, while a long residence time characterizes OBT dynamics [7.1–7.4].

The presence of tritium in soils can result from several processes, mainly atmospheric deposition (including rainfall), addition of labelled POM (see also Sections 6.5.1 and 6.5.3) and the formation of plant litter that evolves into humus. A specific feature of tritium is that it can be 're-emitted' into the atmosphere. The incorporation of tritium into soil organic matter and its subsequent release by 're-emission' is not yet sufficiently understood. A significant proportion of tritium bound to soil organic matter may come from direct deposition or via degradation of the aerial parts of plants (leaf fall and litter formation).

7.2. TRITIUM SOURCE TERMS IN THE TERRESTRIAL ENVIRONMENT

7.2.1. Rainwater

Terrestrial environments are directly affected by precipitation and its associated tritium contents. At the global scale, interstitial waters in soils are mainly expected to reflect the tritium content in precipitation, although evaporation and exchange processes with the atmosphere and exchanges with underground waters (see Chapter 6) are expected to play a potentially modifying role.

7.2.2. Lithogenic or geogenic production

A small part of natural tritium is produced at the surface of the Earth during neutron capture by the small amounts of ^{6}Li contained in rocks. Neutrons involved in such reactions are produced by the spontaneous fission of ^{238}U or by reactions induced by alpha radiation from uranium and thorium, although there will also be a significant cosmogenic contribution. Tritium is produced by neutron absorption of lithium according to the reaction $^{6}Li + n \rightarrow {}^{3}H + {}^{4}He$. Such tritium is termed lithogenic or geogenic and tends to be locked in mineral structures. This type of tritium is generally negligible compared with cosmogenic tritium production in the atmosphere.

7.2.3. Atmospheric production

Tritium is mainly produced in the atmosphere by nuclear reactions between nuclei of the main atmospheric atoms (nitrogen, oxygen) and neutrons from cosmic radiation [7.5, 7.6]. The following reactions lead to the formation of tritium [7.7, 7.8]:

$$^{14}_{7}N + ^{1}_{0}n \rightarrow ^{12}_{6}C + ^{3}_{1}H \tag{7.1}$$

$$^{14}_{7}N + ^{1}_{0}n \rightarrow 3^{4}_{2}He + ^{3}_{1}H \tag{7.2}$$

$$^{16}_{8}O + ^{1}_{0}n \rightarrow ^{14}_{6}C + ^{3}_{1}H \tag{7.3}$$

7.2.4. Atmospheric fallout from nuclear weapon testing

Widespread atmospheric nuclear fallout occurred after more than 500 above ground nuclear weapon tests from 1945 to 1980 [7.9], with most falling in the Northern Hemisphere. Each nuclear weapon test produced a radioactive cloud with a rate and scale of deposition depending on the power yield of the device, the dispersion of the cloud and local weather conditions (e.g. the occurrence of rainfall and wind directions). Background concentrations varied in space and time. Fallout typically contained hundreds of different radionuclides (mostly fission and activation products). Some of these radionuclides — for example, ^{90}Sr, ^{137}Cs and ^{3}H — have medium length half-lives (in the range of years to decades) and are still detectable today. Regarding tritium, approximately 186 000 PBq was cumulatively released and dispersed, but this activity has significantly reduced through radioactive decay [7.9] (see Chapter 3). The Radiation Protection Bureau of Health of Canada has conducted and maintained a nationwide fallout monitoring programme since 1958. In Canada, monitoring stations are situated from coast to coast and in the high Arctic [7.10].

7.2.5. Atmospheric releases of tritium from nuclear facilities

In the atmosphere, tritium is present mostly in the form of tritiated water (HTO) vapours and, to a much smaller extent, in gaseous forms of tritium, altogether representing about 10% to the total amount of tritium in the environment [7.11]. Tritium released from nuclear facilities in the gaseous state is mostly in the form of tritiated water vapour but can also be found as tritiated hydrogen (HT) and tritiated methane (CH_3T). Owing to oxidation processes, most tritium in the gaseous form transforms into HTO and joins the water cycle [7.7].

The main production route of tritium for all reactor types, except HWRs, takes place within the fuel, being formed as a ternary fission product from the fission of certain isotopes of uranium and plutonium. However, it remains largely confined/isolated in the fuel cladding (e.g. in the form of zirconium hydride) and therefore contributes very little to direct releases to the environment [7.7, 7.12]. For most of the tritium released, its production is driven by neutron activation reactions with lithium and boron isotopes in the reactor coolant [7.7, 7.12]. Until recently, HWRs (e.g. CANDU reactors) have been responsible for most production and release of tritium into the environment. In CANDU reactors, most environmental tritium is produced via neutron activation of the deuterium component of heavy water, forming DTO, TTO, DT and TT.

The current international interest in creating fusion reactors will ultimately lead to additional contributions of tritium to the environment, but significant developments are in progress that aim to capture and recycle the largest part of any fusion plant-related tritium.

In 2009–2011, the IAEA organized EMRAS II [7.13], which focused on the improvement of environmental transfer models and the development of reference approaches to estimate the radiological impacts on humans, as well as on flora and fauna, arising from radionuclides in the environment. Reference [7.3] is a report of Working Group 7 (Tritium Accidents) of EMRAS II.

7.2.6. Liquid releases of tritium from nuclear facilities

Tritium is released into the environment mainly as HTO. As tritium is an isotope of hydrogen, its circulation in its liquid form in the environment follows the same pathways as H_2O (see Fig. 7.1), although the different dimensions of the tritium and protium atoms result in some differences in the kinetics of these pathways.

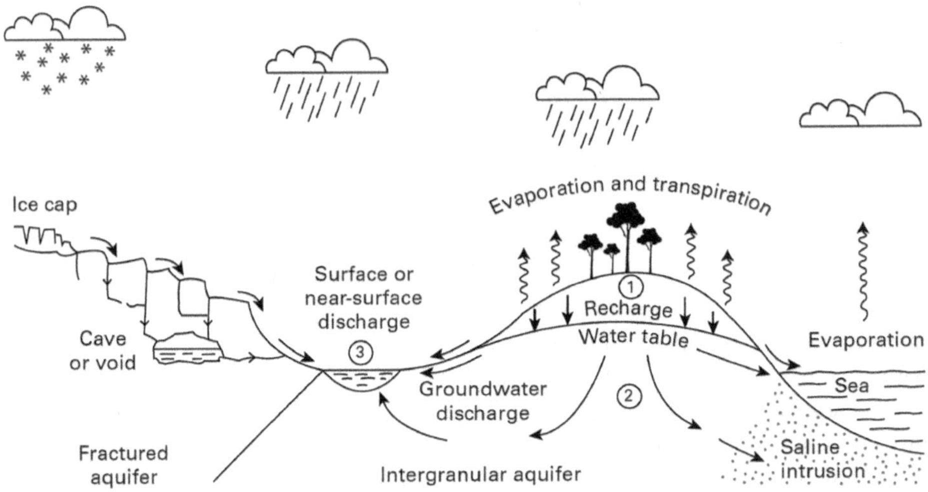

FIG. 7.1. Geochemical water cycle in a temperate oceanic environment with circulation and transfer in terrestrial compartments (reproduced from Ref. [7.14] with permission courtesy of Springer Nature).

Tritiated water, in liquid or gaseous form, is rapidly integrated into the water cycle and dilutes very quickly in the various terrestrial compartments. The gaseous forms of tritium (HT and CH_3T) are not very water-soluble and are carried away and dispersed until exchanges occur with surface molecules (e.g. in water vapour, plants, soil). Exchanges between aquatic compartments and the atmosphere are governed by natural processes of water phase change (e.g. evaporation, condensation) and precipitation. The average tritium concentration found in the environment after releases is reduced to very low levels (below 100 Bq/L in water) by the high dilution capacities of the atmosphere and hydrosphere [7.15, 7.16]. The dilution factor then increases with distance from the initial release point. Transfer of tritium to terrestrial living plants by irrigation with river waters and/or surface waters originating from the vicinity of nuclear reactor sites is another mechanism that explains tritium dispersion into the terrestrial environment.

7.3. BIOINTEGRATION OF TRITIUM IN PLANTS — METABOLIC PROCESSES

Plants are at the base of the food chain and are the main contributors to the formation of OBT. The formation, distribution and resilience of OBT are important for understanding its behaviour in the environment. The integration and circulation of tritium in plants are conditioned by plant–water interactions and functioning, as well as by photosynthesis and diurnal and nocturnal metabolic processes [7.17, 7.18].

The process of integration and circulation of tritium in plants is summarized below and schematically shown in Fig. 7.2.

(a) Tritiated water is absorbed by the plant through the roots, as tritiated water from the soil, and through the leaves, by exchanges with tritiated water vapour from the atmosphere.
(b) The tritiated water absorbed by the roots is transported in the plant radially from the absorbing hairs to the central stele, where the xylem is located, and through the wall or cytoplasm of the cortex cells.
(c) Most of the tritiated water circulates to the leaves and evaporates by cuticular or stomatal transpiration [7.15]. The rest of the tritiated water joins the intra- and extracellular free water fraction of the plant and rapidly equilibrates with the surrounding hydrogen atoms.
(d) The integration of tritium in plants is conditioned by isotopic equilibria and the concentrations of tritium in the root water remain approximately equal to the average concentrations in the soil.
(e) Root uptake is strongly dependent on the physicochemical conditions of the soil and the physiological parameters of the plant.
(f) Atmospheric tritiated water absorbed by foliage is transferred to plants by diffusive transfer through the stomatal orifices. The rate of uptake is directly related to the leaf area index and to variations in osmotic pressures in the guard cells, governing the opening and closing of stomata. From the stomatal cavity, the absorbed water vapour mixes with the water flow of the plant system.
(g) Foliar uptake is strongly dependent on the climatic conditions of the atmosphere and the physiology of the plant [7.1, 7.18]. The tritium concentration of the water supplied by foliar application is generally higher than that already present from the soil water. This initial imbalance leads to an observable decrease in the residence time of the foliar tritium in the plant [7.7].

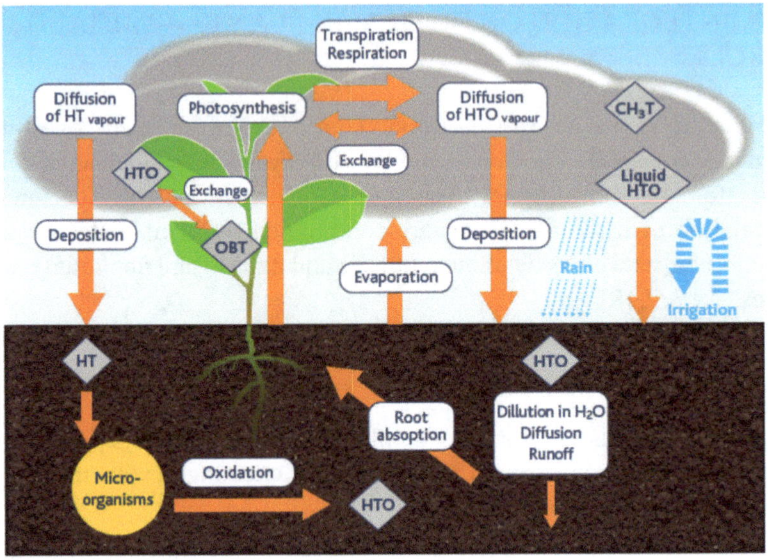

FIG. 7.2. Transfer and circulation of tritium from the environment to plants (reproduced from Ref. [7.19] with permission courtesy of IRSN).

7.3.1. Photosynthesis

During photosynthesis, carbon dioxide (CO_2) and water (H_2O) are taken in from the air and soil. Inside the plant cell, the water is oxidized through electron loss, while the carbon dioxide is reduced by gaining electrons. The overall process transforms the water into oxygen and the carbon dioxide into glucose. The plant then releases the oxygen into the air and stores energy within the glucose molecules.

Small organelles within the plant cell (chloroplast) store the energy of sunlight via thylakoid membranes, which hold the green, light absorbing chlorophyll pigment. During photosynthesis, chlorophyll absorbs energy from blue and red light waves and reflects green light waves, making the plant appear green. While there are many steps involved in the process of photosynthesis, there are two main stages: light dependent reactions and light independent reactions. The light dependent reaction takes place within the thylakoid membrane and requires a steady stream of sunlight. The chlorophyll absorbs energy from the light waves, which is converted into chemical energy in the form of the molecules of adenosine triphosphate (ATP) and nicotinamide adenine dinucleotide phosphate (NADPH). The light independent stage, also known as the Calvin cycle, takes place in the stoma, the space between the thylakoid membranes and the chloroplast membranes, and does not require light. During this stage, energy from the ATP and NADPH molecules is used to assemble carbohydrate molecules, such as glucose, from carbon dioxide.

There are different types of photosynthesis, including C3 photosynthesis and C4 photosynthesis. C3 photosynthesis is used by most plants. It involves producing a three-carbon compound called 3-phosphoglyceric acid during the Calvin cycle, which goes on to become glucose. C4 photosynthesis, on the other hand, produces a four-carbon intermediate compound, which splits into carbon dioxide and a three-carbon compound during the Calvin cycle. A benefit of C4 photosynthesis is that by producing higher levels of carbon, it allows plants to thrive in environments without much light or water. The biomolecules synthesized by photosynthesis are mainly complex carbohydrates, such as cellulose and starch [7.20]. The other products formed belong to families of compounds such as simple carbohydrates, amino acids and organic acids.

Tritium is incorporated into plant organic matter mainly by two distinct processes: isotope exchange and photosynthesis [7.21]. Photosynthesis produces organic molecules in plants from light energy ($h\nu$), H_2O and CO_2 as follows:

$$6\ CO_2 + 12\ H_2O + h\nu \rightarrow C_6H_{12}O_6 + 6\ O_2 + 6\ H_2O \qquad (7.4)$$

or

$$CO_2 + 2\ H_2O + h\nu \rightarrow CH_2O + O_2 + H_2O \qquad (7.5)$$

The tritiated water can therefore become integrated into biomolecules produced by the photosynthetic process and participate in the synthesis of new organic molecules as follows:

$$CO_2 + H_2O + HTO + h\nu \rightarrow CHTO + O_2 + H_2O \qquad (7.6)$$

Photosynthesis is the main process for incorporating tritium into organic matter and forming OBT. However, studies have shown a possible increase in OBT contents in plants under darkness, implying that its formation can also take place through other biochemical processes, such as enzymatic reactions, cellular respiration or the tricarboxylic acid cycle (Krebs cycle) [7.21–7.24].

Isotopic exchange reactions between tissue water and the exchangeable hydrogen positions of already synthesized organic molecules take place in the plant to preserve the isotopic equilibrium defined by Eq. (7.6). Thus, the tritium contained in the cellular or extracellular free water is progressively integrated into organic matter in molecular positions available for isotopic exchange. These positions are involved in weak covalent bonds and the rate of exchange depends on their accessibility, as defined above.

Isotopic fractionation in favour of tritium is observed in mass based OBT formation processes. Indeed, tritium is heavier than protium or deuterium, resulting in a non-negligible isotopic discrimination for certain chemical and biological processes. Thus, enzymatic reactions can lead to the formation of molecules that are potentially depleted in tritium compared with the surrounding concentrations [7.7]. This fractionation can also be found in some evapotranspiration processes of tritiated leaf water, where 9% more tritium is still found in the stomata [7.25].

7.3.2. Evapotranspiration

Evapotranspiration is the transport of water from the surface to the atmosphere and is the sum of direct evaporation of water from the soil and transpiration by plants. Apart from precipitation, evapotranspiration is one of the most significant terms in the water cycle. This process is one of the main consumers of solar energy.

The following factors affect evapotranspiration:

- — Available energy: it takes 600 calories (about 2500 J) to vaporize 1 g of liquid water. As the available energy increases, evapotranspiration increases. Evapotranspiration will therefore vary according to the time of day, the seasons and the latitude.
- — The humidity gradient: the drier the air is, the more water will be vaporized in the atmosphere.
- — The wind: it increases the vaporization of water.
- — Water availability: if there is no water, there is no evapotranspiration.
- — Factors specific to the vegetation under consideration, such as plant growth, height and leaf surface.
- — Stomatal resistance: plants regulate their transpiration by adjusting their opening.
- — Soil characteristics, such as chemistry and albedo.

The behaviour of tritium in the environment is determined by the hydrological cycle. An in-depth understanding and quantification of the processes determining the flow of water through the soil–plant–atmosphere system are therefore essential prerequisites for modelling ^{3}H transfers in terrestrial ecosystems. One of these processes, evapotranspiration, is an important component of the hydrological cycle and affects the water balance of all vegetated landscapes. A better knowledge of evapotranspiration is an important factor in improving management of irrigation or water utilization in agriculture, for example.

7.3.3. Other metabolic processes: Recycling and biodegradation

Biodegradation is the process by which substances are broken down into smaller compounds by microorganisms and involves biologically catalysed reduction in the complexity of organic compounds. Microorganisms, including fungi, bacteria and yeasts, are mostly involved in the process, which can occur aerobically or anaerobically. Algae and protozoa have little involvement in biodegradation. Biodegradation processes vary greatly, but frequently the final product of the degradation is carbon dioxide. Soil type and soil organic matter content affect the potential for adsorption of an organic compound to the surface of a solid.

In the case of longer lived plant species, trees are known to store tritium and can hold records of tritium fallout introduced by atmospheric weapons testing [7.26]. These biotas represent a significant reservoir on a global level that can be recycled into soil following decomposition and then be transferred into river sediments through soil erosion of drainage basins.

In forest ecosystems, litterfall is an important component of the nutrient cycle that regulates the accumulation and development of soil organic matter and hence soil organic carbon. It impacts the input and output of the nutrients, nutrient replenishment, biodiversity conservation and other ecosystem functions. Nutrient cycling is directly linked to productivity in terrestrial ecosystems and is directly involved in plant–soil interaction because it helps to transfer carbon and nutrients via plant matter decomposition into the soil. As hydrogen, and thus tritium, are parts of the soil organic carbon, litter decomposition can reintroduce tritium into the plant and hydrological cycle.

7.4. BEHAVIOUR AND FATE OF TRITIUM IN THE TERRESTRIAL ENVIRONMENT

7.4.1. Exchanges with the atmosphere

Gaseous forms of tritium (e.g. HT, TT, CH_3T) are present in much smaller quantities than vapours of HTO in the atmosphere, where approximately 10% of the global inventory of environmental tritium is found [7.11]. Under the effect of oxidation processes, part of the tritium in the form of HT, TT and CH_3T will convert to HTO and join the water cycle [7.7].

At the soil–atmosphere interface, tritium is exchanged in both directions, from air to soil (deposition) and from soil to air (re-emission by evaporation). The pathways of tritium in the terrestrial environment are schematically shown in Fig. 7.3.

Atmospheric tritium is deposited on the ground by dry or wet (rain or snow) deposition. However, tritiated hydrogen and methane are not easily washed away by rain because of their low solubility (0.002 g of H_2 per kilogram of water and 0.04 g of CH_4 per kilogram of water). Dry deposition of HT into soil is an extremely important process, as the deposition mechanism involves the oxidation of HT (largely by soil microbial action) to the more radiologically hazardous chemical species, HTO. Experiments have shown that direct uptake of HT to vegetation is extremely slow compared with HT uptake by soil, with deposition velocities lower than 10^{-6} m/s per unit area of leaf surface [7.27, 7.28].

Tritium deposited to soil in the form of HTO mixes with soil water and can be re-emitted to the atmosphere through evaporation or taken up through the root systems into vegetation. HTO is deposited from air to soils, vegetation and surface water by the processes of moisture exchange and condensation. Unlike HT, this process does not involve a chemical reaction and is reversible. The deposited HTO can either be re-emitted to the atmosphere or it can migrate to deeper layers of soil or mix into surface waters. Equilibrium between air and soil,

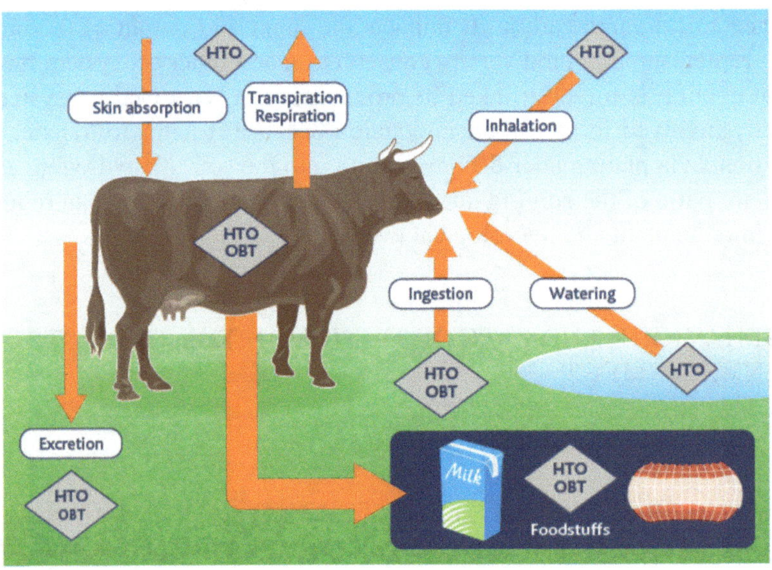

FIG. 7.3. Transfer of tritium in the terrestrial environment at the air–soil–plant interfaces and in animals, including transfer to food products (reproduced from Ref. [7.19] with permission courtesy of IRSN).

vegetation and surface water will be reached when the partial vapour pressure of HTO in the air above the surface is equal to the vapour pressure of the HTO in these compartments. At equilibrium, the net flux of HTO will approach zero.

Tritium from an airborne release plume can be scavenged by precipitation. This 'wet deposition' process is important for two reasons: (a) the plume is depleted owing to preferential removal of HTO compared with ordinary water molecules (H_2O) and (b) wet deposition results in a flux of contaminated water to the surface. The process is termed 'washout' when the precipitation originates above the plume and 'rainout' when the plume is in the region of formation of precipitation. If raindrops are less contaminated than atmospheric water vapour, then uptake of HTO by droplets occurs. A droplet falling through an HTO plume will therefore scavenge HTO from the atmosphere until equilibrium is reached, where the partial pressure of HTO in air is equal to the vapour pressure of HTO in the flux of vapour released by evaporating rain drops on their way to the ground. HT is an isotopic form of hydrogen gas and is expected to have low solubility in water. Precipitation scavenging of HT is therefore often neglected.

7.4.2. Exchanges with underground waters

Tritiated liquid discharges (HTO), by integrating into the hydrological cycle, partly enter the groundwater, where they mix directly with soil water. At the soil–atmosphere interface, tritiated inputs come from the deposition of releases mainly in the form of HT and HTO in precipitation or as vapour. Tritium exchanges are then conditioned by the processes of deposition (from air to soil) and re-emission by evaporation (from soil to air). On contact with the soil, HT diffuses through the pores and is rapidly converted to tritiated water by the oxidizing action of microorganisms. The type of soil and soil water content are key parameters in the rate of deposition of HT and its transformation. These phenomena are conditioned by the temperature of the soil and its moisture content [7.17]. Once converted to tritiated water, HT joins the water fluxes and follows the pathways of tritiated water in liquid or vapour form.

Tritiated water deposition can occur as a dry vapour deposit or as a wet precipitation deposit. Tritiated water vapour is deposited 10–100 times faster than HT on soils. Its diffusion takes place by rapid exchange with water vapour present in the pores of the uppermost layers of the soil [7.7, 7.17]. Most of the tritiated water (up to 90%) deposited in the soil is re-emitted to the atmosphere by evaporation [7.29]. An isotopic equilibrium is established between the soil and the atmosphere, leading to a preferential evaporation of tritiated water molecules [7.30]. With frequent rainfall, HTO in the soil can be diluted and pushed down the soil water profile along with the infiltrating water. Re-emission can be rapid in the first few hours but then decreases progressively with time [7.1]. The tritiated water remaining in the soil can be integrated into the plants by root absorption, which can then participate in the synthesis of biomass by microorganisms. Some will be transferred down to the water table to enter the groundwater domain [7.1].

7.4.3. Organically bound tritium persistence and biodegradation

As previously mentioned, tritium is incorporated into organic matter in the form of OBT, and a large proportion of it (around 70%) appears to be sequestered in a manner that is difficult to reverse (it is not readily exchangeable with the TFWT or the surrounding environment). This is observed in terrestrial plants (grass, lichens, the cellulose of trees, pinecones, leaves, fruit, soybean, peanuts and other vegetables).

OBT has become of increasing interest for environmental tritium research over recent decades because it has a significantly longer residence time in environmental compartments such as soil. There are only limited OBT data available in the literature [7.30–7.36], and many measured OBT activity concentrations in soil samples were obtained by CNL using the method described by Kim and Stuart [7.37].

It thus appears that all compartments of the biosphere act as reservoirs for OBT over time, although the residence time is shortened by the relatively rapid radioactive decay of tritium. The biosphere compartments storing these forms of tritium include terrestrial and oceanic environments that are repositories for living and dead (detrital) organisms, as well as organic soils and sediments. The behaviour of tritium in terrestrial plants mainly depends on their use of water, which determines water absorption by their roots, on photosynthesis and the ensuing (daytime or night time) metabolic processes. Photosynthesis is the main pathway for incorporation of tritium in organic form into vegetation and is the leading source of natural OBT production.

Soil organic matter is composed of a continuum of organic constituents decomposing under the action of microorganisms. These compounds are formed from the degradation of plant biomass and its biopolymers. Thus, OBT can be incorporated into the organic matter of soils through the incorporation of organic molecules of plant origin carrying OBT. While E-OBT is rapidly eliminated from the molecule by isotopic exchange, NE-OBT is incorporated into the following two fractions of soil organic matter:

(a) The ‘labile’ soil organic fraction, which is formed of biomolecules that are very easily metabolized in soils and that can be recycled, such as amino acids, sugars and polysaccharides. In this fraction, NE-OBT can either be progressively removed by degradation of the carrier molecule or join other more stable molecules by incorporation of its molecule into a larger molecular complex of the so-called 'stabilized' fraction.

(b) The 'stabilized' soil organic fraction (POM; see also Sections 6.5.1 and 6.5.3), which is made up of stable detrital materials containing biomolecules that are resistant to the action of soil microorganisms. These include humic substances or other products resulting from the partial degradation of lignin or from molecular reconfigurations. These elements have a recalcitrant (chemically resistant) character mainly due to the diversity of their functional chemical groups and the randomness of their conformation. These substances retain NE-OBT in soils and can persist for decades until reduced via radioactive decay or dispersed through erosion [7.38, 7.39].

On contact with and in the first few centimetres of soil, tritiated hydrogen is rapidly transformed into tritiated water. The conversion rate increases in line with soil temperature and water content, up to 46°C and 25% humidity. This phenomenon is attributed to the presence of microorganisms that can oxidize HT. Tritiated water vapour in the air is deposited between 10 and 100 times more rapidly than HT, by means of a fast process of exchange with the water present in the pores of the surface layers of the soil, at depths of ~2–8 cm [7.27, 7.28]. These exchanges slow down as the soil becomes drier and in the presence of vegetation.

Most of the tritium deposited in soil remains there for only a brief period. A large proportion of the tritiated water (provided that it is in liquid form) is thus re-emitted into the atmosphere by evaporation, at an estimated velocity ranging from 5–10% per minute to 1% per second of the overall amount, depending on environmental conditions. There is uncertainty about the tritium fraction released back into the atmosphere, but it is probable that a part is absorbed by plant roots and a small part quickly migrates into the underlying soil horizons [7.29, 7.40]. Little is also known about this latter fraction, especially as it is subject to significant variation, depending on the hydraulic characteristics of the soil and on climatic conditions (see Chapter 6).

A detailed understanding of the behaviour of OBT in soil remains incomplete at the current time but progress is being slowly made. Isotope discrimination in OBT has been observed in humic and fulvic acids [7.41]. The rates of chemical processes relating to this tritium fraction are slow [7.42, 7.43]. OBT is generally considered to be negligible in cultivated soils with a relatively low humus content and overall, soil is not currently considered a major accumulation compartment for tritium.

7.5. TRITIUM CONTENT IN THE TERRESTRIAL ENVIRONMENT OVER THE NUCLEAR ERA

7.5.1. Tritium under climate change forcing

There is increasing interest in the potential effects of climate change (global warming) on radionuclide mobility and retention in the terrestrial ecosystem. To produce some preliminary models predicting the effects of climate change on HTO and OBT in terrestrial ecosystems, a recent approach is to study key processes across climatically driven changes in temperature, humidity and evapotranspiration. This comparative approach studies ecosystems with elevated tritium levels around nuclear facilities. Some areas have similar soils, plants and aquatic communities but vary greatly in climatic parameters. These sites are representative of large portions of eastern North America and provide a means to directly compare similar ecosystems under different climatic conditions. They also seek to identify key processes and mechanisms from which a predictive framework can be developed to focus detailed research. Further study is planned to assess the cycling of HTO and OBT in relation to plant productivity, litter fall and decomposition and produce estimates of rates and residence times in plants and soil across a climatic gradient.

7.5.1.1. Tritium migration from permafrost melting and wildfires

Since the signing of the Partial Nuclear Test Ban Treaty in 1963, the exponential decline in tritium activity concentration in surface water has been due to a combination of factors including radioactive decay and environmental dispersal processes [7.44]. However, the persistence of tritium in the environment is prolonged if the weapon test era precipitation is isolated by freeze-trapping in snow, ice or permafrost [7.45]. By 2000, tritium activity concentrations in surface waters in Canada had declined exponentially to less than 3 Bq/L owing to mixing and radioactive decay processes. A similar tritium trend has been observed in glacier cores and terrestrial ice meltwaters [7.7, 7.46]. Measurements show that enriched tritium activity concentrations are found in thaw lakes over a regional scale within the discontinuous permafrost zone (permafrost fraction less than 90%; see Fig. 7.4) in the Northwest Territories of Canada. These lakes are fed by sources containing a relatively high level of tritium due to melting of previously frozen precipitations from the 1960s weapon testing era. The thaw of the frozen permafrost, holding about 1000 Bq/L (according to measurements performed at Whitehorse, northern Canada), supplied this formerly sequestered tritium to the lakes [7.45]. The pre-1950s activity concentration of tritium in water was

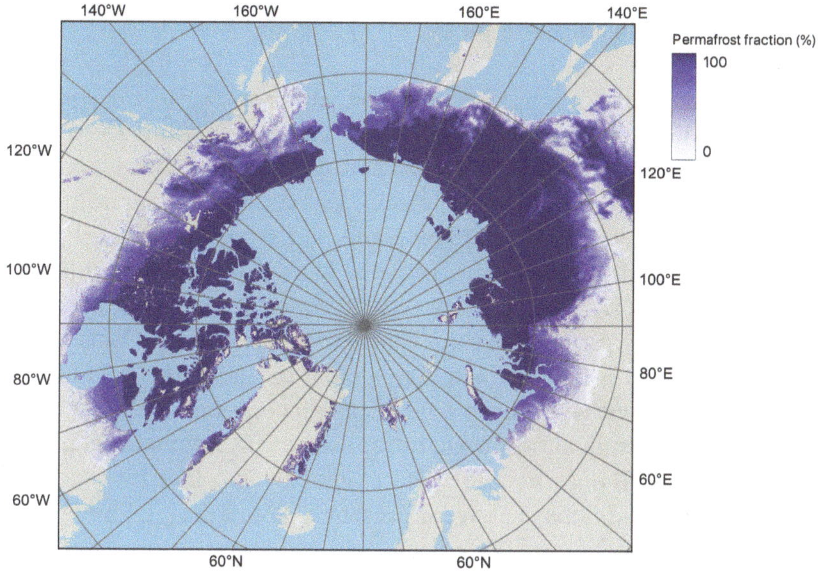

FIG. 7.4. Permafrost areas in Northern Hemisphere (Map: A. Bartsch, b.geos, Korneuburg; data from http://catalogue.ceda.ac.uk/uuid/7479606004d9465bad949671501e5f21/).

about 1 Bq/L, while present day activity concentrations are in the 2–3 Bq/L range across Canada [7.45].

Permafrost melting is now recognized as a major effect of global warming and climate change. Data show that tritium in lakes fed by permafrost thaw in Northern Canada had activity concentrations of about 10 Bq/L [7.46], which were elevated relative to non-thaw lakes. These Northern Canada lakes are therefore receiving tritium loads from formerly sequestered atmospheric fallout from the weapon test era [7.45].

In addition to increasing global temperatures, changing forest fire dynamics also accelerates permafrost degradation [7.47–7.49] and may help to explain elevated tritium activity concentrations in the discontinuous permafrost zone of the Mackenzie River Valley. Observations and descriptive models of fire frequency and severity in Canada suggest that the average area burned per decade will double by 2050 and will increase around 3.5–5.5 times by 2100 owing to a warmer and drier climate [7.50]. The 2014 forest fire season in the Northwest Territories of Canada was the worst in three decades, with more than 380 fires burning a record 3.4 million hectares [7.51], including large fires near the regions sampled in this area [7.49, 7.51]. Indeed, a recent study [7.49] in the Taiga Plains, west of Great Slave Lake (Northwest Territories), found that wildfires had been responsible for about 25% of permafrost thaw over the past 30 years.

7.5.2. Tritium in soils and biota

Once tritium is released into the atmosphere, it can be diluted and transferred/dispersed into the key environmental compartments of the biosphere (e.g. water, soil, biota). About 13% of the precipitation and condensation is re-emitted back into the atmosphere by direct evaporation from plant leaves. About 50% of water is transpired by plants and evaporated by soil [7.1].

Released tritium can diffuse into plant leaves through air moisture, rain, fog and mist as HTO/TFWT. The HTO activity concentration can fluctuate according to the surrounding tritium in air activity concentration and reflects conditions that were present only a few hours or days before the sampling time in plant leaves. During the plant growing season, plants can incorporate tritium in biomolecules (produce OBT) using atmospheric tritium and HTO from soils and groundwater. By contrast, OBT does not fluctuate as quickly in plant leaves, and the OBT levels reflect conditions that were present a few weeks or months before the time of sampling.

In autumn, plant leaves (containing OBT) fall to the ground and are decomposed by microorganisms. Evidence suggests that the OBT in the decomposing plant litter resides in the surface soil for a long time [7.52]. To predict the environmental effects of tritium, the long term behaviour of OBT in soil needs to be quantified, but current data are limited (Table 7.1). One study showed that measured OBT activity concentrations in soil were above 20 Bq/L in Langenburg (Saskatchewan) and Thunder Bay (Ontario) in Canada [7.34], while precipitation values in the two locations averaged 5–10 times lower than the OBT concentration in soil. It is notable that precipitation reflects the current ambient tritium activity concentration in air, including naturally produced tritium, but soil OBT reflects the historical (cumulative) input, which includes the fallout tritium released during the nuclear weapon test period (up to 500 Bq/L).

Soil organic matter is composed of aging plant debris that degrades slowly and reflects OBT in plants at the site. There is little evidence for recycling of soil OBT from soluble organic substances or by metabolic water (HTO) from microbial decomposition. According to recent measurements of the HTO and OBT activity concentrations in plants [7.36], OBT activity concentrations are higher than expected, probably owing to fractionation of the heavier isotope or transfer from soil to plant. Further research focusing on understanding the mechanisms driving OBT activity concentrations in plants and biodegradation is planned, particularly in France and Canada, where there are research groups with long standing interest in environmental tritium.

In Japan, in the early 2000s, studies [7.31] indicated very low OBT activity concentrations. Following the Fukushima Daiichi NPP accident, there was only limited initial concern about terrestrial tritium because the liquid tritium releases

TABLE 7.1. PUBLISHED DATA ON MEASURED SOIL OBT ACTIVITY CONCENTRATIONS

Year	Country	Sample	Soil OBT activity concentration (Bq/L of combustion water)	Reference
			Range (mean ± SD)	
1970	USA	Soil	44–251 (145 ± 59)	[7.29]
1975			92	[7.30]
1976			96	
1997			94	
1985	France	Uncultivated soil	6–26 (13 ± 6)	[7.28]
1991	Japan	Humus, litter, and soil	3	[7.31, 7.32]
2002	France	Grassland soil Cultivated soil	10–16 (13 ± 3)	[7.33]
2003			13–18 (15 ± 2)	
2005			7–11 (9 ± 2)	
2008	Canada	Soil	2–29 (14 ± 10)	[7.34]
2009			5–20 (12 ± 5)	

were much lower than those of radiocaesium [7.53]. The local terrestrial tritium level was slightly increased immediately after the accident but returned to normal values over several months.

Until recently, there has been only limited regulatory or public health interest in the persistence of OBT in soil organic matter because its activity concentrations are very low. The decreasing temporal trend of OBT activity concentrations in soil, from a maximum measured concentration of 250 Bq/L in the 1970s to 20 Bq/L in the 2010s, supported the perception of a low risk.

In certain compartments, it is known that OBT persists longer than HTO in the environment, and the details of retention and recycling and the long term fate of OBT in soil around nuclear sites is of growing interest. At present, OBT activity concentrations in radioactive waste materials and soils around nuclear industries are of limited concern. The availability of further high quality OBT datasets comprising accurate measurements from well designed environmental studies will enable better prediction of the long term behaviour of OBT in soil.

7.5.3. Tritium in plants

Following major nuclear accidents, such as those at the Chernobyl and Fukushima Daiichi NPPs, the fate of radionuclides in the environment was a key focus to ensure the safety of humans and biota. It was recognized that ecological processes of transport, bioaccumulation, biodegradation and physical decay were all important and that some parameters were site specific.

Without any local influences from NPPs, the OBT concentration in living plants has decreased with time since 1963. Data from environmental studies by IRSN [7.54] since the 1990s show that general OBT levels are slowly returning to the natural baseline observed in the 1950s (0.2–0.6 Bq/L in combustion water; see Fig. 7.5).

Over the same period, OBT activity concentrations were also determined for samples collected in the vicinity of several NPPs in France. These showed a clear relationship between the observed level and distance from the NPP (Table 7.2). This effect with distance has also been described around other NPPs elsewhere [7.55, 7.56]. Although the distance effect could be seen, the

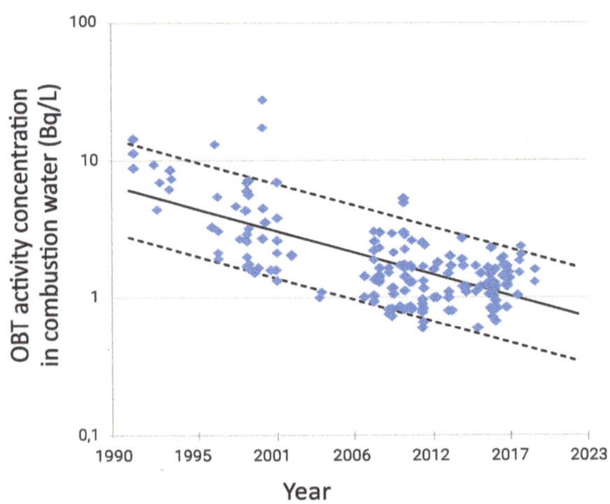

FIG. 7.5. Evolution of OBT activity concentration in combustion water. Data were obtained from grass, leaves, and plant and animal products from 1991 to 2020 (reproduced from Ref. [7.54] with permission courtesy of IRSN).

TABLE 7.2. MEAN OBT ACTIVITY CONCENTRATIONS IN COMBUSTION WATER IN GRASS, LEAVES, PLANT AND ANIMAL PRODUCTS FROM 2017 TO 2020 AS A FUNCTION OF THE DISTANCE FROM THE TRITIUM SOURCE [7.54]

Distance from the NPP (km)	Number of samples	Mean OBT activity concentration (Bq/L)
<2	97	2.3
<5	168	2.1
<10	249	2.0
>10	46	1.4
Background level	—[a]	1.0

[a] —: no data

determined activity concentrations are close to background and remain very low in the vicinity of many NPPs.

Moreover, when considering living plants, such activities of the order of 1 Bq/L of combustion water correspond to activity concentrations below 1 Bq/kg in the plant fresh matter. The activity concentration in dehydrated material is often about half the wet organic content but does vary with different materials; the fresh mass ratio can vary widely from 0.05 in salad to about 0.75 in cereals. HTO, TFWT and OBT have a different environmental behaviour in terrestrial ecosystems and their retention times in the environment may differ. For tritium, three different environmental half-lives (time to eliminate one half of the tritium through physical, biological and ecological processes) are applicable for post-accident recovery phase analysis. Quantification of the integrated half-lives for tritium for the terrestrial environment and component processes in different environment is important.

7.5.3.1. Organically bound tritium in tree rings

The annual nature of tree growth rings has the potential to provide both temporal and spatial records of historical tritium releases. This is because cellulose, hemicellulose and lignin, the main components of wood, are synthesized using soil water and/or atmospheric moisture. Kozák et al. [7.57] reported that single-year tritium fallout anomalies can be resolved by annual growth tree rings. The study showed the reliability of using OBT measurements in tree rings as an indication of historical releases of HT. The viability of using tree sections for high tritium sites with historical HT and HTO releases has been evaluated [7.58]. Two tree ring records from nuclear facilities with high OBT values show values of the order of 5000 Bq/L and 100 Bq/L, respectively (20 to 1000 times higher than background).

Nuclear reprocessing facilities (e.g. in USA, France, Japan) are ideal locations to investigate tritium recorded in tree rings owing to the magnitude and known history of tritium releases over decades from those sites [7.59, 7.60]. Any notable fluctuations in tritium concentration around the sites were expected to be recorded in the growth rings. The results indicated that OBT was retained in tree rings over time and showed the potential benefit of using the technique for identifying historical releases (accidental or intentional) around nuclear facilities [7.61].

7.5.4. Tritium in aquatic and terrestrial living organisms

Unlike the relative ease of collecting environmental samples (e.g. rainwater, air moisture, soil), living organisms in aquatic and terrestrial ecosystems (e.g. game animals and fish around nuclear facilities) can be more difficult because of their limited supply. In France, the environmental monitoring programmes of each NPP include determining tritium in living organisms from both aquatic and terrestrial ecosystems. The data obtained are gathered as part of the French National Environmental Radioactivity Monitoring Network (Réseau National de Mesure de la Radioactivité de l'Environnement), which aims to centralize all environmental radioactivity monitoring data collected in France. For tritium activity concentrations in aquatic and terrestrial living organisms, data are available for fish and farm animals for both TFWT and OBT from 2010 onwards (see Table 7.3).

In France, tritium background levels are of the order of becquerel per litre in rainwaters, surface waters and in water obtained after combustion of plant and animal products (Fig. 7.5) sampled in areas unaffected by NPP releases. In the RNM database, TFWT and OBT are provided in becquerels per kilogram of fresh material, so mean hydrogen percentage of 8% and fresh to mass ratio of 2:1 were chosen to convert from becquerels per litre in combustion water to becquerels per kilogram of fresh sample, leading to an estimated background level of 0.4 Bq/kg of fresh sample. Data gathered from fish and farm animals (Table 7.3) show that 25% of the OBT values have activity concentration levels similar to the background. The higher values observed for TFWT are probably related to a greater influence of the temporal correlation between tritium release

TABLE 7.3. TFWT AND OBT ACTIVITY CONCENTRATIONS IN FRESH SAMPLE FOR FARM ANIMALS AND FISH IN 2010–2020 [7.54]

Tritium fraction	Farm animals (Bq/kg fresh mass)		Fish (Bq/kg fresh mass)	
	TFWT	OBT	TFWT	OBT
Sample size	19	63	197	301
Data > DT	6	63	164	274
Mean	8 .8	2 .0	7 .3	3 .7
Standard deviation	1 .0	2 .2	8 .7	3.3
First quartile	1 .7	0 .4	2 .1	0 .5
Median	2 .5	1 .1	4 .5	0 .9
Third quartile	2 .9	3 .0	7 .8	1 .7
Fourth quartile	8 .8	9 .3	54	540

Note: DT: decision threshold; OBT: organically bound tritium; TFWT: tissue-free water tritium.

from the nuclear site and sample collection. For all data [7.54], the mean is higher than the median and corresponds roughly to the third quartile (75%) of the probability distributions for both TFWT and OBT. For farm animals, this means that the activity concentrations for both TFWT and OBT are equal to or below 3 Bq/kg of fresh sample. For fish, the same trend is observed, but TFWT values are higher than those for OBT, at least for the first three quartiles, with activity concentration below 8 Bq/kg of fresh sample for TFWT and 2 Bq/kg of fresh sample for OBT.

7.5.5. Spatial distribution of OBT around nuclear facilities

Nuclear facilities generally release radioactive materials in a controlled and regulated manner, including tritium in Canada. The CNL CRL site is home to the largest nuclear research complex in Canada. CRL's airborne tritium releases amounted to about 10^{15} Bq/a in 2016. Kim et al. [7.62] conducted a study to characterize the spatial footprint of atmospheric tritium releases in surface soil by the measurement of OBT. Soil OBT activity concentrations were of particular interest because soil represents a long term tritium reservoir that can act as a historical recorder of cumulative tritium releases into the terrestrial environment. Soil samples were collected within the CRL site from 2012 to 2014. Each sample was analysed for HTO and OBT activity concentrations. The highest HTO and OBT measurements obtained during this study were 154.0 ± 7.8 Bq/L and 180.9 ± 37.3 Bq/L, respectively.

The study of the OBT distribution [7.62] (see figure 2 of Ref. [7.62]) concluded that the tritium retained in soil was not related to the distance of the source term (research reactor or waste management areas) but to the prevailing wind direction.

7.5.6. Consequences of accidental releases of tritium on the terrestrial environment

In preparation for the recovery phase that would follow a nuclear accident, there is a need to understand the fate and migration of radioisotopes in the environment to model the large scale remediation of terrestrial and aquatic ecosystems. The best surrogate for this understanding is the global nuclear weapon test fallout data using the radionuclides ^{3}H, ^{14}C, ^{90}Sr and ^{137}Cs.

Tritium released to the terrestrial environment is dispersed and deposited to air, soil and biota, and as HTO, it is highly mobile and is readily assimilated by living organisms. Local residents can be exposed to tritium through many different terrestrial transfer pathways (described in Chapter 3). Typical terrestrial pathways to humans for TFWT and OBT involve aquatic food ingestion, crop ingestion, animal product ingestion and drinking water. It has been established that OBT has a longer retention time in the human body, and this may influence the estimation of radiation dose.

7.5.7. Potential future impacts from fission and fusion energy plants

The fuels used in French NPPs will progressively evolve towards fuels with higher enrichment in ^{235}U, requiring an increase in the boron concentration in the primary reactor coolant circuit to compensate for the excess reactivity of the fuel. Since tritium is a by-product of the neutron activation of ^{10}B, the increase in its concentration in the circuit will lead to an increased production of tritium, leading to greater environmental releases of this isotope. An increase of nearly 90% in annual liquid releases is estimated (0.15 g/a, compared with 0.08 g/a) [7.12].

The ITER facility, under construction in southern France, is one of several thermonuclear reactor projects under development around the world. The operating principle of the fusion reactor is based on the production of energy from the nuclear fusion reaction between deuterium and tritium (D–T reaction). The requirement to recover and recycle most tritium from ITER using advanced detritiation processes is currently estimated to lead to environmental tritium discharges of about 90 TBq/a [7.63].

7.6. CONCLUSIONS

This chapter discusses topics concerning the biointegration of tritium into living plants and the impact of those processes on its chemical speciation, as well as its fate during biodegradation. Also presented are aspects of tritium behaviour in soil and terrestrial biota and the magnitude of background levels at global, regional and local scales. Seasonal and geographical variations of tritium in precipitation following the nuclear weapon test period indicated that the Northern Hemisphere was exposed to higher tritium levels than the Southern Hemisphere. The levels of many fallout radionuclides in precipitation, including tritium, are now negligible, having returned to pre-atmospheric nuclear test fallout levels. The levels fell faster than would have been predicted on the basis of radioactive decay alone, demonstrating the relatively rapid removal/isolation or substantial reduction of contaminants from circulation in the biosphere once the source of contamination is removed [7.10].

OBT research in terrestrial systems has demonstrated the long term residence of tritium in soil and vegetation (see also Chapter 6). The long term retention of OBT activity concentrations in plants explains their similar or higher tritium content compared with current precipitation levels. For example, measured OBT activity concentrations in soil in Canada were up to 20 Bq/L, 2–5 times higher than current tritium activity concentrations in precipitations [7.32].

With a developing impetus for nuclear fusion power development and possible future accidents, there is a continuing need to refine the understanding of the fate and migration of tritium in the environment. By using knowledge of the environmental behaviour of tritium (e.g. impacts of global warming on tritium re-migration from permafrost and glacier tritium repositories), such data can be used to predict what might happen following an accidental or intentional release of tritium to the environment.

REFERENCES TO CHAPTER 7

[7.1] BOYER, C., et al., Tritium in plants: A review of current knowledge, Environ. Exp. Bot. **67** 1 (2009) 34–51, https://doi.org/10.1016/j.envexpbot.2009.06.008

[7.2] CHOI, Y., LIM, K., LEE, W., DIABATE, S., STRACK, S., Tissue free water tritium and organically bound tritium in the rice plant acutely exposed to atmospheric HTO vapor under semi-outdoor conditions, J. Environ. Radioact. **58** 1 (2002) 67–85, https://doi.org/10.1016/S0265-931X(01)00024-8

[7.3] INTERNATIONAL ATOMIC ENERGY AGENCY, Transfer of Tritium in the Environment After Accidental Releases from Nuclear Facilities, IAEA-TECDOC-1738, IAEA, Vienna (2014).

[7.4] KIM, S.B., BAGLAN, N., DAVIS, P.A., Current understanding of organically bound tritium (OBT) in the environment, J. Environ. Radioact. **126** (2013) 83–91, https://doi.org/10.1016/j.jenvrad.2013.07.011

[7.5] GROSSE, A.V., JOHNSTON, W.M., WOLFGANG, R.L., LIBBY, W.F., Tritium in Nature, Science **113** 2923 (1951) 1–2, https://doi.org/10.1126/science.113.2923.1

[7.6] KAUFMAN, S., LIBBY, W.F., The natural distribution of tritium, Phys. Rev. **93** 6 (1954) 1337–1344, https://doi.org/10.1103/PhysRev.93.1337

[7.7] BELOT, Y., ROY, M., MÉTIVIER, H., TUBIANA, M. Le Tritium: De l'Environnement à l'Homme, Institut de Protection et de Sûreté Nucléaire, Les Editions de Physique, Les Ulis (1997).

[7.8] OKADA, S., MOMOSHIMA, N., Overview of tritium: Characteristics, sources, and problems, Health Phys. **65** 6 (1993) 595–609, https://doi.org/10.1097/00004032-199312000-00001

[7.9] BECK, H.L., BENNETT, B.G., Historical overview of atmospheric nuclear weapons testing and estimates of fallout in the continental United States, Health Phys. **82** 5 (2002) 591–608, https://doi.org/10.1097/00004032-200205000-00007

[7.10] LÉTOURNEAU, E.G., MEYERHOF, D.P., AHIER, B., A retrospective of fallout monitoring in Canada, Environ. Int. **20** 5 (1994) 665–673, https://doi.org/10.1016/0160-4120(94)90012-4

[7.11] JACOBS, D.G., Sources of Tritium and Its Behavior Upon Release to the Environment, Technical Report Tl D-24635, US Atomic Energy Commission, Oak Ridge, TN (1968), https://doi.org/10.2172/4799828

[7.12] LE GUEN, B., Impact du tritium autour des centrales nucléaires EDF, Radioprotection **43** 2 (2008) 177–191, https://doi.org/10.1051/radiopro:2008012

[7.13] INTERNATIONAL ATOMIC ENERGY AGENCY, Environmental Modelling for Radiation Safety (EMRAS), IAEA-TECDOC-1678, IAEA, Vienna (2012).

[7.14] DARLING, W.G., BATH, A.H., GIBSON, J.J., ROZANSKI, K., "Isotopes in water", Isotopes in Palaeoenvironmental Research, Springer, Dordrecht (2006) 1–66, https://doi.org/10.1007/1-4020-2504-1_01

[7.15] McFARLANE, J.C., Tritium fractionation in plants, Environmental and Experimental Botany **16** 1 (1976) 9–14, https://doi.org/10.1016/0098-8472(76)90028-9

[7.16] MOMOSHIMA, N., OKAI, T., KAJI, T., TAKASHIMA, Y., Distribution and transformation of various chemical forms of tritium in the environment, Radiochim. Acta **54** 3 (1991) 129–132, https://doi.org/10.1524/ract.1991.54.3.129

[7.17] AUTORITÉ DE SÛRETÉ NUCLÉAIRE, Livre Blanc du Tritium, ASN, Montrouge (2024), https://www.asn.fr/sites/tritium/

[7.18] GALERIU, D., MELINTESCU, A., STRACK, S., ATARASHI-ANDOH, M., KIM, S., An overview of organically bound tritium experiments in plants following a short atmospheric HTO exposure, J. Environ. Radioact. **118** (2013) 40–56, https://doi.org/10.1016/j.jenvrad.2012.11.005

[7.19] INSTITUT DE RADIOPROTECTION ET DE SÛRETÉ NUCLÉAIRE, Tritium and the Environment, IRSN, Fontenay-aux-Roses (2010).

[7.20] KIM, S., WORKMAN, W., KOROLEVYCH, V., DAVIS, P., Field measurements of key parameters associated with nocturnal OBT formation in vegetables grown under Canadian conditions, J. Environ. Radioact. **104** (2012) 94–100, https://doi.org/10.1016/j.jenvrad.2011.08.017

[7.21] GUENOT, J., BELOT, Y., Assimilation of ^{3}H in photosynthesing leaves exposed to HTO, Health Phys. **47** 1 (1984) 849–855, https://doi.org/10.1097/00004032-198412000-00005

[7.22] MELINTESCU, A., GALERIU, D., Uncertainty of current understanding regarding OBT formation in plants, J. Environ. Radioact. **167** (2017) 134–149, https://doi.org/10.1016/j.jenvrad.2016.11.026

[7.23] MOSES, V., CALVIN, M., Photosynthesis studies with tritiated water, Biochim. Biophys. Acta **33** 2 (1959) 297–312, https://doi.org/10.1016/0006-3002(59)90117-9

[7.24] THOMPSON, R.G., NELSONS, C., Photosynthetic assimilation and translocation of ^{3}H-and ^{14}C-organic compounds after ^{3}HHO and $^{14}CO_2$ were simultaneously offered to a primary leaf of soybean, Canad. J. Bot. **49** 5 (1971) 757–766, https://doi.org/10.1139/b71-114

[7.25] GARLAND, J., AMEEN, M., Incorporation of tritium in grain plants, Health Phys. **36** 1 (1979) 35–38, https://doi.org/10.1097/00004032-197901000-00007

[7.26] FUMA, S., INOUE, Y., Simplified and sensitive analysis of organically bound tritium in tree rings to retrospect environmental tritium levels, Appl. Radiat. Isot. **46** 10 (1995) 991–997, https://doi.org/10.1016/0969-8043(95)00201-N

[7.27] NOGUCHI, H., YOKOYAMA, S., Vertical profile of tritium concentration in air during a chronic atmospheric HT release, Health Phys. **84** 3 (2003) 344–353, https://doi.org/10.1097/00004032-200303000-00007

[7.28] YOKOYAMA, S., NOGUCHI, H., ICHIMASA, Y., ICHIMASA, M., Re-emission of heavy water vapour from soil to the atmosphere, J. Environ. Radioact. **71** 3 (2004) 201–213, https://doi.org/10.1016/S0265-931X(03)00169-3

[7.29] CHOI, Y., KIM, S., LIM, K., PARK, H., LEE, W., Incorporation into organically bound tritium and the underground distribution of HTO applied to a simulated rice field, J. Environ. Radioact. **47** 3 (2000) 279–290, https://doi.org/10.1016/S0265-931X(99)00045-4

[7.30] BELOT, Y., CAPUT, C., GAUTHIER, D., Distribution of the organically bound tritium in vegetation exposed to fall-out, Radiat. Prot. Dosim. **16** 1–2 (1986) 111–113, https://doi.org/10.1093/oxfordjournals.rpd.a079724

[7.31] BOGEN, D.C., WELFORD, G.A., "Fallout tritium" distribution in the environment, Health Phys. **30** 2 (1976) 203–208, https://doi.org/10.1097/00004032-197602000-00008

[7.32] BOGEN, D., WELFORD, G., WHITE, C., "Tritium distribution in man and his environment", Behaviour of Tritium in the Environment (San Francisco, 16–20 Oct. 1978), IAEA, Vienna (1979).

[7.33] MOMOSHIMA, N., KAKIUCHI, H., OKAI, T., HISAMATSU, S., MAEDA, Y., Tritium in a pine forest ecosystem: Relation between fresh pine needles, organic materials on a forest floor and atmosphere, J. Radioanal. Nucl. Chem. **243** 2 (2000) 479–482, https://doi.org/10.1023/A:1016050803314

[7.34] HISAMATSU, S., KATSUMATA, T., TAKIZAWA, Y., Tritium concentrations in pine needle, litter and soil samples, J. Radiat. Res. **39** 2 (1998) 129–136, https://doi.org/10.1269/jrr.39.129

[7.35] JEAN-BAPTISTE, P., BAUMIER, D., FOURRÉ, E., DAPOIGNY, A., CLAVEL, B., The distribution of tritium in the terrestrial and aquatic environments of the Creys-Malville nuclear power plant (2002–2005), J. Environ. Radioact. **94** 2 (2007) 107–118, https://doi.org/10.1016/j.jenvrad.2007.01.010

[7.36] THOMPSON, P., KWAMENA, N.-O., ILIN, M., WILK, M., CLARK, I., Levels of tritium in soils and vegetation near Canadian nuclear facilities releasing tritium to the atmosphere: Implications for environmental models, J. Environ. Radioact. **140** (2015) 105–113, https://doi.org/10.1016/j.jenvrad.2014.11.009

[7.37] KIM, S.B., STUART, M., An alternative method of OBT measurement for the limited quantity of environmental samples using a combustion bomb, Appl. Radiat. Isot. **82** (2013) 175–180, https://doi.org/10.1016/j.apradiso.2013.04.037

[7.38] CAMPBELL, C., PAUL, E., RENNIE, D., McCALLUM, K., Applicability of the carbon-dating method of analysis to soil humus studies, Soil Sci **104** 3 (1967) 217–224,
https://doi.org/10.1097/00010694-196709000-00010

[7.39] DOMMERGUES, Y., MANGENOT, F., Ecologie Microbienne du Sol, Masson, Paris (1970).

[7.40] CHOI, Y.-H., et al., Fate of HTO following its acute soil deposition at different growth stages of Chinese cabbage, J. Environ. Radioact. **97** 1 (2007) 20–29,
https://doi.org/10.1016/j.jenvrad.2007.02.007

[7.41] WIERCZINSKI, B., AIGN, J., BAUMGÄRTNER, F., MÜLLEN, G., TÜRLER, A., Proton exchange and tritium enrichment in inorganic systems: Inorganic metal salts, J. Radioanal. Nucl. Chem. **266** 1 (2005) 39–42,
https://doi.org/10.1007/s10967-005-0865-8

[7.42] BALESDENT, J., RECOUS, S., Les temps de résidence du carbone et le potentiel de stockage de carbone dans quelques sols cultivés français, Can. J. Soil Sci. **77** 2 (1997) 187–193,
https://doi.org/10.4141/S96-109

[7.43] MOMOSHIMA, N., KAKIUCHI, H., OKAI, T., MAEDA, Y., Variation of tritium concentration in the course of the degradation of fresh pine needles on a forest floor, J. Radioanal. Nucl. Chem. **242** 1 (1999) 173–175,
https://doi.org/10.1007/BF02345913

[7.44] CLARK, I.D., FRITZ, P., Environmental Isotopes in Hydrogeology, CRC Press, Boca Raton, FL (2013),
https://doi.org/10.1201/9781482242911

[7.45] BOND, M.J., CARR, J., Permafrost thaw and implications for the fate and transport of tritium in the Canadian north, J. Environ. Radioact. **192** (2018) 295–311,
https://doi.org/10.1016/j.jenvrad.2018.07.006

[7.46] GIBSON, J., BIRKS, S., YI, Y., Higher tritium concentrations measured in permafrost thaw lakes in northern Alberta, Hydrol. Process. **30** 2 (2016) 245–249,
https://doi.org/10.1002/hyp.10599

[7.47] JAFAROV, E.E., ROMANOVSKY, V.E., GENET, H., McGUIRE, A.D., MARCHENKO, S.S., The effects of fire on the thermal stability of permafrost in lowland and upland black spruce forests of interior Alaska in a changing climate, Env. Res. Lett. **8** 3 (2013) 035030,
https://doi.org/10.1088/1748-9326/8/3/035030

[7.48] SMITH, S.L., RISEBOROUGH, D.W., BONNAVENTURE, P.P., Eighteen year record of forest fire effects on ground thermal regimes and permafrost in the Central Mackenzie Valley, NWT, Canada, Permafr. Periglac. Process. **26** 4 (2015) 289–303,
https://doi.org/10.1002/ppp.1849

[7.49] GIBSON, C., Long Term Effects of Wildfire on Permafrost Stability and Carbon Cycling in Northern Peatlands, MSc Thesis, Univ. Alberta (2017).

[7.50] BALSHI, M.S., et al., Assessing the response of area burned to changing climate in western boreal North America using a Multivariate Adaptive Regression Splines (MARS) approach, Glob. Change Biol. **15** 3 (2009) 578–600, https://doi.org/10.1111/j.1365-2486.2008.01679.x

[7.51] DARWENT, R., Fire Severity in the 2014 Northwest Territories Fire, Natural Resources Canada, Edmonton (2016).

[7.52] KIM, S., BREDLAW, M., KOROLEVYCH, V., HTO and OBT activity concentrations in soil at the historical atmospheric HT release site (Chalk River Laboratories), J. Environ. Radioact. **103** 1 (2012) 34–40, https://doi.org/10.1016/j.jenvrad.2011.08.013

[7.53] POVINEC, P., et al., Impact of the Fukushima accident on tritium, radiocarbon and radiocesium levels in seawater of the western North Pacific Ocean: A comparison with pre-Fukushima situation, J. Environ. Radioact. **166** (2017) 56–66, https://doi.org/10.1016/j.jenvrad.2016.02.027

[7.54] INSTITUT DE RADIOPROTECTION ET DE SÛRETÉ NUCLÉAIRE, Bilan de l'État Radiologique de l'Environnement Français de 2018 à 2020, IRSN, Fontenay-aux-Roses (2021).

[7.55] BAGLAN, N., ALANIC, G., LE MEIGNEN, R., POINTURIER, F., A follow up of the decrease of non exchangeable organically bound tritium levels in the surroundings of a nuclear research center, J. Environ. Radioact. **102** 7 (2011) 695–702, https://doi.org/10.1016/j.jenvrad.2011.03.014

[7.56] KIM, C.-K., LEE, S.-K., RHO, B.-H., LEE, Y.G., Environmental distribution and behavior of ^{3}H and ^{14}C around Wolsong nuclear power plants, Health Phys. **78** 6 (2000) 693–699, https://doi.org/10.1097/00004032-200006000-00013

[7.57] KOZÁK, K., et al., Retrospective evaluation of tritium fallout by tree-ring analysis, J. Environ. Radioact. **19** 1 (1993) 67–77, https://doi.org/10.1016/0265-931X(93)90059-G

[7.58] RANK, D., A Feasibility Study on the Retrospective Evaluation of Anomalous Local Tritium Fallout by the Analysis of Tree Rings from Selected Districts in Austria and Hungary, Report No. IAEA-R-3452-F, IAEA, Vienna (1987), https://inis.iaea.org/collection/NCLCollectionStore/_Public/18/045/18045560.pdf?r=1

[7.59] DUFF, M.C., et al., Assimilation and transport of organic bound tritium in an irrigated pine forest, Environ. Sci.: Process. Impacts **21** 6 (2019) 938–949, https://doi.org/10.1039/C9EM00165D

[7.60] PETTITT, E.A., DUFF, M.C., VERMEULEN, H., Historical record of tritium from tree cores at the Savannah River Site, Environ. Sci.: Process. Impacts **24** 8 (2022) 1144–1151, https://doi.org/10.1039/D2EM00193D

[7.61] BOURLAT, Y., TAILLADE, J., BAGLAN, N., ALANIC, G., POINTURIER, F., Reconstruction of gaseous tritium waste from the CEA centre facilities at Saclay using rings from two trees over 50 years old, Controle (Paris) **188** (2010) 24–28.

[7.62] KIM, S.B., et al., Organically bound tritium (OBT) activity concentrations in surface soil at the Chalk River Laboratories, Canada, J. Environ. Radioact. **208–209** (2019) 105999,
https://doi.org/10.1016/j.jenvrad.2019.105999

[7.63] INSTITUT DE RADIOPROTECTION ET DE SÛRETÉ NUCLÉAIRE, Le Tritium Dans l'Environnement: Synthèse Critique des Connaissances et Perspectives de Recherches, IRSN/DEI/2009-05, Vol. IRSN/DEI/2009-05, IRSN, Fontenay-aux-Roses (2009).

Chapter 8

DATA AND MODELS FOR DOSIMETRY OF TRITIUM

M.I. BALONOV, A. ULANOWSKI, V. VOSTROTIN

8.1. INTRODUCTION

Tritium is a pure beta emitter with low radiation energy output (the maximal energy of the emitted electrons is $E_{\beta,max}$ = 18.6 keV, with a mean electron energy of 5.682 keV per decay [8.1]). Because of the low energy of its beta emission, the ionizing radiation from the decay of tritium has a short range; for example, the continuous slowing down approximation range for electrons [8.2] emitted by tritium does not exceed 7.3 mm in air and 8 μm in liquid water or in biological tissue. Correspondingly, ionizing radiation from external sources of tritium cannot reach radiosensitive biological tissues of humans, animals or plants, which are normally protected by outer layers of dead skin, fur, scales or bark. External exposure to beta particles from tritium may become of dosimetric relevance and need to be considered only for microscopic entities (e.g. single-cell organisms) or for some aquatic organisms under special exposure conditions. For exposures of human beings and most other animals and plants, external exposure to environmental tritium can be safely ignored, so the only practically significant exposure situations are related to internally distributed tritium. As an isotope of hydrogen, tritium is ubiquitous and can be found across the environment in diverse physical and chemical forms (HT, T_2, HTO, DTO[1], T_2O, $CH_{4-x}T_x$, organic hydrocarbons, other compounds), thus being able to easily enter an organism via respiration of air, ingestion of food and water, and absorption through skin or wounds.

From past radiological experience, environmental tritium is known to result in only low dose exposures of human and biota, thus it is associated with low risks of detrimental effects on human health or biodiversity. However, exposures to high activities of anthropogenic tritium, either accidental or controlled in radiobiological experiments, can result in notable radiobiological effects.

---

[1] Tritium compounds that also include deuterium may become of relevance near nuclear facilities with HWRs. Otherwise, such compounds are rare in the general environment and can be safely ignored from the environmental perspective.

This chapter addresses diverse topics related to exposures to different forms and compounds of tritium and focuses specifically on the following:

(a) Intake pathways and scenarios of tritium entry into the human body from the environment;
(b) The systemic behaviour of tritium compounds in the human body;
(c) The radiobiological effectiveness of tritium radiation in comparison with external exposure due to penetrating, sparsely ionizing photon radiation, with respect to diverse endpoints for humans and biota;
(d) Current ICRP recommendations and dose coefficients for humans and biota;
(e) Outcomes of bioassay studies of tritium internal content for members of the public and occupationally exposed workers.

8.2. HUMAN INTAKE OF TRITIUM FROM THE ENVIRONMENT

Intake of tritium from the environment can occur in three main forms:

(1) Elemental gaseous hydrogen (HT and, to a lesser extent, T_2) in air.
(2) Tritium oxide (tritiated water, HTO, and, to a lesser extent, T_2O or DTO) in the form of water vapour in the air, and in the liquid phase in surface and groundwaters, as well as in sea water. HTO is also present in plant and animal food products, either cultivated or wild.
(3) OBT, formed in the environment through metabolic processes in living organisms. Like tritiated water, OBT is found in food products of plant and animal origin.[2]

The current ICRP recommendations for occupational and environmental exposure of human beings [8.3–8.5] systematically consider various tritium compounds and differentiate their intake and biokinetics in the human body following uptake via inhalation of air, drinking of water and food consumption. Conventionally, the dose coefficients recommended by ICRP are expressed as committed effective dose per becquerel of activity intake.

Tritium in the form of elemental gaseous hydrogen (HT or T_2) enters the human body through the respiratory tract with inhaled air. The ICRP human respiratory tract model [8.6, 8.7], with revisions introduced in Ref. [8.5], is

---

[2] OBT in biological tissues is carbon bound tritium that was originally formed in living systems through natural environmental or biological processes from HTO (or HT via HTO). OBT is not exchangeable with hydrogen in water.

applied to describe the intake of inhaled material in diverse compartments and tissues of the lungs, as well as its deposition, absorption and exhalation.

The integral intake of tritium with gaseous hydrogen (HT and T_2) can be evaluated given the known — either measured or modelled — time dependent activity concentration of HT or T_2 in air and the age dependent inhalation rate [8.8], which depends on the type of physical activity or lifestyle. For a person of given age, sex, occupation and lifestyle, integration of representative parameters yields an estimate of integral intake during the period ΔT (d) as follows:

$$I_{\mathrm{HT}}^{\mathrm{inh}} = \int_{\Delta T} V_{\mathrm{air}}(t) C_{\mathrm{HT}}^{\mathrm{air}}(t) \mathrm{d}t \tag{8.1}$$

where

$I_{\mathrm{HT}}^{\mathrm{inh}}$ is the integral intake of tritium with gaseous hydrogen (HT and T_2) (Bq);

$C_{\mathrm{HT}}^{\mathrm{air}}$ is the time dependent activity concentration of HT or T_2 in air (Bq/m^3);

and $V_{\mathrm{air}}(t)$ is the age dependent inhalation rate (m^3/d).

Inhaled air contains not only elemental gaseous hydrogen but also tritiated water (HTO and T_2O) vapours, either released into the air along with HT or formed in the environment as a result of HT oxidation under the action of soil microorganisms [8.9].

Tritium oxide or tritiated water (HTO) is the most common form of tritium in the environment. Tritiated water vapours in air enter the body during breathing, and all inhaled tritium activity is almost instantaneously absorbed in the respiratory tract. Given monitoring data of the concentration of tritiated water in the air, its integral intake into the body can be estimated similarly to Eq. (8.1), taking into account person specific data.

Vapours of tritiated water in air may also penetrate the surface of human skin and enter the body at a rate of about half the inhalation rate [8.10, 8.11]. The rate of intake of HTO through the skin depends on the air temperature and the protective properties of the clothing.

The intake of HTO into the body with drinking water depends on the source of water supply; for example, underground or surface water bodies or bottled water. The activity concentration of HTO in the water of surface water bodies can be in dynamic equilibrium with water vapour in the air.

The specific activity of HTO in food products is also determined by the place and technology of production. In irrigated crop production, the content of

HTO in products is largely determined by the HTO activity concentration in the irrigation water. The content of HTO in livestock products depends on the activity concentration of HTO in drinking water and feed. Assessment of the HTO intake by members of the public with foods is complicated by the various origins of the foods, including imported ones, and their culinary processing. Therefore, for such an assessment, it is preferable to use monitoring data collected by the duplicate diet or market basket sampling methods. If monitoring data are available, then the annual ingestion intake rate of HTO in the body, $I_{\mathrm{HTO}}^{\mathrm{ing}}$, can be estimated by summing up intake with food and drinking water as follows:

$$I_{\mathrm{HTO}}^{\mathrm{ing}} = M_{\mathrm{f}}\, S_{\mathrm{HTO}}^{\mathrm{f}} + V_{\mathrm{dw}}\, C_{\mathrm{HTO}}^{\mathrm{dw}} \tag{8.2}$$

where

$I_{\mathrm{HTO}}^{\mathrm{ing}}$	is the annual ingestion intake rate of HTO in the body (Bq/a);
M_{f}	is the average annual food consumption rate (kg/a) [8.8];
$S_{\mathrm{HTO}}^{\mathrm{f}}$	is the annual average specific activity of HTO in food (Bq/kg);
V_{dw}	is the annual consumption rate of drinking water (L/a) [8.8];

and $C_{\mathrm{HTO}}^{\mathrm{dw}}$ is the annual average activity concentration of HTO in drinking water (Bq/L).

In the human body, all the ingested HTO is known to be quickly and fully absorbed from the gastrointestinal tract [8.10].

OBT enters the human body with food produced in a tritium-containing environment. Since a person can consume food and drinking water coming from sources with different tritium content, the ratio of equilibrium activity concentrations of HTO and OBT in the body tissues may vary from that resulting from intake of tritium either as HTO or OBT. It is accepted that all OBT activity ingested with food is fully absorbed in the human body, so the alimentary tract transfer factor is set to unity: $f_{\mathrm{A}} = 1$ [8.10].

The actual levels of tritium intake into the human body for members of the public may vary considerably with the distance from potential tritium sources, such as nuclear industry objects (HWRs, tritium production facilities, thermonuclear installations, other industries operating with large amounts of tritium) [8.12] (see also Section 6.5.2). Historically, an increased intake of tritium into the human body for members of the public occurred worldwide in the early 1960s and shortly thereafter owing to the testing of thermonuclear weapons in the atmosphere [8.13].

8.3. BIODISTRIBUTION AND BEHAVIOUR OF TRITIUM COMPOUNDS IN THE HUMAN BODY

The biodistribution and behaviour of tritium in the human body is considered here in relation to the intake of tritium from the environment and is therefore limited to the three main forms of tritium: HT, HTO and OBT. The biokinetic patterns of their turnover in the body have common elements.

For any route of intake —inhalation, ingestion or percutaneous intake — HTO is rapidly absorbed through tissue barriers and enters the bloodstream. For organs with a high blood supply (e.g. lungs, liver, kidneys), HTO is rapidly (in minutes) absorbed into fluids with the blood flow, while its absorption in organs with a lower blood supply (e.g. muscles, connective and adipose tissue, bones) is slower. The equilibrium concentration of HTO in bodily fluids is achieved in a time from few hours to one day. In the body tissues, a certain fraction of HTO is involved in metabolic processes, and tritium atoms are gradually included in the OBT.

HTO is excreted from the body with water, for example, with urine, sweat and exhaled vapour. Measurements from 16 studies including about 400 adults [8.10, 8.12] found the mean biological half-life of HTO to be equal to 9.5 days, while individual values ranged from 4 to 18 days, depending on climate, season, workload and other factors.

A small part of tritium converted from HTO to OBT in the body — about 1% according to Refs [8.12, 8.14] — can then be excreted from the body with HTO owing to the catabolism of biomolecules at slower rates, with a mean half-life of 59 days, ranging from 23 to 104 days.

A simple first-order two-compartment model reflecting the described processes (Fig. 8.1) was first proposed by Snyder et al. [8.15] and was further used in some publications [8.12]. The model established relationships between the HTO intake rate, $I(t)$ (Bq/d); the tritium activity in the body as HTO, $W(t)$ (Bq); the tritium activity in the body as OBT, $B(t)$ (Bq); and the tritium activity excreted from the body as HTO, $U(t)$ (Bq/d). Transfer rates (d^{-1}) between the HTO and OBT compartments in the body and the excretion rate (d^{-1}) are denoted in the figure as α, β and γ, respectively.

The contribution of OBT to total lean body tissue doses after intake of HTO was estimated to be equal to 1.8–4.6% (mean 3.0 ± 0.9%), using 17 available individual datasets [8.14]. ICRP currently uses a more complex conservative model for radiation protection purposes [8.10], which treats blood and extravascular water in the body separately and includes two separate compartments for OBT.

Following inhalation of gaseous hydrogen HT, most of the HT is exhaled, and only a small fraction (about 0.01%) of tritium is dissolved in body fluids

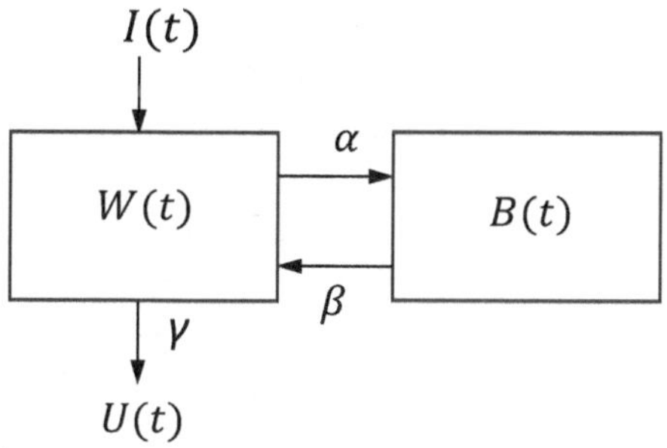

FIG. 8.1. Schematic of two-compartment recurrent model [8.12, 8.15] of tritium biokinetics in the human body following HTO uptake.

and oxidized to HTO [8.16, 8.17]. Conversion from HT to HTO presumably results from microbial action in the large intestine of humans. The resulting HTO is transferred with the bloodstream to organs and tissues and becomes the predominant contributor to the committed effective dose from HT intake. HT is not significantly absorbed through the skin. Irradiation of the lungs by inhaled HT does not significantly increase the committed effective dose [8.10], because of the short range of the beta particles from tritium in lung tissue. The dose coefficient for inhalation of HT is therefore about 0.01% of the dose coefficient used for inhalation of HTO.

In comparison with elemental HT, inhalation of gaseous tritiated methane $CH_{4-x}T_x$ results in higher, up to 0.3%, deposited fraction in the human respiratory tract, which can be assumed as instantaneously absorbed [8.10]. Following absorption, the tritium activity can be further treated as described by the systemic model for HTO.

OBT enters the human body with ingested food — namely, predominantly with proteins, fats and carbohydrates — and participates in the processes of digestion of these macromolecules in the gastrointestinal tract and in further metabolism. In these processes, part of the tritium from the OBT composition is released from C–H bonds and passes into the body water as HTO. Another part of tritium in metabolic processes remains in the C–H bonds of the breakdown products of macromolecules and is subsequently absorbed from the gastrointestinal tract into the blood and enters the tissue cells. In this way, tritiated sugars, amino acids and fatty acids and the products of digestion in the gut are transferred through the gut wall into the bloodstream and transported around the body. The share of OBT retained in C–H bonds is largely determined by the chemical properties and elemental composition of the foodstuffs; correspondingly, this proportion varies widely. Experiments on mammals have shown that the fraction of bound tritium in body tissues after intake of OBT with food is up to several tens of per cent of intake. On the basis of these data, the systemic model of ICRP for OBT intake [8.10] effectively assumes for human radiation protection purposes that, following intake, 50% of tritium is directly distributed to OBT in tissues and the remaining part enters the blood and circulates in the body as described by the HTO systemic model. More detailed insights into the metabolism of tritium compounds in mammals can be provided by biologically plausible models of tritium kinetics [8.18–8.20].

It is believed that OBT from decomposition of food further metabolizes in the same way as OBT formed in the human body from HTO. Accordingly, tritium is excreted from the body with HTO owing to the catabolism of biomolecules and in the composition of organic molecules with a period of several tens of days [8.12].

8.4. BIOLOGICAL AND HEALTH EFFECTS FROM EXPOSURE TO TRITIUM

Over the decades since the beginning of large scale use of tritium in military and industrial nuclear activities and its creation as a by-product of weapon tests or of the operation of nuclear facilities, tritium concentrations in the environment have remained at low levels. These could not lead to human exposures with doses exceeding the range of 'low doses' [8.21]. Correspondingly, no deterministic effects (tissue reactions), whose severity depends on radiation dose, are addressed in this section and the main focus is instead given to radiobiological endpoints of a stochastic nature, whose probability depends on the dose, the dose rate and radiobiological effectiveness of the given radiation type [8.22].

Compared with the radiobiological consequences of an acute exposure to external photon radiation, radiobiological effects in mammals due to prolonged internal exposure to tritium are affected by at least three biophysical factors:

(a) The radiobiological 'quality' of the ionizing radiation, namely, the relative ability of radiation of various types to cause harmful damage to biological tissue. Because of the ionization density created by the low energy beta radiation of tritium (mean energy ~5.7 keV, maximum energy 18.6 keV), the quality of the ionizing radiation from tritium exceeds that of the reference gamma radiation.
(b) The low dose rate of beta radiation from tritium during its retention in the body of mammals, namely, from a week to few months.
(c) The inhomogeneity of the spatial distribution of tritium and its radiation within tissues and their cells.

The appropriate quantity used for assessment of the effectiveness of different radiations in causing detrimental effects is the relative biological effectiveness (RBE)[3]. The RBE is an empirical quantity that depends on the biological system and endpoints, the dose and the experimental conditions, and it generally correlates with the linear energy transfer (LET) of ionizing radiation. In Ref. [8.23], ICRU showed that the LET of tritium beta particles is significantly higher than that of the reference photon radiation (for example, ^{60}Co or X rays with a peak voltage of 200 kVp) but is comparable to the LET of X rays with a peak voltage of 50 kVp and is an order of magnitude lower than that of 5.3 MeV alpha radiation (see Table 8.1 [8.23, 8.24]). This LET value, which is higher than that commonly assumed for electrons [8.22], may result in higher effectiveness in causing biological and adverse health effects, including cancer and inheritable effects.

Because of the higher LET of tritium β radiation compared with that conventionally assumed for all electrons, in recent decades, several experiments have been conducted using mammals (mostly mice) and their cells to determine the RBE for tritium as HTO under various experimental conditions. Such studies included endpoints related to the development of the embryo or foetus, carcinogenic, heritable and reproductive effects, as well as cell death. These biological effects have been studied either after single administration of HTO in

TABLE 8.1. TRACK AVERAGE LET, $\overline{L_\Delta}$, IN WATER FOR VARIOUS RADIATIONS BASED ON A CUT-OFF ENERGY OF 100 eV [8.23]

Radiation	$\overline{L_\Delta}$ (keV/mm)
^{60}Co gamma rays	0.22
200 kVp X rays	1.7
^{3}H beta rays	4.7
50 kVp X rays	6.3
5.3 MeV alpha particles	43

[3] The RBE is defined as the ratio of the absorbed dose of a reference radiation needed to cause a specific biological response to the absorbed dose of the radiation of interest that causes the same response [8.22].

the body or under conditions of chronic exposure to HTO, for example, through daily drinking water. To account for possible reduction of biological effects by the administration of a low dose rate, in some experiments the dose rate of the reference photon radiation changed with time, along with the dynamics of the dose rate of tritium β radiation.

The total number of experimental RBE studies with HTO approached 50 [8.12, 8.25], with 28 of them being conducted in vivo on mammals. In summary (Table 8.2), experimental estimates of the RBE values for the tritium β radiation in mammals have been reported to be in the range 1.0–3.9 (mean 2.5) and 0.4–2.3 (mean 1.2) using a reference radiation of gamma rays and orthovoltage X rays, respectively. There was a tendency for RBE values to increase with decreasing doses. RBE values derived from stochastic effect studies are only slightly higher (mean 3.0 compared with prolonged gamma radiation) than those obtained from studies of deterministic effects. Only three experimental studies directly addressed carcinogenic effects in mammals. In general, the set of RBE values obtained from experimental studies with HTO on animals is consistent with microdosimetric considerations [8.12, 8.26].

Bellamy et al. [8.27] estimated the RBE for electrons and photons derived from Monte Carlo simulations of electron tracks in water as a tissue substitute medium. They assumed that the low energy electrons with high LET can produce more damage to biological tissue, so the fraction of the total dose from electrons or photons that is produced by secondary low energy electrons can represent the

TABLE 8.2. SUMMARY OF TRITIUM BETA RADIATION RBE VALUES FROM EXPERIMENTAL IN VIVO STUDIES ON MAMMALS USING DIFFERENT BIOLOGICAL ENDPOINTS AND DIFFERENT REFERENCE RADIATION [8.12]

Reference radiation	All studies				Studies of stochastic effects with prolonged reference exposure			
	Number of studies	RBE			Number of studies	RBE		
		Mean	Median	Range		Mean	Median	Range
Prolonged gamma	21	2.5	1.1	1.0–3.9	9	3.0	3.0	1.8–3.9
X rays	7	2.5	1.2	0.4–2.3	4	1.2	1.2	1.1–1.3

Note: RBE: relative biological effectiveness.

radiobiological effectiveness of the type of source radiation. With these results, they systematically simulated energy deposition spectra due to source electrons and photons and estimated the RBE for internal exposure from 1070 radionuclides from Ref. [8.28]. They found RBE $\approx$ 2 for tritium beta particles compared with photons with high energy of $\geq$1 MeV. This RBE estimate for tritium agrees with radiobiological experience (see Table 8.2).

Another specific feature of tritium as a 'soft' beta emitter is that almost all its radiation is absorbed inside the tissues and cells of the tissues where decay has occurred. Therefore, the uneven spatial distribution of tritium atoms in human tissues and in tissue cells can lead to uneven irradiation of their radiosensitive parts. This effect is typical for cases when tritiated precursors of macromolecules (e.g. DNA, proteins) enter the body, as a rule, under occupational conditions [8.12, 8.29]. For members of the public, tritium entering the human body in the form of HT, HTO and OBT from the environment practically does not affect the spatial distribution of internal radiation dose.

UNSCEAR [8.12] also analysed and summarized numerous radiation epidemiological studies, performed in countries with advanced nuclear technologies, of cancer incidence and mortality among workers and the general population in relation to their exposure to tritium radiation. Most epidemiological studies of tritium workers have not used estimates of tritium specific doses in their analyses, which limits the inferences that may be derived from these studies about the tritium specific risk. It is also unlikely that epidemiological studies of exposure of the public to environmental tritium will produce meaningful tritium specific risk estimates, because the associated radiation doses are very low.

Based on the existing radiobiological and epidemiological data, ICRP decided, for the purpose of human radiation protection, to apply a radiation weighting factor for tritium β radiation equal to be unity, as is used for other low LET radiations [8.30]. For the purposes of environmental radiological protection, ICRP [8.24] indicated that the observed RBE values show no clear pattern of differences between species and suggested application of the same radiation weighting factor for all organisms, which for low LET radiations, including beta particles of tritium, was recommended to be equal to one. However, a caveat was included that if exposure to tritium beta particles (or other low energy, low LET radiations) results in absorbed dose rates within or close to the recommended reference levels for an organism, then additional review and possible application of radiation weighting might be warranted.

8.5. METHODS OF DOSE ASSESSMENT

Radiation protection of workers or members of the public requires dose assessment for specific exposure situations [8.22, 8.31]. For situations of existing or planned exposures from environmentally distributed radionuclides, including tritium, a commonly accepted approach is to estimate the committed dose of internal exposure due to the annual intake of radionuclides in the human body. Correspondingly, for workers and members of the public, the annual internal dose is assessed using the estimated annual intake I (see Section 8.2) and the corresponding dose coefficient, which is the committed effective dose per unit intake e, as follows:

$$E = I\,e \tag{8.3}$$

where

E is the annual internal dose (Sv);
I is the estimated annual intake (Bq);

and e is the dose coefficient (Sv/Bq).

The value of the effective dose coefficient, e, is determined by the physical properties of the radionuclide, the physical and chemical properties of its compound, the intake pathway, the personal age, the adopted biokinetic models and other factors. By definition [8.22], the effective dose coefficient e is a sex-independent weighted average of organ equivalent doses following acute intake and integrated up to age 70 for members of public, assuming an adult of age 20. For workers, the dose coefficient is computed for a period of 50 years, thus being compatible with the dose coefficients for adult members of the public. During the past decade, ICRP has published a series of reports [8.5, 8.10, 8.32–8.34] presenting effective dose coefficients for workers from occupational intake of radionuclides, using the current ICRP 2007 Recommendations [8.22]. These effective dose coefficients have replaced those based on the previous ICRP 1990 Recommendations [8.35], presented for workers in Refs [8.35–8.37]. However, the dose coefficients for members of the public have not been updated for consistency with the current ICRP recommendations, and the values from the most recent compendium [8.35] of the dose coefficients [8.3, 8.4, 8.38–8.40] are used here for members of the public internally exposed owing to intake of tritium distributed in the environment. ICRP has also developed dose coefficients for internal irradiation of embryos and foetuses [8.41] and for feeding of infants with mother's milk [8.42].

To calculate the dose coefficients, ICRP uses sophisticated models describing deposition of radionuclides in the input compartments (organs of respiratory and digestive tracts), changes in their physicochemical forms and due to radioactive decay, absorption and transfer within the body and excretion. In most models, also including the model for tritium, ICRP applies first-order (exponential) multi-compartmental biokinetic models.

For dosimetry of tritium in members of the public of different ages, ICRP developed two main systemic models: for HTO and for OBT [8.4, 8.38, 8.43]. In the model for tritiated water, it is assumed that following intake of HTO by any route, tritium activity instantaneously enters the body water (97% of the activity) and the body tissues as OBT (3% of the activity). Subsequently, tritium activity is excreted as HTO, following an exponential law with a half-life ranging from 3 days for infants to 10 days for adults, and as OBT with a half-life ranging from 8 days in infants to 40 days in adults. The exchange between HTO and OBT was not considered.

For the intake of OBT with food, 50% of the tritium activity goes to the body water compartment and 50% to the body OBT compartment. The subsequent excretion of each component from the body is the same as described above for the intake of HTO.

ICRP dose coefficients for HTO, OBT, as well as for HT and $CH_{4-x}T_x$, are given in Table 8.3. Currently recommended by ICRP [8.5, 8.10, 8.32–8.34], the dose coefficients for occupational intake of radionuclides are based on the updated physiologically based systemic models, dosimetric data and detriment-based tissue weighting factors. For tritium, the revised dose coefficients for intake by adult workers [8.10] have been increased for HTO by 5% and 11% for ingestion and inhalation, respectively, and for OBT by about 20%, as shown in Table 8.3.

Assessment of doses resulting from exposures to tritium of a particular person using data on tritium transfer in the environment and its intake with air, water and food may be associated with significant uncertainties due to several factors. These include environmental variability, uncertainties of the assumed intake and estimated content in the body or organs, variability of monitoring data, and uncertainties associated with the reference biokinetic models and data.

An alternative method of dose estimation is based on bioassay techniques, namely, on measurements of tritium activity concentration in body fluids excreted from the human body. Most often, activity concentrations are measured in urine samples of workers or residents. This method allows the exclusion of uncertainties related to environmental processes and estimation of intake in the human body. The activity concentration of the radionuclide already present in the human body is measured. However, interpretation of the measurement results and estimation of the effective dose are conditional on the knowledge or the

TABLE 8.3. DOSE COEFFICIENTS FOR HTO, OBT, HT AND $CH_{4-X}T_X$ RECOMMENDED BY ICRP FOR THE MEMBERS OF PUBLIC [8.4, 8.43], FOR PRENATAL EXPOSURE [8.41], EXPOSURE DURING NURSING [8.42] AND PERSONS EXPOSED OCCUPATIONALLY [8.10]

Age	Committed effective dose per unit intake (pSv/Bq)				
	HTO		OBT	HT	$CH_{4-x}T_x$
	Ingestion	Inhalation	Ingestion	Inhalation	Inhalation
	Member of the public [8.4, 8.41–8.43]				
Prenatal[a]	36	36	76	—[b]	—
Prenatal[c]	31	31	63	—	—
Nursing infant[a]	22	22	35	—	—
Nursing infant[c]	20	20	30	—	—
0.25 years	63	64	120	0.0064	0.64
1 year	48	48	120	0.0048	0.48
5 years	31	31	73	0.0031	0.31
10 years	23	23	57	0.0023	0.23
15 years	18	18	42	0.0018	0.18
20 years	18	18	42	0.0018	0.18
	Worker [8.10]				
20 years	19	20	51	0.002	0.059

[a] Dose coefficients per unit acute maternal intake at the end of week 10 of pregnancy.
[b] —: no data.
[c] Dose coefficients per unit chronic maternal intakes for the whole duration of pregnancy.

plausibility of assumptions regarding the form of tritium and dynamics of its intake into the body.

The simplest estimation of the committed effective dose using measurements of tritium activity concentration in urine is for a situation of pulse intake of the radionuclide; for example, following inhalation during passage of a cloud carrying tritium in the form of HTO, HT or CH_2T, or a single consumption of drinks or food with increased tritium content. The measured tritium activity concentration in a urine sample, $u(t)$, at time t following an acute intake can be recalculated back to the initial activity concentration of tritium in the urine, $u(0)$, at the time of the expected intake, assuming simple exponential retention in the body:

$$u(0)=u(t)e^{\frac{\ln 2}{T_{½}}t} \tag{8.4}$$

where $T_{1/2}$ is the biological half-life of HTO in the body of Reference Man [8.4, 8.8, 8.38, 8.43, 8.44], and the activity concentrations $u(t)$, $u(0)$ are expressed in Bq/L.

This simple approach is appropriate for a short time since acute intake, typically $t \leq 100$ d. For time periods exceeding 100 days for adults (e.g. workers), the additional longer term excretion component might need to be taken into account.

Accordingly, the initial intake of HTO, $I(0)$ (Bq), is equal to the HTO content in body water, $W(0)$, which is equal to the product of $u(0)$ by the volume of water in the body of Reference Man, BW, which is 42 L in an adult [8.8, 8.44].

$$I(0)=u(0)\mathrm{BW} \tag{8.5}$$

Then, the committed effective dose from a single intake of HTO into the body is calculated from $I(0)$ according to Eq. (8.3).

An alternative dynamic of tritium intake can be a chronic intake of tritium from the environment with air, water and food. This scenario can be found to be relevant for residents of areas with elevated concentration of tritium, for example, areas in proximity to nuclear power or reprocessing plants, industrial or military activities and waste disposal facilities. In this situation, chronic intake of tritium from the environment can be assumed with a constant intake rate with air, drinking water and/or food. Assuming the equilibrium content of HTO in body water, the daily intake for an adult can be estimated as the product of the measured activity concentration in urine and the daily water intake of Reference Man:

$$i_{eq} = u_{eq}\mathrm{WI} \tag{8.6}$$

where

i_{eq} is the daily intake for an adult (Bq/d);
u_{eq} is the measured activity concentration in urine (Bq);

and WI is the daily water intake of Reference Man, which is equal to 2.9 L/d [8.8, 8.44].

Accordingly, the annual intake is $I_{an} = 365 \times i_{eq}$ and the expected effective dose from the annual intake is calculated from I_{an} using Eq. (8.3).

This approach and Eqs (8.4–8.6) are developed for intake of tritiated water and are not applicable in the unlikely case of a predominant intake of tritium in the form of OBT with food. In addition, it was assumed above that all the tritium in the urine samples is in aqueous form or the sample was distilled before the activity was measured. Procedures for bioassay of tritium need to account for this possibility and appropriate checks need to be implemented during monitoring.

8.6. RESULTS OF BIOASSAY OF TRITIUM FOR MEMBERS OF THE PUBLIC

Systematic assessment of the tritium activity concentration in the body of residents living close to nuclear facilities that discharge significant activities of the radionuclide into the air and water bodies is an important element of radiological protection of the population. For comparison, the same measurements can be also made away from the nuclear facilities or before their commissioning [8.45–8.49]. In most of the considered cases, a bioassay was implemented via measurement of tritium activity concentration in urine samples collected from residents, followed by internal dose assessment, determination of the statistical properties of its distribution and of the associated uncertainties. Assuming equilibrium conditions for tritium in the body, the annual internal effective dose was calculated from the measured activity concentration in urine. To evaluate tritium intake with water, its activity concentration in drinking water sources was measured in parallel. To improve the dose estimates, in some studies the concentration of organic tritium compounds in the urine of residents was also measured.

A review of several studies conducted in Canada [8.45, 8.46], the Republic of Korea [8.47, 8.48] and the Russian Federation [8.49] was presented by UNSCEAR in Ref. [8.12]. Table 8.4 summarizes outcomes and data from studies published since 2015 on internal tritium exposure of individuals living both near and far from operating nuclear facilities or before their commissioning.

One study performed in Finland [8.50], where no local tritium discharge sources exist, presented data on tritium found in 2005 in urine collected from 227 randomly selected Finnish adults. The ages of the people ranged from 18 to 65 and the mean activity concentration of HTO in urine was 2.5 Bq/L, with a maximum of 18.3 Bq/L. The detection limit varied from 1.5 to 2.3 Bq/L. The uncertainties of the reported results varied from 5% to 31%. For the significant part of the data, 64 samples (28%) had tritium activity concentration below the detection limit. Most of the measurement results (92%) were below 4 Bq/L. Only for two persons were the tritium activity concentrations found to be higher than 10 Bq/L. No significant difference was found between males and females. In 2005, activity concentration of tritium in drinking water in Finland was determined in the range from 0.8 ± 2 to 1.7 ± 2 Bq/L. In 58% of the urine samples, the activity concentrations were higher than that in the water. No regional differences or differences caused by the origin of water supplies were found in the results.

In Türkiye, several studies [8.51–8.53] had been completed before the construction of NPPs. Urine samples from 100 persons aged from 18 to 66 years (65 males and 35 females) were selected randomly from a pilot region, Rize, and analysed [8.51]. The average tritium activity concentration in the urine samples was found as 4.7 ± 1.9 Bq/L, with a maximum of 28 Bq/L. The activity concentration of tritium in 33 samples (33%) was below the detection limit, which was equal to 2.4 Bq/L. In the Giresun region, the average activity concentration of tritium in urine was 14.2 ± 2.2 Bq/L (102 adults) and the maximum was 46 Bq/L [8.52]. Studies conducted later in the Trabzon region showed similar results, with the average activity concentration of tritium in urine being 13.4 ± 2.3 Bq/L (for 100 adults) and a maximum value of 35 Bq/L [8.53].

A recent study was conducted in Sweden in the city of Lund, where the European Spallation Source has been under construction [8.54]. The average tritium activity concentration in tap water in Lund, measured between October 2018 and January 2019, was found to be 1.5 ± 0.6 Bq/L (the detection limit equals 1.2 Bq/L). The average urinary HTO activity concentration found in the public was below the detection limit of 2.1 Bq/L. Only one value from a member of the public was slightly above the detection limit, at 2.2 ± 1.3 Bq/L, which was not significantly different from the tritium activity concentration of Lund's tap water.

A study in China [8.55] collected and analysed urine samples from 34 adults living near the operating Qinshan NPP. The average activity concentration in the urine of participants living at distances of 2, 10 and 22 km from the power plant were found as 3.5 ± 1.6, 2.1 ± 0.6 and 3.0 ± 0.8 Bq/L, respectively. Thus, no

TABLE 8.4. MEASUREMENTS OF TRITIUM ACTIVITY CONCENTRATION IN THE URINE OF MEMBERS OF THE PUBLIC AND IN SOURCES OF DRINKING WATER SUPPLY AT THE BEGINNING OF THE 21st CENTURY

Country	Year	Studied population		Number of persons	Tritium activity concentration in urine (Bq/L)		Tritium activity concentration in water (Bq/L)	Reference
					Mean ± standard error or range of means	Range of individual values	Mean ± standard error or range	
General public								
Finland	2005	Randomly selected adults		227	2.5	<1.5–13.7 (males) <1.5–18.3 (females)	0.8–1.7	[8.50]
Korea, Republic of	2013	General public		50	2.8 ± 1.4	<1.8–5.6	—[a]	[8.47, 8.48]
Türkiye	2015	Randomly selected adults from regions	Rize	100	4.7 ± 1.9	<2.4–28	—	[8.51]
	2018		Giresun	102	14.4 ± 2.2	<2.6–46	—	[8.52]
	2020		Trabzon	100	13.3 ± 2.3	<2.6–33	—	[8.53]

TABLE 8.4. MEASUREMENTS OF TRITIUM ACTIVITY CONCENTRATION IN THE URINE OF MEMBERS OF THE PUBLIC AND IN SOURCES OF DRINKING WATER SUPPLY AT THE BEGINNING OF THE 21st CENTURY (cont.)

Country	Year	Studied population	Number of persons	Tritium activity concentration in urine (Bq/L)		Tritium activity concentration in water (Bq/L)	Reference
				Mean ± standard error or range of means	Range of individual values	Mean ± standard error or range	
Sweden	2018–2019	Volunteers living in Lund	37	<2.1	<2.1–2.3	1.5 ± 0.6	[8.54]
Population in the vicinity of nuclear facilities							
Canada	2001	Near NPP with HWRs	3	18.2	6.5–32.2	—	[8.45, 8.46]
Russian Federation	2008	Near Mayak PA	45	100–800	—	—	[8.49]
	2010	Near Mayak PA, Ozersk	—	72 ± 20	—	—	
China	2014	2 km off NPP Qinshan 10 km off NPP Qinshan 22 km off NPP Qinshan	34	3.5 ± 1.6 2.1 ± 0.6 3.0 ± 0.8	1.3–6.7 1.3–3.1 2.2–3.8	—	[8.55]

TABLE 8.4. MEASUREMENTS OF TRITIUM ACTIVITY CONCENTRATION IN THE URINE OF MEMBERS OF THE PUBLIC AND IN SOURCES OF DRINKING WATER SUPPLY AT THE BEGINNING OF THE 21st CENTURY (cont.)

Country	Year	Studied population	Number of persons	Tritium activity concentration in urine (Bq/L)		Tritium activity concentration in water (Bq/L)	Reference
				Mean ± standard error or range of means	Range of individual values	Mean ± standard error or range	
Russian Federation	2015	Near NPP Beloyarskaya	50	39	12–65	15–55	[8.56]
Russian Federation	2016	Near Mayak PA, Ozersk	30	12.6 ± 2.3 (children) 18.1 ± 1.1 (adults)	<8–33	25–35	[8.57]

dependence was found of the tritium activity concentration in urine samples from members of the public from the distance to the Qinshan NPP.

Another recent study was conducted in the vicinity of the operating Beloyarskaya NPP in the Russian Federation [8.56]. This study included 50 people including adolescents and young adults aged 15–18 years and adults aged 20–25, 27–39 and 45–57 years. The mean tritium activity concentration in the urine samples varied with age group from 12 to 65 Bq/L, with a total sample mean of 39 Bq/L. The difference in the mean tritium activity concentration in the urine samples of men and women was statistically insignificant. During the study period, the activity concentration of tritium in drinking water was found to vary in the range 15–55 Bq/L.

Another study in the Russian Federation was conducted in the territory close to the large reprocessing plant Mayak PA [8.49]. Urine samples were collected by clinical laboratories in 2008 from residents of five settlements: Ozersk, Tatysh, Kyshtym, Metlino and Kasli, located within 5–20 km of the plant. The studied cohort included 45 persons of both sexes and different ages and occupations. The tritium activity concentration in the urine of people living near Mayak PA during the study period varied from 100 to 800 Bq/L, depending on the distance from the facility. The tritium activity concentration in the urine of residents of Ozersk in 2010 was 72 ± 20 Bq/L.

In 2016, another study [8.57] in the vicinity of the Mayak PA plant was completed. Thirty residents of Ozersk, including 22 adults (aged 27–75 years) not occupationally associated with Mayak PA and 8 children (aged 6–15 years), participated in the study. The detection limit of the tritium activity concentration in urine samples was 8 Bq/L. The mean values of tritium activity concentration in the urine of children and adults were 12.6 ± 2.3 Bq/L and 18.1 ± 1.1 Bq/L, respectively. The tritium concentration in the water of Lake Irtyash, which is the main source of water supply of Ozersk, was in the range 25–35 Bq/L from 2014 to 2018. Five children and three adults, in addition to tap water, consumed imported bottled water. The analysis showed that the amount of tritium in adults and children was more dependent on the source of drinking water supply than on age.

The data summarized in Table 8.4 [8.45–8.57] show that at the beginning of the 21st century, in the four studied areas in Europe and Asia, the mean tritium activity concentrations in the urine of members of public resulted from the natural production of tritium in the atmosphere and from thermonuclear weapon tests conducted in the mid-20th century, and were in the range <2.1–4.7 Bq/L. These mean values were reported for 414 people, for whom individual values of tritium activity concentration in urine were found to range from <1.5 to 29 Bq/L. For members of public residing near nuclear facilities, the tritium activity concentrations were found to be generally higher. Specifically, for members of the public living in the four studies areas in Asia and North America, contemporary values of the mean tritium activity concentration in urine from 162 persons varied from 2.1 to 39 Bq/L, with individual variability in the range 1.3–800 Bq/L. In the area affected by operations at Mayak PA, the activity concentration of tritium in urine had reduced by a factor of more than ten in the period 2008–2016, thus indicating significant reduction of tritium releases from this plant.

For members of the public — whether those residing close to operating nuclear facilities or nuclear facilities under construction or those residing far from those — the estimated annual effective doses from exposure to tritium did not exceed 10^{-3} of the annual effective dose from natural background radiation or the dose limit set for the public [8.22, 8.31].

8.7. TRITIUM DOSIMETRY FOR ANIMALS AND PLANTS

The system of radiological protection of the environment [8.24, 8.28, 8.58–8.61] differs from that for human beings in many aspects. Historically, the main focus of radiation protection was on humans and the pertinent biokinetic and dosimetric models. Even data originating from animal experiments were originally formulated for the purpose of human radiation protection.

Biological endpoints for the protection of non-human organisms are formulated as those relevant and expressed at the level of populations, not individual organisms. Hence, the relevant biological endpoints are represented by a combination of deterministic and stochastic harmful effects of radiation [8.24, 8.61]. Protection criteria have been formulated by ICRP in terms of dose rate bands (DCRLs), thus being implicitly assigned for a scenario of chronic exposure typical of situations of planned or existing exposures. The nuclear accident at the Fukushima Daiichi NPP demonstrated [8.62, 8.63] that such reference levels may appear impractical at the early stage of an accidental situation with rapidly changing ambient dose rates and non-equilibrated environmental contamination of the environmental media.

Another outstanding challenge for environmental dosimetry is due to the enormous diversity of animals and plants in relation to their habitats, behaviour, size, anatomy, physiology and biokinetics. Correspondingly, the current dosimetric approaches for animals and plants are based on using simple models approximating their bodies with basic shapes and assuming homogeneous volumes with uniformly distributed sources of radiation. The dose coefficients for non-human biota [8.28, 8.61] are formulated in terms of the absorbed dose averaged for the total body. No biokinetic behaviour is currently considered and instantaneous intake from the environment is estimated using CRs [8.59], which relate activity concentrations to the body of the organism and in the environmental media (air, water, soil). Ignoring the effects of biokinetic circulation and excretion from an organism brings conservatism to the recommended internal dose coefficients, conditional on the half-life and biokinetic behaviour of a radionuclide and the lifetime of an organism. Because of extensive biodiversity and the absence of comprehensive biokinetic models for non-human organisms, the dose coefficients for animals and plants are conventionally expressed as ratios of absorbed dose rates per unit radionuclide activity concentration inside the organism's body or in the environment (air, water, soil). For example, the dose coefficients for internal exposure are expressed in units of $\mu Gy \cdot h^{-1} \cdot Bq^{-1} \cdot kg$; the dose coefficients for external exposure of aquatic organisms in units of $\mu Gy \cdot h^{-1} \cdot Bq^{-1} \cdot L$; and the dose coefficients for terrestrial biota in units of $\mu Gy \cdot h^{-1} \cdot Bq^{-1} \cdot kg$ for radionuclides uniformly distributed within the soil, in units of $\mu Gy \cdot h^{-1} \cdot Bq^{-1} \cdot m^2$ for radionuclides distributed in the upper soil and approximated by a planar source or in units of $\mu Gy \cdot h^{-1} \cdot Bq^{-1} \cdot m^3$ for radionuclides distributed in the ambient air.

To facilitate radiation protection for environmental species, ICRP defined a set of Reference Animals and Plants [8.28] covering aquatic and terrestrial organisms, for which the dose coefficients had been systematically developed [8.28, 8.61] for various cases of exposure to radiation. Reference [8.61], which presents the updated dose coefficients for non-human species, was complemented by the online tool BiotaDC[4], which allows 'on the fly' computation of the dose coefficients for organisms of arbitrary ellipsoidal shapes with masses in range 1 mg–1 t exposed to radionuclides and their radioactive progeny in aquatic or terrestrial environments. The computations can be done for any of 1252 radionuclides of 97 elements included in Ref. [8.64], using various options to account for the effect of radioactive progeny.

Under these conditions, the dose coefficients for internal exposure to tritium, with its low energy beta particles, are the same for all Reference Animals and Plants and equal to 3.3 $pGy \cdot h^{-1} \cdot Bq^{-1} \cdot kg$ [8.61]. Doses of external exposures to tritium are shown only for aquatic organisms and are very small, of the order of 10^{-18}–10^{-20} $Gy \cdot h^{-1} \cdot Bq^{-1} \cdot L$. For terrestrial organisms covered by protective layers of skin, fur, feathers or scales, external exposure to tritium radiation is ignored, and the dose coefficients are equal to zero.

In some situations, more accurate or detailed dose estimates can be required, for example, for laboratory animals exposed in controlled conditions or for studies dealing with specific organisms in a specific environment. For such dose assessments, the inherent conservatism of the ICRP recommended dose coefficients might not be acceptable and biokinetic considerations might become important. For longer lived radionuclides, such as tritium, ignoring the effect of excretion after acute intakes may result in significant overestimation

[4] http://biotadc.icrp.org

of the dose estimates. In this situation, specific ad hoc models need to be developed considering time dependent retention of a radionuclide in the body of an organism.

Known or assumed dynamics of intake, in combination with an organism and radionuclide specific retention function, may substantially reduce uncertainties of the dose estimates for non-human organisms. Commonly used approaches [8.65, 8.66] are based on generic single compartment, first-order kinetic models, with retention of a radionuclide characterized by a single exponential with biological half-life T_{bio}. The latter quantity is known to vary largely across the diverse biota in diverse ecosystems. On the other hand, animals and plants express similarities and common properties that can be used to derive organism specific parameters for a given organism. For this purpose, the so-called allometric laws [8.67–8.70] are widely used to express essential biological or physiological properties of organisms as a function of their sizes or body masses. For the biological half-life T_{bio} of a radionuclide, the allometric equation is commonly expressed as a function of body mass:

$$T_{\text{bio}} = a\, M^{b} \tag{8.7}$$

where

T_{bio}	is the biological half-life of the radionuclide (d);
M	is the body mass of the organism (g);

and a (d) and b (unitless) are approximation parameters.

Recently, generalized allometric relationships were introduced [8.61, 8.71] and suggested for improved characterization of the basal metabolic rate and ventilation rate of mammals. These generalized relationships provided a better description of the observed experimental data and allowed the estimation of uncertainty due to interspecies' variability, which was found to be characterized by a geometric standard deviation of ~1.5.

For tritium turnover in mammals, parameters of the allometric parameterization of Eq. (8.7) for the biological half-life of tritium in mammals, which first appeared in a monograph by Whicker and Schultz [8.72], were incorrect; unfortunately, these parameters were quoted and reproduced in numerous subsequent publications on the subject. Küppers et al. [8.73] had analysed the parameters of the allometric equation for tritium before the publication of Ref. [8.72] and, using a weighted least squares fit of the original data from Ref. [8.74], found that an appropriate approximation can be achieved with $a = 1.4$ and $b = 0.14$, if in Eq. (8.7) the body mass is expressed in grams and the resulting half-life in days.

8.8. CONCLUSIONS

As a radionuclide emitting low energy beta particles with negligible penetrating ability, tritium is insignificant in external exposure scenarios. Radiological hazards from tritium are related to internal exposures, where this radioactive isotope of ubiquitous hydrogen enters metabolic and biokinetic processes in various chemical forms ranging from simple water to complex organic compounds. Systemic biokinetic models describing tritium behaviour in the human body have been recommended and used by ICRP for developing recommendations on radiological protection of humans [8.5, 8.6, 8.7, 8.43]. The ICRP radiological protection framework for non-human species still relies on simplified approaches [8.28, 8.61], designed to provide conservative dose estimates suitable for comparison with the recommended DCRLs for animals and plants [8.28, 8.60].

The low energy of the emitted radiation leads to a low radiological hazard of tritium, which is reflected by small internal dose coefficients [8.10, 8.35] and high values of exemption and clearance levels [8.31].

The low energy electrons emitted in the radioactive decay of tritium produce higher ionization density than other sparsely ionizing radiations such as high energy electrons and photons. This results in higher radiobiological effectiveness for tritium radiation, with mean values for various biological endpoints ranging from 2 to 3 compared with gamma radiation [8.12, 8.22, 8.24]. However, for practical purposes of radiological protection, the radiation weighting factor for tritium is assumed to be the same as that of other sparsely ionizing radiations and set equal to 1 [8.22].

Section 8.5 contains descriptions of methodology for human dose assessment based on either estimated intake of tritium as HTO, HT or OBT or on tritium activity concentration measured in the body water (urine), both for a single intake and for chronic intake of tritium from the environment with air, water and food.

For members of the public — whether they reside close to operating nuclear facilities or nuclear facilities under construction or far from those — the estimated annual effective doses from exposure to tritium have not or will not exceed 10^{-3} of the annual effective dose from natural background radiation or the dose limit set for the public [8.22, 8.31].

REFERENCES TO CHAPTER 8

[8.1] KONDEV, F.G., WANG, M., HUANG, W.J., NAIMI, S., AUDI, G., The NUBASE2020 evaluation of nuclear physics properties, Chin. Phys. C **45** 3 (2021) 030001,
https://doi.org/10.1088/1674-1137/abddae

[8.2] SELTZER, S., et al., Key Data for Ionizing-Radiation Dosimetry: Measurement Standards and Applications, ICRU Report 90 (2014),
https://doi.org/10.1093/jicru_ndw043

[8.3] INTERNATIONAL COMMISSION ON RADIOLOGICAL PROTECTION, Age-dependent Doses to Members of the Public from Intake of Radionuclides: Part 2 Ingestion Dose Coefficients, ICRP Publication 67, Pergamon Press, Oxford and New York (1993).

[8.4] INTERNATIONAL COMMISSION ON RADIOLOGICAL PROTECTION, Age-dependent Doses to Members of the Public from Intake of Radionuclides: Part 4 Inhalation Dose Coefficients, ICRP Publication 71, Pergamon Press, Oxford and New York (1995),
https://doi.org/10.1016/S0146-6453(00)80008-1

[8.5] INTERNATIONAL COMMISSION ON RADIOLOGICAL PROTECTION, Occupational Intakes of Radionuclides: Part 1, ICRP Publication 130, Pergamon Press, Oxford and New York (2015),
https://doi.org/10.1177/0146645315577539

[8.6] INTERNATIONAL COMMISSION ON RADIOLOGICAL PROTECTION, Human Respiratory Tract Model for Radiological Protection, ICRP Publication 66, Pergamon Press, Oxford and New York (1994),
https://doi.org/10.1016/0146-6453(94)90018-3

[8.7] INTERNATIONAL COMMISSION ON RADIOLOGICAL PROTECTION, Guide for the Practical Application of the ICRP Human Respiratory Tract Model, ICRP Supporting Guidance 3 (2002),
https://doi.org/10.1016/S0146-6453(03)00011-3

[8.8] INTERNATIONAL COMMISSION ON RADIOLOGICAL PROTECTION, Basic Anatomical and Physiological Data for Use in Radiological Protection Reference Values, ICRP Publication 89 (2002),
https://doi.org/10.1016/S0146-6453(03)00002-2

[8.9] AMANO, H., et al., Formation of organically bound tritium in plants during the 1994 chronic HT release experiment at Chalk River, Fusion Technol. **28** 3P1 (1995) 803–808,
https://doi.org/10.13182/FST95-A30503

[8.10] INTERNATIONAL COMMISSION ON RADIOLOGICAL PROTECTION, Occupational Intakes of Radionuclides: Part 2, ICRP Publication 134, Pergamon Press, Oxford and New York (2016),
https://doi.org/10.1177/0146645316670045

[8.11] OSBORNE, R., Absorption of tritiated water vapour by people, Health Phys. **12** 11 (1966) 1527–1538, https://doi.org/10.1097/00004032-196611000-00002

[8.12] UNITED NATIONS, Biological Effects of Selected Internal Emitters — Tritium, 2016 Report of the United Nations Scientific Committee on the Effects of Atomic Radiation (UNSCEAR) to the General Assembly, Annex C, UN, New York (2017).

[8.13] UNITED NATIONS, Sources and Effects of Ionizing Radiation, Vol. I: Sources, 2000 Report of the United Nations Scientific Committee on the Effects of Atomic Radiation (UNSCEAR) to the General Assembly, UN, New York (2000).

[8.14] BALONOV, M., CHIPIGA, L., Dose assessment for intake of tritiated water in humans: Role of tritium incorporation in organic matter, Radiat. Hyg. **9** 4 (2017) 16–25 (in Russian), https://doi.org/10.21514/1998-426X-2016-9-4-16-25

[8.15] SNYDER, W.S., FISH, B.R., BERNARD, S.R., FORD, M.R., MUIR, J.R., Urinary excretion of tritium following exposure of man to HTO — a two exponential model, Phys. Med. Biol. **13** 4 (1968) 547–559, https://doi.org/10.1088/0031-9155/13/4/305

[8.16] PINSON, E.A., LANGHAM, W.H., Physiology and toxicology of tritium in man, J. Appl. Physiol. **10** 1 (1957) 108–126, https://doi.org/10.1152/jappl.1957.10.1.108

[8.17] PETERMAN, B., JOHNSON, J., McELROY, R., HT/HTO conversion in mammals, Fusion Technol. **8** 2P2 (1985) 2557–2563, https://doi.org/10.13182/FST85-A24664

[8.18] GALERIU, D., A. BERESFORD, N., TAKEDA, H., MELINTESCU, A., CROUT, N.M.J., Towards a model for the dynamic transfer of tritium and carbon in mammals, Radiat. Prot. Dosimetry **105** 1–4 (2003) 387–390, https://doi.org/10.1093/oxfordjournals.rpd.a006264

[8.19] PRIEST, N.D., et al., Tritium (-H) retention in mice: Administered as HTO, DTO or as ^{3}H-Labeled amino-acids, Health Phys. **112** 5 (2017) 439–444, https://doi.org/10.1097/HP.0000000000000637

[8.20] MASUDA, T., YOSHIOKA, T., Estimation of radiation dose from ingested tritium in humans by administration of deuterium-labelled compounds and food, Sci. Rep. **11** 1 (2021) 2816, https://doi.org/10.1038/s41598-021-82460-5

[8.21] UNITED NATIONS, Summary of Low-Dose Radiation Effects on Health, 2010 Report of the United Nations Scientific Committee on the Effects of Atomic Radiation to the General Assembly, UN, New York (2011).

[8.22] INTERNATIONAL COMMISSION ON RADIOLOGICAL PROTECTION, The 2007 Recommendations of the International Commission on Radiological Protection, ICRP Publication 103, Pergamon Press, Oxford and New York (2007).

[8.23] INTERNATIONAL COMMISSION ON RADIATION UNITS AND MEASUREMENTS, Linear Energy Transfer, ICRU Rep. 16, ICRU, Washington, D.C. (1970).

[8.24] INTERNATIONAL COMMISSION ON RADIOLOGICAL PROTECTION, Radiation Weighting for Reference Animals and Plants, ICRP Publication 148, Pergamon Press, Oxford and New York (2021), https://doi.org/10.1177/0146645319896548

[8.25] STRAUME, T., CARSTEN, A., Tritium radiobiology and relative biological effectiveness, Health Phys. **65** 6 (1993) 657–672, https://doi.org/10.1097/00004032-199312000-00005

[8.26] MORSTIN, K., KOPEC, M., OLKO, P., SCHMITZ, T., FEINENDEGEN, L., Microdosimetry of tritium., Health Phys. **65** 6 (1993) 648–656, https://doi.org/10.1097/00004032-199312000-00004

[8.27] BELLAMY, M., PUSKIN, J., HERTEL, N., ECKERMAN, K., An empirical method for deriving RBE values associated with electrons, photons and radionuclides, Radiat. Prot. Dosim. **167** 4 (2015) 664–670, https://doi.org/10.1093/rpd/ncu358

[8.28] INTERNATIONAL COMMISSION ON RADIOLOGICAL PROTECTION, Environmental Protection, The Concept and Use of Reference Animals and Plants, ICRP Publication 108, Pergamon Press, Oxford and New York (2008).

[8.29] HILL, R.L., JOHNSON, J.R., Metabolism and dosimetry of tritium., Health Phys. **65** 6 (1993) 628–647, https://doi.org/10.1097/00004032-199312000-00003

[8.30] COULOT, J., et al., Internal dosimetry and tritium — the ICRP position, J. Radiol. Prot. **28** (2008) 131–135, https://doi.org/10.1088/0952-4746/28/2/E02

[8.31] EUROPEAN COMMISSION, FOOD AND AGRICULTURE ORGANIZATION OF THE UNITED NATIONS, INTERNATIONAL ATOMIC ENERGY AGENCY, INTERNATIONAL LABOUR ORGANIZATION, OECD NUCLEAR ENERGY AGENCY, PAN AMERICAN HEALTH ORGANIZATION, UNITED NATIONS ENVIRONMENT PROGRAMME, WORLD HEALTH ORGANIZATION, Radiation Protection and Safety of Radiation Sources: International Basic Safety Standards, IAEA Safety Standards Series No. GSR Part 3, IAEA, Vienna (2014), https://doi.org/10.61092/iaea.u2pu-60vm

[8.32] INTERNATIONAL COMMISSION ON RADIOLOGICAL PROTECTION, Occupational Intakes of Radionuclides: Part 3, ICRP Publication 137, Pergamon Press, Oxford and New York (2017), https://doi.org/10.1177/0146645317734963

[8.33] INTERNATIONAL COMMISSION ON RADIOLOGICAL PROTECTION, Occupational Intakes of Radionuclides: Part 4, ICRP Publication 141, Pergamon Press, Oxford and New York (2019), https://doi.org/10.1177/0146645319834139

[8.34] INTERNATIONAL COMMISSION ON RADIOLOGICAL PROTECTION, Occupational Intakes of Radionuclides: Part 5, ICRP Publication 151, Pergamon Press, Oxford and New York (2022), https://doi.org/10.1177/01466453211028755

[8.35] INTERNATIONAL COMMISSION ON RADIOLOGICAL PROTECTION, Compendium of Dose Coefficients Based on ICRP Publication 60, ICRP Publication 119, Pergamon Press, Oxford and New York (2012), https://doi.org/10.1016/j.icrp.2012.06.038

[8.36] INTERNATIONAL COMMISSION ON RADIOLOGICAL PROTECTION, Dose Coefficients for Intakes of Radionuclides by Workers, ICRP Publication 68, Pergamon Press, Oxford and New York (1994).

[8.37] INTERNATIONAL COMMISSION ON RADIOLOGICAL PROTECTION, Conversion Coefficients for Use in Radiological Protection Against External Radiation, ICRP Publication 74, Pergamon Press, Oxford and New York (1996), https://doi.org/10.1016/S0146-6453(96)90003-2

[8.38] INTERNATIONAL COMMISSION ON RADIOLOGICAL PROTECTION, Age-dependent Doses to Members of the Public from Intake of Radionuclides: Part 1: A report of a Task Group Committee of the International Commission on Radiological Protection, ICRP Publication 56, Pergamon Press, Oxford and New York (1990).

[8.39] INTERNATIONAL COMMISSION ON RADIOLOGICAL PROTECTION, Age-dependent Doses to Members of the Public from Intake of Radionuclides: Part 3: Ingestion Dose Coefficients, ICRP Publication 69, Pergamon Press, Oxford and New York (1995).

[8.40] INTERNATIONAL COMMISSION ON RADIOLOGICAL PROTECTION, Age-dependent Doses to the Members of the Public from Intake of Radionuclides: Part 5: Compilation of Ingestion and Inhalation Coefficients, ICRP Publication 72, Pergamon Press, Oxford and New York (1995), https://doi.org/10.1016/S0146-6453(00)89192-7

[8.41] INTERNATIONAL COMMISSION ON RADIOLOGICAL PROTECTION, Doses to the Embryo and Fetus from Intakes of Radionuclides by the Mother, ICRP Publication 88, Pergamon Press, Oxford and New York (2001), https://doi.org/10.1016/S0146-6453(01)00022-7

[8.42] INTERNATIONAL COMMISSION ON RADIOLOGICAL PROTECTION, Doses to Infants from Ingestion of Radionuclides in Mothers' Milk, ICRP Publication 95, Pergamon Press, Oxford and New York (2004), https://doi.org/10.1016/j.icrp.2005.02.001

[8.43] INTERNATIONAL COMMISSION ON RADIOLOGICAL PROTECTION, Human Alimentary Tract Model for Radiological Protection, ICRP Publication 100, Pergamon Press, Oxford and New York (2006), https://doi.org/10.1016/j.icrp.2006.03.001

[8.44] INTERNATIONAL COMMISSION ON RADIOLOGICAL PROTECTION, Report on the Task Group on Reference Man, ICRP Publication 23, Pergamon Press, Oxford and New York (1975).

[8.45] CANADIAN NUCLEAR SAFETY COMMISSION, Tritium Releases and Dose Consequences in Canada in 2006, CNSC, Ottawa (2014).

[8.46] KOTZER, T., TRIVEDI, A., Dosimetric implications of atmospheric dispersal of tritium near a heavy-water research reactor facility, Radiat. Prot. Dosimetry **93** 1 (2001) 61–66, https://doi.org/10.1093/oxfordjournals.rpd.a006414

[8.47] KIM, C.-K., HAN, M.-J., Dose assessment and behavior of tritium in environmental samples around Wolsong nuclear power plant, Appl. Radiat. Isot. **50** 4 (1999) 783–791, https://doi.org/10.1016/S0969-8043(98)00075-X

[8.48] YOON, S., HA, W.-H., LEE, S.-S., Tritium analysis of urine samples from the general Korean public, Appl. Radiat. Isot. **81** (2013) 276–278, https://doi.org/10.1016/j.apradiso.2013.03.011

[8.49] CHEBOTINA, M.Ya., NIKOLIN, O.A., The current tritium concentrations in human urine in the area of nuclear fuel cycle facilities, Dokl. Biol. Sci. **447** 1 (2012) 390–391, https://doi.org/10.1134/S0012496612060117

[8.50] PUHAKAINEN, M., HEIKKINEN, T., Tritium in the urine in Finnish people, Radiat. Prot. Dosimetry **128** 2 (2008) 254–257, https://doi.org/10.1093/rpd/ncm334

[8.51] DIZMAN, S., YILMAZ, A., KESER, R., Determination of tritium concentrations in humans before the development of a nuclear power plant in Turkey, Isot. Environ. Health Stud. **51** 3 (2015) 478–484, https://doi.org/10.1080/10256016.2015.1056182

[8.52] DIZMAN, S., KESER, R., YILMAZ, A., ÇAKIR, B., Tritium concentrations in people living in Giresun Province of Turkey, NWSA **13** 1 (2018) 1–6 (in Turkish), https://www.doi.org/10.12739/NWSA.2018.13.1.4A0055

[8.53] DİZMAN, S., Türkiye'de nükleer enerji öncesi Trabzon ilinde yaşayan bazi insanlarda trityum konsantrasyonlari, J. Inst. Sci. Technol. **10** 2 (2020) 998–1005, https://doi.org/10.21597/jist.651114

[8.54] PÉDEHONTAA-HIAA, G., HOLSTEIN, H., MATTSSON, S., RÄÄF, C.L., STENSTRÖM, K.E., Tritium in urine from members of the general public and occupationally exposed workers in Lund, Sweden, prior to operation of the European Spallation Source, J. Environ. Radioact. **213** (2020) 106141, https://doi.org/10.1016/j.jenvrad.2019.106141

[8.55] SHEN, B.-M., et al., Determination of total tritium in urine from residents living in the vicinity of nuclear power plants in Qinshan, China, Int. J. Environ. Res. Public. Health **12** 1 (2015) 888–894, https://doi.org/10.3390/ijerph120100888

[8.56] CHEBOTINA, My., NIKOLIN, O., BONDAREVA, L., RAKITSKY, V., Tritium in urine of people living in the area of influence of the Beloyarskaya NPP, Radiat. Hyg. **9** 4 (2016) 87–92 (in Russian), https://doi.org/10.21514/1998-426X-2016-9-4-87-92

[8.57] FINASHOV, L., VOSTROTIN, V., YANOV, A.Y., Tririum in urine in residents of Ozyorsk, the Chelyabinsk region in 2016, Radiat. Hyg. **12** 3 (2019) 42–49, https://doi.org/10.21514/1998-426X-2019-12-3-42-49

[8.58] INTERNATIONAL COMMISSION ON RADIOLOGICAL PROTECTION, 1990 Recommendations of the International Commission on Radiological Protection, Ann. ICRP **21** 1–3 (1991).

[8.59] INTERNATIONAL COMMISSION ON RADIOLOGICAL PROTECTION, Environmental Protection: Transfer Parameters for Reference Animals and Plants, ICRP Publication 114, Pergamon Press, Oxford and New York (2009), https://doi.org/10.1016/j.icrp.2011.08.009

[8.60] INTERNATIONAL COMMISSION ON RADIOLOGICAL PROTECTION, Protection of the Environment Under Different Exposure Situations, ICRP Publication 124, Pergamon Press, Oxford and New York (2014), https://doi.org/10.1177/0146645313497456

[8.61] INTERNATIONAL COMMISSION ON RADIOLOGICAL PROTECTION, Dose Coefficients for Non-human Biota Environmentally Exposed to Radiation, ICRP Publication 136, Pergamon Press, Oxford and New York (2017), https://doi.org/10.1177/0146645317728022

[8.62] UNITED NATIONS, Levels and Effects of Radiation Exposure Due to the Accident at the Fukushima Daiichi Nuclear Power Station: Implications of Information Published Since the UNSCEAR 2013 Report, Scientific Annex B, 2020/2021 Report of the United Nations Scientific Committee on the Effects of Atomic Radiation (UNSCEAR) to the General Assembly, Vol. II, UN, New York (2022).

[8.63] INTERNATIONAL ATOMIC ENERGY AGENCY, Environmental Transfer of Radionuclides in Japan Following the Accident at the Fukushima Daiichi Nuclear Power Plant, IAEA-TECDOC-1927, IAEA, Vienna (2020).

[8.64] ECKERMAN, K., ENDO, A., Nuclear decay data for dosimetric calculations, ICRP Publication 107, Pergamon Press, Oxford and New York (2008), https://doi.org/10.1016/j.icrp.2008.10.005

[8.65] HIGLEY, K., DOMOTOR, S., ANTONIO, E., A kinetic-allometric approach to predicting tissue radionuclide concentrations for biota, J. Environ. Radioact. **66** 1–2 (2003) 61–74, https://doi.org/10.1016/S0265-931X(02)00116-9

[8.66] HIGLEY, K., BYTWERK, D., Generic approaches to transfer, J. Environ. Radioact. **98** 1–2 (2007) 4–23, https://doi.org/10.1016/j.jenvrad.2007.02.013

[8.67] RUBNER, M., Über den Einfluss der Körpergrösse auf Stoff- und Kaftwechsel, Z. Biol. **19** (1883) 535–562.

[8.68] KLEIBER, M., Body size and metabolic rate, Physiol. Rev. **27** 4 (1947) 511–541, https://doi.org/10.1152/physrev.1947.27.4.511

[8.69] MARQUET, P.A., et al., Scaling and power-laws in ecological systems, J. Exp. Biol. **208** 9 (2005) 1749–1769, https://doi.org/10.1242/jeb.01588

[8.70] WEST, G.B., BROWN, J.H., The origin of allometric scaling laws in biology from genomes to ecosystems: towards a quantitative unifying theory of biological structure and organization, J. Exp. Biol. **208** 9 (2005) 1575–1592, https://doi.org/10.1242/jeb.01589

[8.71] ULANOVSKY, A., Dosimetry for animals and plants: Contending with biota diversity, Ann. ICRP **45** (2016) 225–238, https://doi.org/10.1177/0146645316630710

[8.72] WHICKER, F.W., SCHULTZ, V., Radioecology: Nuclear Energy and the Environment, Vol. 1, CRC Press, Boca Raton, FL (1982).

[8.73] KÜPPERS, C., USTOHALOVA, V., ULANOVSKY, A., Systematische Untersuchung der Exposition von Flora und Fauna bei Einhaltung der Grenzwerte der StrlSchV für den Menschen -Vorhaben 3609S70006, Bericht BfS-RESFOR-62/12, Bundesamt für Strahlenschutz, Salzgitter (2012), http://nbn-resolving.de/urn:nbn:de:0221-201203227814

[8.74] RICHMOND, C., LANGHAM, W., TRUJILLO, T., Comparative metabolism of tritiated water by mammals, J. Cell. Comp. Physiol. **59** (1962) 45–53, https://doi.org/10.1002/jcp.1030590106

Chapter 9

MEASUREMENT OF TRITIUM IN ENVIRONMENTAL MATERIALS

I.W. CROUDACE, P.E. WARWICK, J. SÜLTENFUSS, M. GRÖNING

9.1. INTRODUCTION

9.1.1. Forms of tritium

Tritium can replace hydrogen in a wide range of inorganic and organic molecules. In the environment, tritium may exist in simple molecules, including HT, HTO and CH_3T. In environmental waters, most tritium will likely exist in the form of HTO. Where organic materials contain some ^3H in free water (HTO) this is referred to as TFWT. Alternatively, tritium may substitute for hydrogen in organic molecules, where the tritium may be in an exchangeable form (e.g. associated with O–H, N–H or S–H bonds) or a more strongly bound, non-exchangeable form (e.g. via a C–H bond). Where the tritium substitution occurs via biological processes, the tritium is referred to as OBT. It is then further subdivided into either E-OBT or NE-OBT. If the tritium has substituted for hydrogen in a synthesized organic molecule via an artificial process (e.g. radiolabelling), the form is known as T-TOM [9.1, 9.2]. Of less significance environmentally, but of considerable importance for waste characterization during nuclear facility decommissioning, is tritium associated with inorganic substances. In such a case, ^3H has either substituted for hydrogen via exchange or has been produced in situ from neutron capture by B or Li components in various materials (e.g. concrete containing clay minerals or graphite). This form of tritium is referred to as either exchangeable or non-exchangeable tritiated inorganic matter (Fig. 9.1).

The analytical procedure adopted will depend on the type of sample being analysed and on whether measurement of total ^3H activity concentrations are sufficient or additional information on the ^3H distribution and speciation is required. The choice of measurement technique will also be determined by the required detection limit.[1]

[1] This chapter describes analytical techniques used to measure tritium content and discusses specific types of instrument implemented for this purpose. Mentioning of specific brand names or instruments serves solely informational purposes and does not imply endorsement by the IAEA.

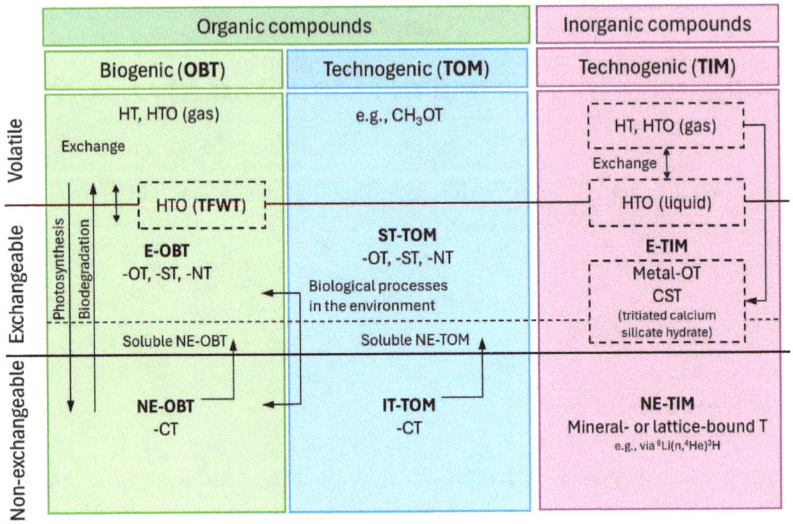

FIG. 9.1 Chemical forms of T for organic biogenic (OBT), organic technogenic (TOM) and inorganic technogenic (TIM) compounds, stratified according to their mobility. Shown in the figure are tissue-free water tritium (TFWT), exchangeable OBT (E-OBT), non-exchangeable OBT (NE-OBT), soluble TOM (ST-TOM), insoluble TOM (IT-TOM), exchangeable TIM (E-TIM), non-exchangeable TIM (NE-TIM), tritiated metal hydroxide (Metal-OT) and tritiated calcium silicate hydrate (CST).

9.1.2. Sample preservation

Given the volatility of certain ^3H species and the potential for ^3H migration between TFWT, E-OBT and environmental H_2O, specific precautions are required to ensure that the sample submitted for analysis is representative of the original material. For environmental samples, considerable care is required during sampling and sample storage to prevent exchange between the TFWT/E-OBT components in the sample and the ambient environment. Typically, at the point of sampling, the samples are sealed in polythene bags in plastic boxes or are double-bagged. Samples are transferred to freezers at −20°C within 24 h of receipt at the laboratory [9.3, 9.4].

9.1.3. Contamination and cross-contamination

Contamination control is critical during low level ^3H analysis and where TFWT is being determined. Contamination may arise from atmospheric tritium released from nuclear fission or fusion reactor facilities, radiopharmaceutical manufacture, luminous dial manufacture, research facilities, etc. In addition, local contamination may arise from the introduction of ^{3}H-containing products into the laboratory environment. These include tritium powered light sources incorporated into luminous watches and instrument dials, emergency signs and vintage telephones. Examples include the Trimphone (a model of telephone designed in the late 1960s in the UK that included a tritium based luminous dial) and novelty trinket-like items, such as luminous key chains.

Contamination from the environment can be minimized by limiting exposure of the sample to the atmosphere and by drying the sample under an inert gas or in a vacuum oven. Environmental ^{3}H activities in the laboratory may be monitored to provide warning of increases in background tritium concentrations in ambient atmosphere and corresponding increase in risk of ^{3}H contamination. This is of relevance when the laboratory is located close to nuclear facilities or is mainly analysing tritium at very low environmental levels. All glassware used in the preparation and analysis of samples is generally dried to remove residual ^{3}H contamination [9.5]. Le Goff et al. [9.6] noted that atmospheric ^{3}H generated significant contamination of water samples in contact with the laboratory atmosphere, reaching equilibrium specific activities with atmospheric ^{3}H at a rate of 1% per hour. Elevated atmospheric ^{3}H activities also resulted in significant biases in specific activity measurements when free water was extracted by freeze-drying. This was due to the ingress of atmospheric moisture into the system and subsequent co-condensation with the free water of the sublimed sample. One simple approach to monitor ambient atmospheric ^{3}H activity concentrations is to expose ~200 mL water to the laboratory atmosphere over a period of two weeks, whereby isotopic exchange occurs. After the exposure period, the ^{3}H content in the water is measured by LSC and the ambient atmospheric ^{3}H activities are inferred. Another method established at the IAEA laboratories in Vienna is to immerse thin aluminium plates into liquid nitrogen, which after cooling will freeze-trap any moisture in the laboratory air. The collected frozen water can be quickly directly assessed by LSC. Other important aspects of assessing contamination are covered by Morgenstern and Taylor in Ref. [9.7].

Where solid samples are first combusted, blank tritium-free samples are usually processed with the combustion process to clean the combustion glassware and to provide a measure of the procedural blank. The magnitude of any memory effects is evaluated and will depend on the analytical extraction method used (e.g. thermal oxidation process). A memory effect of 0.1% has been reported for combustion furnace based procedures [9.5].

9.1.4. Sample preparation

Sample preparation approaches are dependent on the material to be analysed and the purpose of the analysis. For environmental surveillance programmes, and to determine the dose arising from consumption of foodstuffs, the edible fraction is isolated and reserved for analysis. Samples are routinely freeze-dried or oven-dried to remove free water and then stored under vacuum prior to further analysis [9.8]. Aqueous samples such as sea waters and groundwaters are typically filtered through a 0.45 μm filter prior to analysis of ^3H in the filtrate.

9.1.5. Measurement of ^3H

The most appropriate measurement technique (Fig. 9.2) for the determination of ^3H in environmental materials will depend on the following factors:

(a) The matrix being analysed (solid, liquid or gaseous);
(b) The chemical speciation of the ^3H;
(c) The required detection limit;
(d) Whether total ^3H activity concentration is sufficient or if ^3H speciation is required.

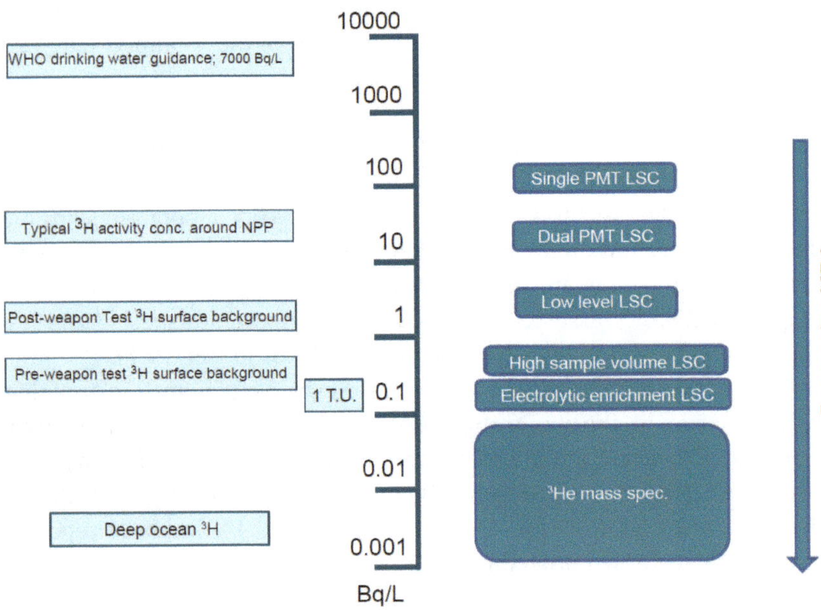

FIG. 9.2. Generalized levels of ^3H activity concentrations in the environment and relative sensitivity of various measurement techniques. LSC: liquid scintillation counting; MDA: minimum detectable activity; PMT: photomultiplier tube.

In most instances, ^3H needs to first be separated from the sample matrix and from any interfering species and converted to a chemical form suitable for measurement. Given the volatility of ^3H, most separation techniques utilize thermal desorption, distillation or vacuum extraction. Where required, the speciation of the ^3H can be determined through selective extraction of the ^3H-labelled species of interest.

Measurement of ^3H activity is often performed radiometrically using techniques capable of detecting its low energy beta emission. Examples include LSC or gas flow proportional counting (see Section 9.4). For low level, sensitive radiometric measurements, the ^3H concentration can be enhanced relative to ^1H using electrolytic enrichment (see Section 9.2.6). Ultralow level measurements of ^3H generally require the more sensitive mass spectrometry technique by measuring the ingrown progeny ^3He. If ^3He ingrowth mass spectrometry is being used, then the sample will need to be handled and stored according to the requirements of the technique (see Section 9.5).

9.2. SAMPLE TREATMENT PROCEDURES

9.2.1. Common tritium extraction techniques — solids

9.2.1.1. Freeze-drying

Tritiated water may be isolated from wet solids such as sediments and biota using freeze-drying, also termed lyophilization. The sample is frozen and then placed under vacuum in the freeze-drying apparatus. The sample is allowed to gradually warm under vacuum, causing the ice to sublime. The resulting water vapour diffuses or is pumped away from the sample vessel and is subsequently isolated in a separate cold trap. Under these conditions, any $^1H/^3H$ mass fractionation is negligible, and the resulting water may be directly measured by LSC with no further treatment. As many commercial freeze-drying units are designed to collect water from multiple samples in one condensing receptacle, a custom-built apparatus is required to permit segregated collection of water from samples.

9.2.1.2. Azeotropic distillation

Like freeze-drying, azeotropic distillation [9.9] has been used to isolate tritiated water from wet solids whilst reducing the fractionation that accompanies distillation. The sample is mixed with an organic solvent (see Table 9.1) and the mixture is warmed. Water and solvent combine to produce an azeotrope

TABLE 9.1. COMMON SOLVENTS AND DISTILLATION TEMPERATURES USED IN AZEOTROPIC DISTILLATION [9.10]

Solvent	Distillation temperature (°C)
Benzene	69
Xylene	94.5
Cyclohexane	81.4

that distils over at a temperature below the normal boiling point of water. The distillate is then condensed, and the two phases are separated. The water may then be recovered and the ^3H activity determined by LSC.

9.2.1.3. Microwave drying

In this technique, proposed by Akata et al. [9.11], samples of pine needles are heated using microwave radiation. Recoveries of up to 97% of the free water can be quickly achieved but consideration needs to be given to possible losses of volatile compounds containing tritium.

9.2.1.4. Wet oxidation

This method has been employed for the decomposition of organic materials, whereby the solid sample is placed in a flask along with a strong oxidant such as $K_2Cr_2O_7 + H_2SO_4$ [9.12], $FeSO_4 + H_2O_2$ [9.13], $K_2S_2O_8 + AgNO_3 + H_2SO_4$ [9.14] or $KMnO_4 + H_2SO_4$. The mixture is heated to oxidize tritiated species to HTO, which is subsequently isolated by distillation.

9.2.1.5. Tube furnace combustion

This method is usually applied to extract total OBT or analyse solid samples for tritium (Fig. 9.3) [9.1, 9.15–9.20]. Extraction of ^3H is achieved using a dual or triple-zone tube furnace, enabling independent heating of the sample and catalyst/oxidant. Typically, the sample is placed into a quartz or alumina sample boat, which is transferred into a quartz work tube. The sample is heated in a flow of air or a mixture of N_2 and O_2 using a predefined heating profile comprising a series of heating ramps and dwells at key temperatures. Routinely, samples are initially heated to 110°C, liberating tritiated water and volatile organic compounds.

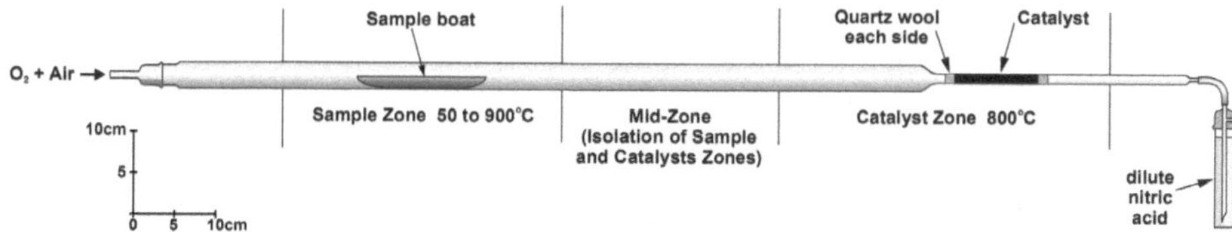

FIG. 9.3. Schematic diagram of a tube furnace used in the analysis of ^{3}H and ^{14}C in environmental samples (see also Ref. [9.17]).

The temperature of the sample zone is then increased to around 600°C or higher to thermally oxidize the previously dehydrated sample and to liberate any OBT. Oxygen is often passed over the sample to aid complete oxidation. The water vapour and combustion products resulting from these heating stages is passed over a heated oxidant (e.g. CuO) or catalyst (e.g. Pt/alumina) to ensure that any ^{3}H species are converted to HTO, which is subsequently trapped in a cooling trap or water bubbler.

The choice of heating profile is important to ensure controlled oxidation of the sample and quantitative recovery of ^{3}H. Thermal ramp rates (typically 3–5°C/min) are optimized to ensure that the sample does not combust in an uncontrolled manner, which could result in a rapid pulse of combustion products passing through the catalyst zone, leading to incomplete oxidation. Any rapid pulse of gases could also result in an overpressure in the system (potentially compromising gas seals). Such an event is often followed by an underpressure as the system relaxes, leading to a 'suck back' of trapping solution into the hot work tube. Anti-suck-back designs of bubblers are available to prevent the back flow of trapping solution and minimize the impact of such pressure excursions. Target temperatures and dwell times are selected at key transition temperatures to ensure that reactions have reached completion prior to commencement of the next heating stage (Fig. 9.4).

Oxygen is typically introduced towards the end of the run to ensure complete oxidation of intractable organic material. Early introduction of oxygen can result in uncontrolled combustion. In certain designs, oxygen is introduced downstream from the sample ahead of catalyst to ensure a sufficient O_2 supply to the catalyst and efficient oxidation. In many instances, ^{3}H is trapped in a water bubbler that is slightly acidified with HNO_3 to prevent any CO_2 (potentially containing ^{14}C) from co-trapping with the ^{3}H. Trapping of HTO in the bubbler system is usually quantitative, even with relatively simple bubbler designs. However, evaporative loss of water from the bubbler occurs throughout the run and this needs to be corrected for. Evaporative loss may be minimized by cooling the bubbler solution or mitigated by using two bubblers in series. Alternatively,

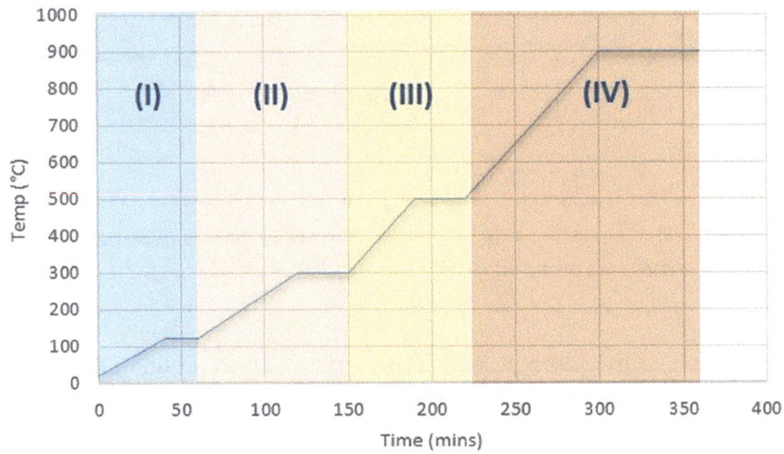

FIG. 9.4. Typical stages of heating of the tube furnace. (I) Extraction of water; (II) decomposition of readily oxidizable organic species and oxyhydroxides; (III) decomposition of remaining organics and oxyhydroxides; (IV) decomposition of intractable species.

the water vapour may be trapped on silica gel and subsequently recovered via freeze-drying [9.18]. For organic-rich samples, where combustion waters are generated in sufficient quantities, the water may be condensed and then purified by distillation following pH adjustment with Na_2O_2 [9.7].

Among flame oxidizers, the Perkin Elmer 307 Flame Oxidizer was well established during the period up to 2000 but has progressively become less available and less suited to running relatively large mass environmental samples. It was better suited for the processing of small, medium-activity samples and semiautomatic handling of samples.

Oxygen bomb combustion — or, more correctly, pressurized oxygen combustion — has been used for the rapid decomposition and oxidation of solid organic materials [9.4, 9.21–9.23]. The sample (up to 20 g and pelletized) is sealed into a purpose designed and pressure tested heavy stainless steel chamber. It is then pressurized with oxygen at 10–20 bar and then ignited electrically. During the sample ignition process, the chamber rapidly heats, which assists in the later extraction of the HTO during the controlled release of vessel pressure. One efficient method collects the combustion water from the vessel by careful release of vapour pressure through a cryotrapping system (−110°C). Any condensed combustion water remaining in the vessel is extracted through the same cryotraps by a vacuum pumping stage [9.22].

9.2.2. Common tritium extraction techniques — liquids

9.2.2.1. Direct counting

Direct counting of water samples permits rapid assessment of ^{3}H activity and is suitable for many applications. The water sample is typically filtered prior to subsampling and an aliquot of the sample is mixed with a commercial scintillation cocktail. The presence of dissolved solids will impact the miscibility of the sample with the scintillation cocktail and the detection efficiency achievable.

9.2.2.2. Distillation

Distillation has been routinely used to separate HTO from aqueous solutions. In its simplest form, distillation can be achieved by placing the sample in a Petri dish and leaving the dish on a warm surface. Water evaporates from the sample and condenses on the underside of the Petri dish lid, from where it can be recovered. The volume of water that may be collected in this way is limited. A development to this simple procedure is to use a concave lid and a central collection reservoir [9.24]. The condensed water runs to the lowest point of the concave surface and drops into the collector. In this way, sufficient water may be collected to permit the measurement of environmental levels of ^{3}H.

More conventionally, aqueous samples are purified by distillation prior to measurement. Two standard procedures for ^{3}H purification by distillation have been reported in the literature. In the ISO 9698:2019 standard [9.25], the sample is made alkaline with NaOH. An oxidant, such as $KMnO_4$, is added to oxidize any organic tritium species that may be present in the water to HTO during the distillation, and to ensure that potentially volatile Pb and Bi are in their highest, least volatile, oxidation state. The APHA-AWWA-WEF standard [9.26] recommends the addition of sodium carbonate to render the sample alkaline and the addition of sodium thiosulfate to convert iodine species to iodide to prevent co-distillation. The aqueous sample is heated in a flask and the resulting water vapour is condensed using a Liebig condenser and collected in a second flask. In all instances, $^{1}H/^{3}H$ fractionation during the distillation procedure has to be accounted for. Fractionation can be overcome by only collecting the middle, constant boiling point, fraction of the distillate. Discarding the first 10 mL of the distillate also removes any volatile organics. Alternatively, the total volume of the aqueous sample can be distilled to dryness to avoid fractionation effects.

In the IAEA's Isotope Hydrology Laboratory, where approximately 1000 natural water samples are processed annually to determine tritium concentrations [9.27], electrolytic enrichment of water samples is used after a prior vacuum distillation. In the early years of the laboratory development, tritium analysis mostly involved LSC measurements but now the more sensitive He ingrowth mass spectrometry technique is largely used. The LSC method is also widely adopted by many IAEA partner laboratories worldwide. The vacuum distillation system consists of 20 individual distillation lines, arranged in groups of five in four compartments. A 3 m long water-cooling bath kept at about 2°C supports all lines. The 20 distillation units are operated in parallel to reduce throughput bottlenecks, resulting in an improved overall workflow. A gentle vacuum distillation is carried out over six hours at 60°C, with water moisture passing from the heated round flask through the column and glass wool diaphragm into the receiving flask, kept at about 2°C. A stable heating temperature in the system is maintained by heating each round-bottomed flask in its own heating mantle. The elevated temperature avoids condensation of water except in the receiving bottles. Sea water can be distilled down to an electric conductivity of 10 µS/cm in one run, avoiding the double distillation of samples, which was necessary with the earlier system. Evacuation of the distillation system through a capillary ensures insignificant water loss of less than 0.2 g for 250 g distilled water, which permits stable isotope measurements on the distillate without significant fractionation effects.

9.2.2.3. Ambient temperature distillations

Ambient temperature distillations were developed for the determination of ^{3}H in temperature sensitive materials [9.9]. A sample is placed in an evaporator chamber that is connected to a condenser chamber (consisting of the scintillation vial and a modified cap cooled in an alcohol ice bath) via a heated glass tube. A vacuum pump is connected to the off-gas outlet of the condenser chamber. The sample is warmed to 20–25°C and the apparatus is evacuated of air. Under this reduced air pressure, the water will evaporate at low temperatures, pass through the glass tube and be condensed.

9.2.2.4. Reverse osmosis

Reverse osmosis has been utilized for the purification of water samples prior to liquid scintillation measurement of ^{3}H [9.28]. Processing times of <1 h for typical sample sizes of 16 L were significantly quicker than for conventional distillation based techniques. Samples with conductivities ranging from 6.98 µS/cm to 919 µS/cm could be treated in a single pass, whereas samples with higher conductivities of up to 1744 µS/cm required a double pass to achieve a target conductivity in the purified fraction of <5 µS/cm. In all instances, turbidity values were <1.0 NTU[2]. Measurable memory effects were observed following the processing of water samples containing 50 Bq/L of ^3H.

9.2.2.5. Solid phase extraction purification

Commercially available solid phase extraction columns have been produced for use in rapid analysis of HTO in aqueous samples. These columns are based on a mixture of anion exchange resin, cation exchange resin and organic extractant (a polyacrylamide based resin). The aqueous sample is passed through the column, which removes most radioactive interferents whilst HTO passes through into the eluent. The effectiveness and application of this approach has not been fully assessed, and it needs to be noted that OBT is likely to be retained on the column.

9.2.2.6. Conversion to hydrogen and methane

Water samples may be converted to hydrogen or methane, which is subsequently blended with argon prior to measurement of ^3H by gas flow proportional counting. The method was first described by Anand and Lal [9.29] and Povinec [9.30] and subsequently optimized for small sample volumes by Moran et al. [9.31]. Water is first converted to hydrogen via reaction with zinc as follows:

$$Zn + H_2O \rightarrow ZnO + H_2$$

The resulting H_2 is combined with CO_2 over a Ru catalyst to produce CH_4:

$$CO_2 + 4H_2 \rightarrow CH_4 + 2H_2O$$

This procedure allows for 120 mg of water to be converted to 70 cm^3 of CH_4, which is sufficient for proportional counting. Grabczak et al. [9.32] reduced 6 mL of water to hydrogen gas using Mg metal, with the reduction taking approximately 1 h. The resulting gas was mixed with P10 gas and counted for 1000 min using gas flow proportional counting with sensitivity approximately two times better than direct counting by LSC.

---

[2] NTU: nephelometric turbidity unit. The unit is used to express measurements of turbidity done with a nephelometer, which detects amount of light scattered by particles in a liquid.

9.2.3. Common tritium extraction techniques — gases

Tritium is present in air as either water vapour (HTO) or, to a lesser extent, HT. HTO in air may be sampled by passing air at a known flow rate through water bubblers and then measuring the ^3H activity of the water. Alternatively, water vapour may be isolated by passing the air through silica gel, a molecular sieve or even a cryogenically cooled aluminium surface or glass trap (see Section 9.1.3). Tritium present as HT is usually oxidized to HTO using a Pd catalyst and then collected as HTO by condensation. A combined HTO/HT concentrator was used by Brown et al. [9.33], who first removed HTO on a molecular sieve (type MS 4A). The HT was then passed through a Pd catalyst coated onto a molecular sieve (type MS 4A), whereby the HT was first oxidized to HTO and then immediately retained on the molecular sieve support. Approximately 10^{-4} of the initial HT was oxidized and retained on the first molecular sieve, along with the atmospheric water vapour. The HTO was recovered from the first molecular sieve by heating the molecular sieve in a vacuum line at 520°C and collecting the liberated HTO in liquid N_2 cooled traps.

Uda et al. [9.34] described a development of this approach to include the oxidation of methane (see Fig. 9.5). After trapping water vapour on a type MS 3A molecular sieve (which effectively traps HTO, but not HT or CH_3T), HT was oxidized using a Pt catalyst loaded onto a Fe–Cr–Al metal honeycomb at temperatures <100°C. Performing HT oxidation at low temperatures prevents the CH_3T from also being oxidized. A Pt/alumina catalyst was not suitable, because at low temperatures, HTO can adsorb onto the alumina, resulting in lower recoveries and memory effects. CH_3T was oxidized using a Pd/alumina catalyst at 350°C. For both HT and CH_3T oxidation, ^3H-free H_2 and CH_4 were introduced as carrier gases prior to the oxidation stages, and the resulting HTO was trapped on type MS 3A molecular sieves.

Atmospheric CH_3T can also be isolated and then blended with Ar to produce a counting gas suitable for proportional counting. Such an approach has been utilized for monitoring of landfill CH_3T [9.35].

9.2.4. Common tritium extraction techniques — aerosols

In the environment, tritium may be present and associated with atmospheric aerosols arising from resuspension of labelled soil particles, bioaerosols or aerosols derived from industrial processes [9.36]. Measurement of ^3H in air filters can be achieved either via direct LSC (ensuring that no other natural radionuclides, such as Rn progeny, are present in the sample) or following combustion of the air filter and LSC of the resulting combustion water.

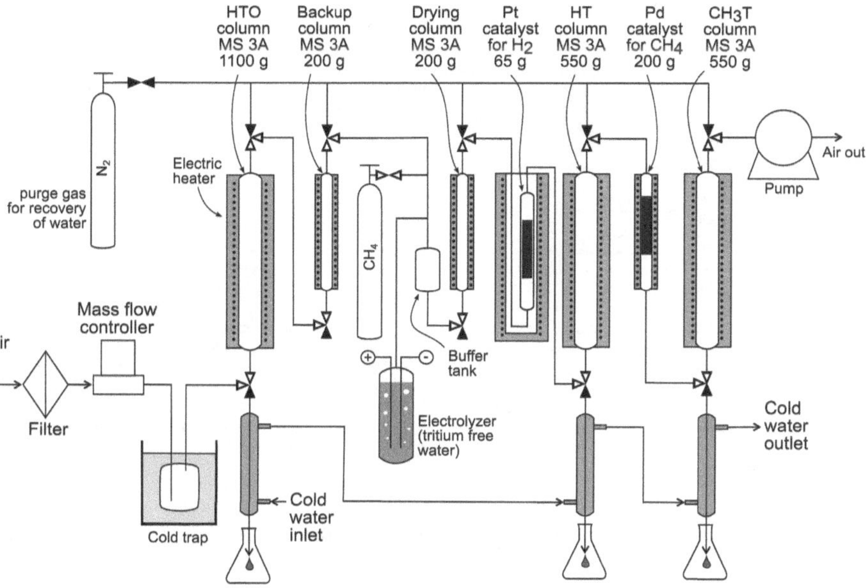

FIG. 9.5. System for collecting atmospheric HTO, HT and CH_3T (after Ref. [9.34])

9.2.5. Electrolytic enrichment of ^3H prior to liquid scintillation counting

Low level measurements of tritium in water by LSC are limited by the volume of water that may be mixed with a scintillant. Any improvement on the detection limit requires increasing the concentration of tritium in the water sample. However, as HTO is chemically identical to H_2O, such concentration is difficult to achieve. Concentration of ^3H by electrolytic enrichment was first reported by Kaufman and Libby [9.37]. Electrolytic enrichment employs the electrolysis of a diluted alkaline solution to degrade ordinary water molecules preferentially to HTO molecules, which have a slightly higher binding energy. As electrolysis continues, the water volume decreases, and the resulting concentrate has a higher level of HTO compared with the original samples. Concentration factors of 15–60 are achievable, depending on the electrolysis conditions, although a routine enrichment factor of 75 was reported by Morgenstern and Taylor [9.7]. More recently, enrichment factors >100 were reported by Kumar et al. [9.38].

For the electrolytic enrichment technique, initial sample pre-treatment is important. Other radionuclides present in the initial water fraction will also concentrate in the aqueous phase during electrolysis. The first stage in the separation is therefore to distil the sample to obtain a pure water sample, free

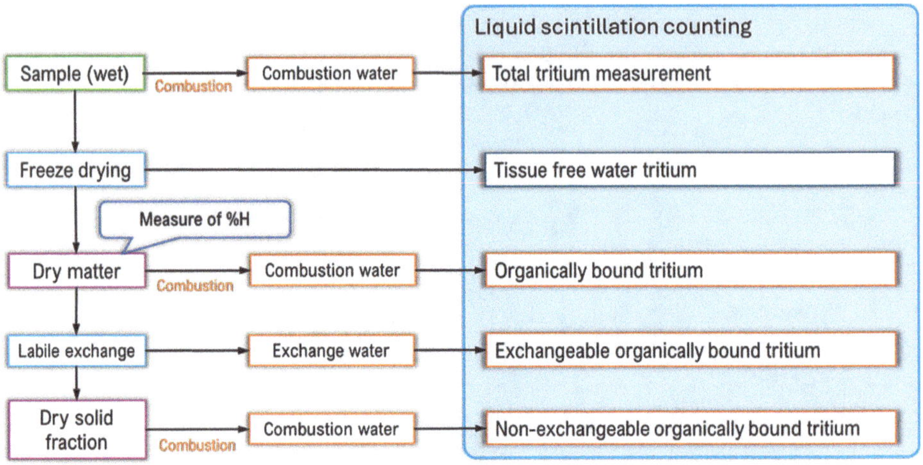

FIG. 9.6. Procedure for determining the speciation of tritium in environmental media [9.44].

from contamination (e.g. Fig. 9.6). Sodium carbonate is often added to the sample prior to distillation to prevent volatile radionuclides (e.g. ^{14}C in CO_2) from distilling over.

9.2.5.1. The electrolysis cell

The electrolysis cell is designed to provide the optimum conditions for enrichment of tritium whilst preventing contamination of the sample from atmospheric moisture containing tritium. One such cell designed by Allen et al. [9.39] consists of a glass cylinder tapered at the base to permit the efficient collection of the concentrate. Evolved gases pass through a silica gel trap and a U-tube containing silicone oil. The inclusion of a U-tube produces a small positive pressure that prevents atmospheric moisture reaching the sample in the cell. PTFE (polytetrafluoroethylene) baffles at the top of the cell to prevent spray carry-over during the electrolysis. The splash head at the top of the cell also incorporates an alternative vent that permits the introduction of more sample if volumes greater than 100 mL are required for electrolysis. The periodic addition of sample during the electrolysis allows greater concentration factors to be obtained. However, this system requires considerable supervision to determine the correct time to add the sample. A modified electrolysis cell incorporating a feed reservoir was developed by Metson [9.40]. In this modified design, the tapered glass cylinder was replaced with a commercially available tapered measuring cylinder. It was found that the continuous feed approach resulted in more reproducible results and less operator intervention compared with periodic addition of sample.

9.2.5.2. *The electrodes*

The cathode is typically composed of high purity iron or stainless steel [9.40]. The anode is composed of nickel or a nickel–iron alloy, which is more resistant to chemical attack. Initial cleaning of the electrodes is crucial. Electrodes are normally degreased and then washed in an alkaline solution of sodium phosphate and sodium hydroxide. The electrodes are then treated with a solution of 25 g/L oxalic acid, 13 g/L hydrogen peroxide and 0.1 g/L sulfuric acid. This chemically polishes the iron electrodes and prevents corrosion during the electrolysis. Treatment of steel electrodes by immersion into 5N phosphoric acid at 50°C for 30 min was found to improve tritium recovery and reproducibility [9.40]. The spacing of the electrodes is crucial to obtaining reproducible results. Spacing is usually maintained using PTFE spacers. Tritium recovery is also found to be lower when new electrodes are used and 'running in' of the electrodes is usually required before reproducible tritium recoveries are obtained [9.27].

9.2.5.3. *The electrolysis stage*

For electrolysis to proceed, the initial distillate should be made alkaline using sodium hydroxide or sodium peroxide. Addition of sodium peroxide has the advantage that no hydrogen is being added to the sample, and this limits the possibility of ^{3}H contamination. Approximately 0.5% by mass of sodium peroxide is added. The sample is then electrolysed with a current of 6–10 A. The current is chosen so that the electrolysis is achieved in such a way that ohmic heating of the sample is limited. The electrolysis cells are placed in a cooling bath at around 2°C to enhance tritium enrichment and prevent overheating and vapour production. Ethylene glycol has been suggested as a suitable coolant. Electrolytic enrichment of 100 mL of sample to 5 mL, using an initial current of 6 A that reduces continuously to 3 A, requires 60 h. At the end of the electrolysis, some of the tritium is present as hydroxyl ions. Lead nitrate may be added to the sample, producing lead hydroxide. This decomposes at 150°C, producing lead oxide and water. The electrolytically concentrated sample, along with lead hydroxide, is transferred to a distillation rig operating under reduced pressure of approximately 5 mmHg (667 Pa). The sample is heated at 160°C and the distillate is collected. Alternatively, the sodium hydroxide may be neutralized with CO_2 prior to distillation, again releasing any hydroxyl-bound tritium to the unstable carbonic acid, which decomposes to CO_2 and water. The distillate is mixed with a suitable scintillant and counted on a low level liquid scintillation counter.

9.2.5.4. Calibration of electrolytic enrichment using deuterium

The greatest contribution to the overall measurement uncertainty in electrolytic enrichment measurements derives from the tritium recovery determination. The tritium recovery is usually determined through the electrolysis of a limited number of reference standards and the determined recovery is applied to all unknown samples. This assumes that the sample and standard cells are operating under identical conditions and that no gross variation in tritium recoveries has occurred. If a gross variation does occur in a sample cell this would go undetected. Thus, the determination of the tritium recovery by analysis of the deuterium enrichment for each cell using a microliter amount of sample eliminates this problem [9.41].

9.2.6. Electrolytic enrichment at the IAEA's Isotope Hydrology Laboratory

The IAEA's Isotope Hydrology Laboratory has been a global centre for tritium analysis of natural waters for well over 50 years. They have continuously developed and improved their methodologies and use them to guide IAEA partner laboratories worldwide [9.27]. Improvements have included a simple method for pretreatment of electrolytic cells to ensure a high tritium separation factor, an upgraded design of the exhaust system for explosive gases, and a vacuum distillation line for faster initial preparation of water samples for electrolytic enrichment and for tritium analysis. Achievements included the reduction of the variation of individual enrichment parameters of all cells to better than 1% and an improvement of 50% of the stability of the mean background. This led to improved detection limits and measurement precisions. Such a capability was crucial for ultralow tritium measurements using LSC and gas flow proportional counting. The overall improvements and optimizations led to better performance, higher operational safety standards and increased protection against tritium contamination during sample preparation.

The operation of the tritium electrolysis system used by the IAEA's Isotope Hydrology Laboratory allows processing of 24 (or 48) samples and generates large amounts of potentially explosive hydrogen–oxygen gas mixtures. Management of this risk has involved the introduction of the following critical safety and security features at the IAEA facility:

- Stopping the electrolysis after a predefined number of ampere-hours to prevent the electrolytic cell from running dry, which might lead to a spark when the electric current is interrupted.

— Controlling upper and lower temperature limits for the refrigerator system, with shutdown of the electrolysis in case of an exceedance of the temperature limits. An excessive sample evaporation rate or too low a temperature could lead to a spark ignition when the electric current is interrupted.
— Shutting down in case of failure of the cooling compressor.
— Providing an over-voltage protection circuit, with shutdown of electrolysis in case of the supplied electrolysis voltage exceeding 80 V. This avoids excessive voltage on the current regulated source due to bad or broken contacts. An over-voltage could indicate that cells are approaching dryness (routine operation voltage: about 70 V for 24 cells in series).
— Having a remote emergency power isolator at the laboratory entrance (in case of dealing with a fire or other emergency).
— Using glass bubblers filled with silicone oil for each of the 24 (or 48) electrolytic cell exhaust lines to enable visual checking of system tightness and bubbling rate and to prevent cross-contamination between samples in case of a shutdown.
— Employing an explosion-proof, high volume gas exhaust system to manage all gaseous exhausts from the 24 (or 48) electrolytic cells that pass into the transparent wall mounted cabinet. All exhaust tubes from the bubblers are connected to one main exhaust line and an associated main gas bubbler, which vents the gas through the explosion proof gas exhaust system.

Recent advances at the IAEA's Isotope Hydrology Laboratory include the development of a new electrolytic enrichment system (using solid polymer electrolyte (SPE) technology; see Section 9.2.7) with full layout and construction information made available[3]. Further information is freely available via the open tritium data processing tool TRIMS[4] [9.42].

9.2.7. Electrolytic enrichment using solid polymer electrolyte

In conventional alkaline electrolytic enrichment of tritium in waters, NaOH or Na_2O_2 is typically added as an electrolyte. The concentration of the electrolyte solution increases, making it difficult to proceed with electrolysis at a CR of 13 times or more. In addition, O and H generated during electrolysis could, in principle, explode. SPE electrolysis was developed by General Electric Inc. in the early 1970s and enabled pure water electrolysis (see also Ref. [9.43]). The method uses an ion exchange membrane (an SPE) as the electrolyte. A membrane with a thickness of 0.1–0.3 mm is used for this electrolysis to suppress the

[3] https://nucleus.iaea.org/sites/ihn/Pages/InstallationCompactTEU.aspx
[4] https://nucleus.iaea.org/sites/ihn/Pages/Software.aspx

TABLE 9.2. ADVANTAGES AND DISADVANTAGES OF ELECTROLYTIC ENRICHMENT USING SOLID POLYMER ELECTROLYTE

Advantages	Disadvantages
No danger of hydrogen explosion	Only one sample can be processed at a time
Easy to operate	^{3}H concentration factors typically half of those achievable by electrolytic enrichment
No need to add acid or base electrolytes	Temperature control is difficult
No restriction on the concentration ratio	
Current density is large and electrolysis is fast	
No need for complicated electrode cleaning after electrolysis	

amount of hydrogen and oxygen diffusing through the membrane. One example of a ^{3}H concentrator using an SPE is Tripure XZ014 (De Nora Permelec Ltd). The system comprises a porous cathode made of 316 stainless steel and a porous anode made of iridium oxide supported on titanium. This SPE membrane is a strongly acidic cation exchange membrane (e.g. DuPont Nafion). The apparatus uses these electrodes with SPE installed on both sides. Since the SPE behaves as an electrolyte in this system, electrolysis proceeds with pure water alone, and gas is generated from the gaps in the metal fibres. Hydrogen ions generated from water at the cathode move through the SPE and reach the cathode, and hydrogen gas is generated at the anode; oxygen gas is generated at the cathode. Since oxygen gas and hydrogen gas are separated by the SPE membrane, the risk of a gas explosion due to mixing is most unlikely.

Positive and negative features of the SPE method are summarized in Table 9.2.

9.2.8. Tritium speciation determination

In environmental monitoring programmes, it is often necessary to differentiate between ^{3}H associated with TFWT and with OBT (E-OBT or NE-OBT) [9.16]. However, such distinction is very dependent on the analytical approach used to quantify the various ^{3}H forms. Typically (Fig. 9.6), the sample is first dehydrated to remove the TWFT using azeotropic distillation (although organic residues can interfere with subsequent analytical steps), freeze-drying or oven-drying (if the TFWT fraction is not being collected). This removes the ^{3}H component associated with free water. Where freeze-drying is used to recover the TFWT, the residue is typically then oven-dried to ensure that all TFWT has been removed prior to combustion. The sample is then combusted to decompose OBT (combined E-OBT and NE-OBT) and convert it into combustion water, which is collected and measured for ^{3}H activity concentration. Alternatively, the ^{3}H content in the solid fraction can be determined using a ^{3}He ingrowth based technique (see Section 9.6)

E-OBT and NE-OBT can be differentiated by introducing an intermediate step [9.2, 9.4, 9.5, 9.16], whereby the dried, ground sample is soaked in ^{3}H-free water prior to combustion. In some instances, the water is warmed to 40°C to increase the rate of exchange, with an exchange rate half-time (time taken to extract 50% of the exchangeable ^{3}H) of ~30 min and quantitative recovery of exchangeable ^{3}H within four hours [9.5]. Exchangeable ^{3}H will equilibrate with the water, which is then recovered by filtration under dry air (often in a glove box) and freeze-drying the solid fraction to remove traces of water. The E-OBT can be quantified by measuring the quantity of ^{3}H associated with the filtered water. The sample is then dried and combusted to determine the residual NE-OBT fraction, or the solid fraction retained for ^{3}H measurement via ^{3}He ingrowth.

Nivesse et al. [9.8] have noted that the measurement of E-OBT via an intrusive liquid exchange approach can be biased by solubilization of NE-OBT molecules and degradation of organic molecules in contact with water. The researchers detailed an alternative, non-invasive, technique whereby the sample is equilibrated with water vapour under controlled temperature and humidity conditions to determine the E-OBT component more reliably.

Le Goff et al. [9.6] undertook a systematic study of fractionation effects associated with the processing of milk samples for speciated ^{3}H measurement. Milk samples were dehydrated using either freeze-drying or distillation (using a distillation bridge or rotary evaporator, in both cases under reduced pressure) to determine the free water ^{3}H content. The authors noted a significant bias associated with sample mass, method of water removal and degree of dehydration. Elevated ^{3}H specific activities in free water obtained by freeze-drying were found to result from a greater degree of contamination from ^{3}H-enhanced atmospheric water vapour for samples extracted by freeze-drying compared with those extracted by distillation. Equipment dead volumes significantly impacted the measurement of per cent free water and specific activities when performing distillations under reduced pressures. Significant fractionation of ^{3}H was also observed during the dehydration process if the drying stage was not taken to completeness. The ^{3}H specific activity in the distillate obtained from 20 mL of milk using a rotary evaporator with a 7 mL dead volume was found to be 91% of the true value if 10% of the water remained in the solid phase — clearly a significant bias.

9.3. RADIOMETRIC MEASUREMENT OF ^{3}H BY LIQUID SCINTILLATION COUNTING

Tritium, as a low energy beta emitting radionuclide (E_{max} = 18.6 keV), can be measured radiometrically using either windowless proportional counters or, more commonly, by LSC techniques. Most commercial liquid scintillation counters can accommodate a total maximum volume of about 20 mL, thus limiting the amount of sample that can be counted. Large volume liquid scintillation counters (e.g. Hitachi ALOKA AccuFLEX LB7) are commercially available; these allow larger sample sizes to be measured, which can reduce the limit of detection achievable (see Section 9.3.1). Other ultralow level LSC manufacturers rely on other technological novelties to enhance sensitivity (e.g. Hidex ULLA). Apart from LSC instrument design, it is also possible to enhance sensitivity by electrolysing the sample prior to counting. Electrolytic enrichment increases the concentration of ^{3}H relative to ^{1}H, which results in a reduction in the limit of detection. Alternatively, non-radiometric techniques can be used to measure the ingrowth of ^{3}He arising from the beta decay of ^{3}H using He ingrowth mass spectrometry. This approach results in the lowest limits of detection (see Fig. 9.2 and Section 9.5).

In LSC, a liquid sample fraction containing the ^{3}H (usually water) is mixed with a cocktail comprising an aromatic solvent, primary and secondary scintillators, emulsifiers and stabilizers. The beta particle emitted during the decay of ^{3}H excites the solvent molecule. The excitation energy is transferred to the primary and secondary scintillators, ultimately resulting in the emission of light photons, which are detected by the photomultiplier tubes (PMTs) of the liquid scintillation counter (Fig. 9.7). The emulsifier enables intimate mixing of the aqueous (sample) and organic (solvent/scintillator) phases through micelle formation.

Robust and sensitive measurement of ^{3}H by LSC requires the optimization of a number of factors, including choice of scintillation cocktail, aqueous/organic ratio, source preparation, quench reduction and background reduction.

9.3.1. Source preparation and choice of scintillation cocktail

For LSC measurement of aqueous samples (distillates, combustion waters or bubbler trap solutions), an aliquot of the sample is mixed with a scintillation cocktail in a vial. Aqueous to organic scintillant ratios are optimized to maximize

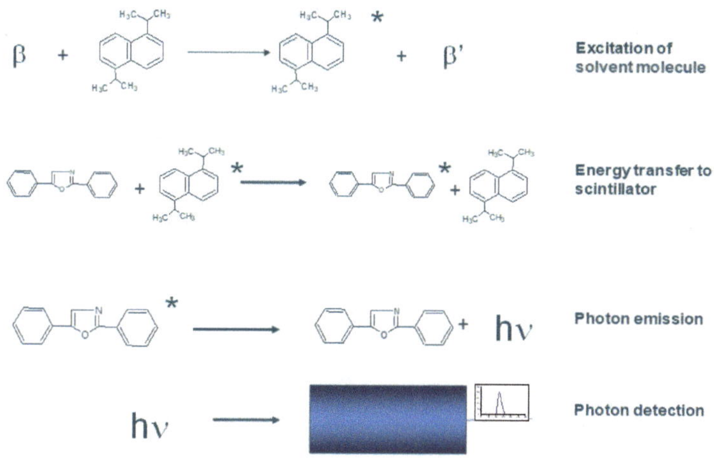

FIG. 9.7. Stages of scintillation process in a liquid scintillation counter.

the measurement figure of merit, which in turn depends on the volume of sample being counted, the ^3H detection efficiency and the background. In addition, the mixture has to be stable (miscible) for sufficient time to permit counting. To assist optimization, the figure of merit, FOM, can be calculated as follows:

$$\mathrm{FOM} = \frac{\varepsilon^2 V^2}{B}$$

where

ε	is the ^3H detection efficiency;
V	is the volume of aqueous sample counted;

and B is the background count rate.

In general, increasing the sample volume counted will decrease the counting efficiency, and there is a balance to be achieved between the volume counted and the impact that this has on the efficiency (Fig. 9.8).

Specific commercially available cocktails are formulated to maximize the volume of the aqueous sample that can be counted whilst maintaining relatively high counting efficiencies. The presence of dissolved solids and the pH will affect the miscibility of the aqueous sample with the scintillation cocktail and will also impact the ^3H detection efficiency.

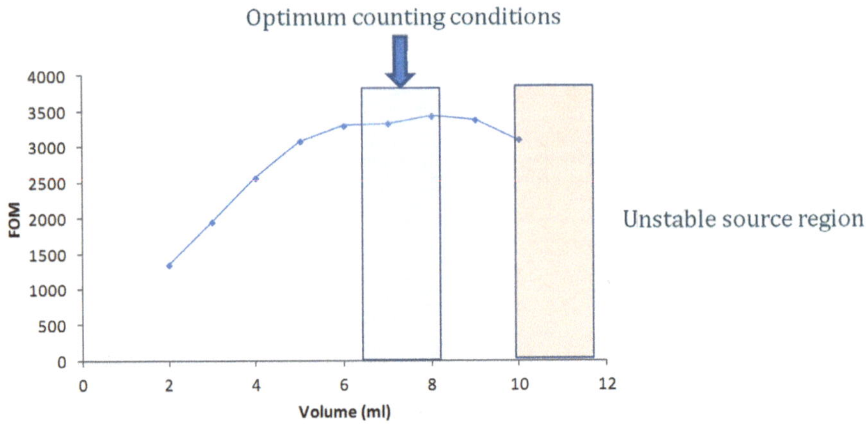

FIG. 9.8. Effect of increasing sample volume on the figure of merit (20 mL total volume, Gold Star scintillation cocktail, polythene vial).

9.3.2. Counting efficiency and quench

Counting efficiency is dependent on the beta energy of the radionuclide. The spectrum of beta particles emitted by ^{3}H extends to 18.6 keV, with the mean energy equal to 5.7 keV. It is estimated that only 4% of the beta particle energy can be converted to light, so most PMTs will detect only 25% of the photons emitted. For the mean energy of beta particles, it is estimated that only 20–25 scintillations can be produced for each single decay event. A significant proportion of ^{3}H beta emissions will be at energies below the threshold for detecting two photons required for coincidence counting, thus limiting the maximum counting efficiency of ^{3}H to ~60%. However, detection efficiencies will be further reduced by any process that competes with the production and detection of the light photons.

The detection of light photons produced from the interaction of ionizing radiation with the scintillant cocktail can be adversely affected by the photons interacting with other compounds, contaminants, etc., that might be present prior to reaching the PMTs. This reduction in efficiency of the detection system is known as 'quench'. The effect of quench is to decrease the number of decay events that are registered by the PMTs (decrease in counting efficiency) and to shift the energy spectrum of a nuclide to a lower energy region, as fewer photons are registered for a given particle energy than in an unquenched sample (see the example spectrum in Fig. 9.9). Quenching may also occur when the light photon is produced but is prevented from reaching the detector. The light photon is either adsorbed by a molecule dissolved in the sample–cocktail mixture (colour quench)

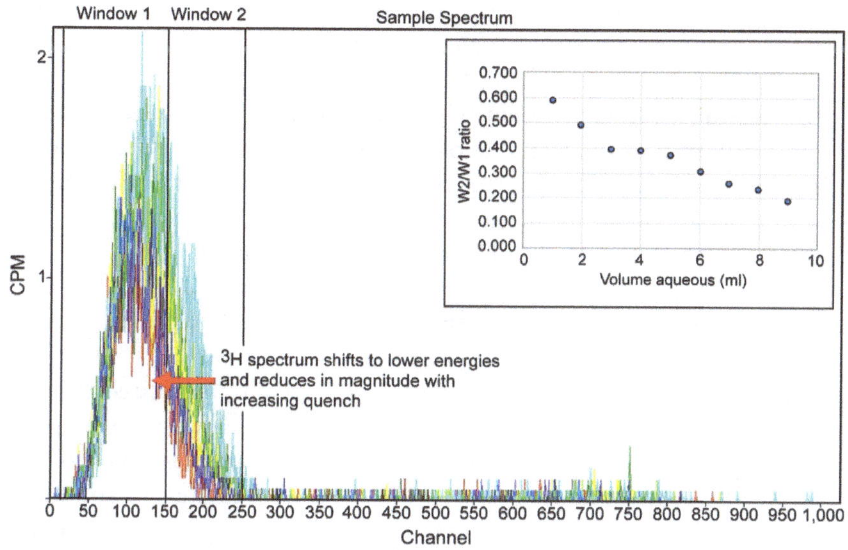

FIG. 9.9. Shift in ^3H pulse height spectrum with quench and correction using the sample channels ratio method. CPM: counts per minute.

or is physically prevented from reaching the detector by an obstruction such as particulate matter suspended in the mixture (physical quench).

9.3.3. Chemical quench

The chemical reactions that transfer the energy of the beta particle to produce a light photon can be inhibited. This is known as chemical quench. The excited solvent or primary scintillant molecule interacts with a molecule other than the secondary scintillant. The solvent or primary scintillant is de-excited and the second molecule is excited. This excited molecule then de-excites, emitting the excess energy as heat instead of a light photon. No signal is therefore produced by the PMTs and the event is not detected.

A wide range of compounds can induce chemical quench. These include dissolved oxygen, chlorinated organic compounds, acetone and water. The chemical quench process can be subdivided as follows:

(a) Acid quenching. Protonation of the primary or secondary scintillator modifies the excitation energy levels of the scintillator, creating a mismatch for solvent–scintillator energy transfer.

(b) Concentration quenching. The concentration of a component within the scintillator mixture reaches such a level as to interfere with the scintillation process.
(c) Dilution quenching. Dilution of the sample with any molecule that does not participate in the scintillation process increases the effective distance between solvent molecules and reduces the efficient transfer of energy from the ionizing radiation to the scintillant molecule.
(d) Dipole–dipole quenching. Certain materials interact with the excited solvent molecule, with subsequent transfer of energy via dipole–dipole interactions and dissipation of energy via non-radiative processes. Oxygen and nitromethane quench in this fashion.
(e) Capture of secondary electrons. Part of the excitation process involves the emission of secondary electrons by the solvent molecule after interaction with the ionizing radiation. These secondary electrons can go on and excite further solvent molecules. Any molecule that captures these secondary electrons will effectively quench the scintillation process. Halogenated molecules quench effectively via this process.

Chemical quenching has a marked effect on the pulse height spectrum. As quenching of the sample increases, the total number of observed events decrease as the counting efficiency falls. The pulse height spectrum shifts to lower energy channels as fewer photons are produced for a given energy of ionizing radiation. The magnitude of the spectral maximum also decreases (Fig. 9.9). The extent of chemical quench for a given quantity of quenching agent depends on the cocktail being used. Chemical quench due to dissolved oxygen can be removed by bubbling nitrogen through the sample. However, care needs to be taken when using after-pulsing corrections[5], as a reduction, instead of an increase, in counting efficiency may be observed. In most other cases, quenching is limited by sample purification prior to liquid scintillation measurement and by correction of the final measurement for the inevitable quenching that still occurs.

In colour quench, the quenching agent has an absorbance band at the same wavelength as the emission band of the scintillator. The light photons are absorbed and hence do not reach the PMTs for detection. As scintillators usually emit in the blue region of the visible spectrum, the greatest quenching is observed with red compounds. Colour quench is an absorbance process and therefore obeys the Beer–Lambert law. In the Beer–Lambert law, the path length of the process (in this case, the distance between the photon production and the edge of the vial)

---

[5] A technique adopted in time resolved LSC that uses the number of pulses detected within a short time period immediately after the main pulse to distinguish between beta and alpha particles or cosmic radiation events.

is critical to the degree of absorbance observed. For this reason, colour quench differs from chemical quench in that it is dependent on the vial size. In practice, this results in broader pulse height spectra with a lower maximum compared with a sample chemically quenched to the same counting efficiency.

9.3.4. Physical quench

In physical quenching, the light photons produced are prevented by a physical barrier from reaching the PMTs. Any dispersed particulate matter in the vial will drastically quench the sample. Physical quench also occurs if the water droplet micelle size is significant compared with the range of the ^{3}H beta particle or if the sample adsorbs onto the surface of the vial. The emitted beta particle may not interact with the solvent under these conditions. Furthermore, in the case of sample adsorbed onto the vial surface, even if the adsorbed sample is in intimate contact with the scintillant, the counting efficiency will still be significantly reduced, as only the radiation emitted towards the contents of the vial will excite the solvent molecule, whereas that emitted towards the vial wall are lost.

9.3.5. Quench correction and calibration of measurements

Physical and colour quenching are usually mitigated through adoption of robust sample preparation procedures prior to liquid scintillation measurement. However, although chemical quenching can be minimized in the sample preparation stages, it cannot be totally avoided, and some form of quench correction is typically required.

In all instances, the aim of quench correction is to determine the counting efficiency for the sample, accounting for the composition of the sample and its impact on the quench. Quench correction falls into three main categories:

(a) Measurement of a source of comparable composition containing a known quantity of ^{3}H. The source may either be prepared separately from the samples (a matrix-matched standard) or be produced by spiking one of the samples with a known activity of ^{3}H (internal standardization).
   (i) Matrix-matched standards. Standards are prepared with comparable chemical compositions to the samples but with a known quantity of ^{3}H. The standards are measured alongside the samples, and the instrument response to the standard is used to determine the counting efficiency. The approach is relatively straightforward but requires that standards are prepared and counted with each batch of samples. The approach is also dependent on the standard compositions being representative of the samples.

(ii) Internal standardization. The counting efficiency for a sample is determined by measuring the increase in measured counts following the addition of a known activity of ^{3}H to the sample. The sample is first counted to determine the count rate of the ^{3}H in the sample. The sample is then spiked with a ^{3}H standard of similar composition and the mixture is recounted. The approach determines the specific counting efficiency for each sample and therefore permits correction for inter-sample variability in the quench level. However, the method requires each sample to be counted twice (increasing measurement time and reducing instrument throughput) and increases the quantity of radioactive organic waste generated. Moreover, the addition of the ^{3}H spike may change the quench level of the sample, although the effect is usually small.

(b) Using the shift in the pulse height spectrum to assess the degree of quench and to compare this with independently measured quench calibration curves to estimate the counting efficiency. This may be achieved using either the sample ^{3}H beta spectrum (sample channels ratio and internal quench parameter measurement) or the Compton spectrum generated from an external gamma source (external channels ratio and external quench parameter measurement). More complex approaches use a spectral fitting algorithm to determine the best fit to a particular quench level and hence counting efficiency (total disintegration per minute approach).

(i) Sample channels ratio method. This approach utilizes the shift to lower energies in the pulse height spectrum, which is observed with increasing quench. The measurement window is divided into two counting regions: a low energy window, W1, and a high energy window, W2. The boundary between the two windows bisects the pulse height spectrum. As the quench level of the sample increases, the proportion of the total counts registered in the low energy counting window (W1) increases whilst that registered in the high energy window decreases. The ratio of measured counts in the low and high energy measurement windows (W1/W2) provides an indicator of the quench level of the sample. A series of standards covering a range of quench levels are prepared and counted under identical counting conditions to provide a calibration between the W1/W2 ratio and the ^{3}H counting efficiency. This approach provides a sample specific quench correction that requires only one measurement per sample. Another benefit is that it does not generate additional radioactive organic wastes. However, the approach does rely on there being a sufficient ^{3}H activity in the sample to precisely measure count rates in the two measurement windows and is therefore not appropriate for low activity measurements. A similar approach can be used with an external standard (ECR) although more robust approaches tend to be preferred.

(ii) Internal quench parameter measurement. This approach is a development of the sample channels ratio approach. Rather than using two measurement windows, the position of the ^{3}H pulse height spectrum is defined by either the channel number associated with the spectrum centre of gravity (Spectral Index Sample, SIS) or its endpoint (Spectral Quench Parameter – Internal, SQP(I)), depending on the instrument manufacturer. In practice, the spectrum endpoint is difficult to define, as the low count rates measured at this point are difficult to distinguish from the background. In practice, a position (in terms of channel number) is defined whereby a set percentage of the spectrum lies below this point. Measurement of a set of standards covering a quench range under identical counting conditions provides a calibration of counting efficiency based on peak position. As with the sample channels ratio, this approach provides sample specific quench correction and requires a single measurement but again can only be applied to samples containing significant ^{3}H activities.

(iii) External quench parameter measurement. This approach utilizes a gamma induced Compton spectrum to determine the quench level in the sample. A moveable, external gamma emitting source (^{226}Ra, ^{133}Ba or ^{152}Eu, depending on the instrument manufacturer) is positioned adjacent to the sample during part of the measurement process. The secondary Compton electrons will interact with the scintillation cocktail in the same manner as a beta particle, producing a characteristic spectrum. The first measurement will be a combination of the Compton spectrum and the sample spectrum. A second measurement is performed in the absence of the gamma source to determine the sample spectrum, which is subtracted from the first measurement to produce a Compton spectrum independent of the sample activity. The spectrum will be shifted to lower energies depending on the quench level of the sample. The position of the spectrum is determined using one of several parameters, including the centre of gravity (Spectral Index External, SIE), point of inflection (Horrocks number, H#) or channel number at which a defined percentage of the spectrum lies below this point (Spectral Quench Parameter-External, SQP(E)). A calibration curve is used to relate the external quench parameter to the ^{3}H counting efficiency. This approach has the benefits associated with the Internal Quench Parameter approach but does not rely on the presence of appreciable ^{3}H activity in the sample itself. Many of the external quench parameter approaches are affected by the 'wall effect', whereby scintillation cocktail penetrates into the vial wall, resulting in a distortion of the Compton spectrum. A modification of the Reverse Spectrum Transform method was developed and is currently used in instruments of PerkinElmer, Inc., which uses a more robust parameter, called the Transformed Spectral Index External (tSIE), for correction of the distortion.

(c) Triple to double coincidence ratio. Triple to double coincidence ratio (TDCR) counting is an absolute approach for determining detection efficiency. The technique was originally developed for metrological applications [9.45, 9.46], and it is now well established [9.47] in commercially available instruments (e.g. HIDEX 600 SL Super Low Level and ULLA). The technique utilizes the ratio of the coincidence signals recorded in pairs of PMTs to that recorded in three PMTs counting in coincidence to determine the measurement efficiency and hence the activity of the source. The technique is applicable to radionuclides emitting electrons via beta decay or internal conversion and corrects for both chemical and colour quench. Chemiluminescence contributions are also eliminated.

9.3.6. Preparation of quench curves

For many of the quench correction approaches, a set of standards of known ^{3}H activity, covering the quench range expected for the samples, needs to be counted. The measurement of these calibration standards permits the correlation between the selected quench parameter and the ^{3}H counting efficiency to be determined. The quench level of the sample can then be determined, and the appropriate correction for the counting efficiency applied, using the calibration curve. The quench level in the standards can be varied by either changing the ratio of aqueous phase to scintillant (as water is likely to be the main quenching agent when counting purified aqueous samples) or by the addition of increasing quantities of a quenching agent, such as nitromethane. Care needs to be taken when changing the ratio of aqueous to scintillant phases to ensure that the capacity of the scintillation cocktail is not exceeded. Ideally, the quenching agent is selected to represent one most likely to be present in the sample. Quench standards are counted using identical instrumental parameters as those chosen for the samples. The measured count rate and quench parameter for each standard is then used along with the known ^{3}H activity to determine the counting efficiency at a given quench parameter value and draw a graph to interpolate for the range of quenches.

9.3.7. Background reduction in liquid scintillation counting

Background contributions derive from many sources, including: (a) natural background radiation associated with cosmic radiation, building materials and the local environment; (b) the materials used to manufacture the instrument components; (c) the vial and scintillation cocktail; (d) electronic noise from the PMTs; (e) non-radiation-induced scintillation events (Fig. 9.10).

Background contributions arising from cosmic radiation and the local environment are minimized through a combination of active and passive shielding. Active shielding is provided by a guard detector located close to the sampling chamber that counts in anticoincidence with the sample PMT circuit. The guard detector can be a mineral scintillant filled chamber or a solid scintillator such as bismuth germanium oxide (BGO). Cosmic and environmental radiation will trigger both the guard chamber and sample chamber PMTs and will be rejected by the anticoincidence circuit. Passive background reduction is achieved through the use of graded shielding comprising layers of low background Pb, Cd and Cu. Background contributions arising from inherent radioactivity in instrument components is minimized through careful selection of construction materials and use of low potassium glass in PMT windows.

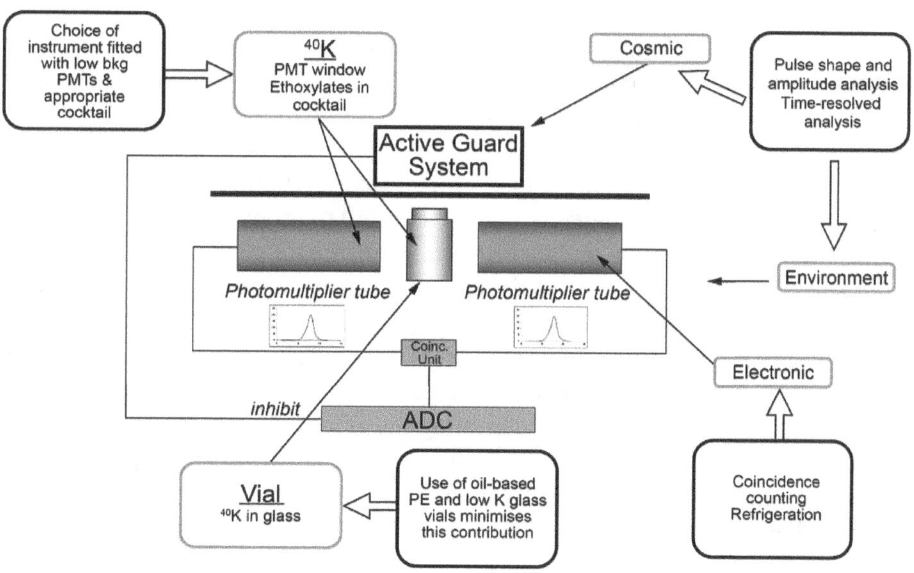

FIG. 9.10. Contributions to background signal on liquid scintillation counters. ADC: analogue to digital converter; bkg: background; PE: polyethylene; PMT: photomultiplier tube.

Two types of counting vial are routinely used for ^{3}H measurement. Low potassium glass vials provide good transparency and are impermeable to higher vapour pressure solvents, such as benzene, toluene and xylene, that were historically used in scintillation cocktails. Polythene vials offer significantly lower backgrounds compared with glass vials and are still impermeable to the newer, environmentally friendly cocktails. Teflon coated polythene vials have also been developed to minimize solvent diffusion into plastic vials, which can result in a wall effect affecting the detection efficiency and source stability. Polyethene vials must be manufactured from oil derived polyethylene to avoid incorporation of ^{14}C, which will increase background.

Electronic noise from the PMTs can give rise to significantly elevated background signals but is readily overcome by measuring the sample with two PMTs counting in coincidence. Higher backgrounds are observed for single-PMT systems, which are not suitable for low level ^{3}H measurement. Background signals arising from buildup of electrostatic charge on the vials (particularly polythene vials) is routinely overcome through the incorporation of a static eliminator into the liquid scintillation counter.

Electrostatic charge on plastic vials can cause significant interference during LSC. Electrostatic charge may be caused by friction generated when two different materials rub together. The combination of human skin and the polyethylene of the vials is likely to produce high levels of electrostatic charge, especially with the moisture present on the skin surface, which is conducive to the formation of electrostatic charge. In addition, the construction material of liquid scintillation counters increases the likelihood of static charge buildup. The level of electrostatic charge can be tens of thousands of volts. Static charge is variable and changes with time — an effect that makes it even more difficult to correct for. Most current scintillation counters therefore incorporate a static eliminator, comprising a radioactive source or an electric field to produce a counter-ion field around the vial, thus reducing the electrostatic charge.

Transient elevated background count rates can arise from chemiluminescence and photoluminescence, whereby light production is initiated through a chemically induced or photoinduced process, rather than from radioactive decay. Chemiluminescence and photoluminescence are detected within the low energy region of the energy spectrum and overlap significantly with the 3H spectrum. Chemiluminescence differs from radiation induced scintillation processes in that it is a single-photon event. To trigger two PMTs counting in coincidence, multiple chemiluminescence events are required to occur within the coincidence gate time. Chemiluminescence is relatively short lived and may be overcome by dark-adapting the sample prior to counting to permit chemiluminescence reactions to subside. Jakonić et al. [9.48] estimated the chemiluminescence half-life for Optiphase Hisafe 3 cocktail with distilled water to be approximately 16.4 min.

Photoluminescence appeared to decay more rapidly, and both interferences could be overcome through dark adaption of the sample for approximately 3 h. Chemiluminescence may also be corrected for using delayed coincidence counting (e.g. installed on the widely distributed Quantulus 1220 liquid scintillation counters). The chemiluminescence process results in the production of a single photon. For chemiluminescence to be detected by a two-PMT system counting in coincidence, the rate of chemiluminescence-derived photon production needs to be sufficiently high so that multiple events will occur within the coincidence circuit time gate, with each PMT detecting photons from separate, uncorrelated events. The probability of this occurring is the same even if a time delay is applied to one of the circuits, given the lack of correlation between the events producing the photons. However, for a genuine beta decay, multiple photons are produced from a single event, resulting in simultaneous triggering of both PMTs. If a time delay is introduced on one of the PMTs, the signal will no longer be in coincidence. By comparing the standard coincidence count rate with the delayed coincidence rate, it is possible to determine the contribution from chemiluminescence.

Background measurements need to be performed with each batch of samples counted. The background sample is prepared in an identical manner to the samples, using the same cocktail, vial and aqueous/scintillant ratio. Surface waters will contain ^3H associated with cosmogenic ^3H and ^3H derived from atmospheric nuclear weapon testing. Therefore, blank samples need to be prepared using deep borehole water (sometimes called 'dead water') that has been isolated from the atmosphere for sufficient time to permit the decay of cosmogenic ^3H and which has not been contaminated with fallout derived ^3H. Typically, $Na_2S_2O_3$ and Na_2CO_3 are added to the borehole water, which is then distilled to remove dissolved salts that would impact the detection efficiency [9.49].

9.3.8. Liquid scintillation counting instrument stability

Over long counting periods, the stability of the counting electronics is important. With improvements in electronics, it has become possible to construct counting equipment that is stable over considerable time periods. For example, the long term stability of the first generation of Quantulus 1220 ultralow level liquid scintillation counter, which is still widely employed by many tritium analysis laboratories, has been assessed over a period of 10 years [9.50]. An average decrease in efficiency of 0.54%/a was determined for ^3H. Despite this long term stability, monitoring of instrument performance is still a requirement for robust measurement. Counting background and standard samples at the beginning of each run and interspersed with the samples is one way of checking that the instrument set-up has not altered in any way during a count sequence. Regular checks on the instrument calibrations by counting a set of standards is also good practice.

9.4. RADIOMETRIC MEASUREMENT OF GASEOUS ^{3}H BY GAS FLOW PROPORTIONAL COUNTING

Gas flow proportional counting, with its associated high counting efficiency and low background, offers a sensitive approach for measurement of ^{3}H in the gaseous phase. Measurement of ^{3}H in air has been achieved using multi-wire proportional counters, whereby the anode–cathode distance is minimized, reducing primary electron attachment to atmospheric oxygen and subsequent signal suppression [9.51]. This approach was developed into a gas flow proportional counter, incorporating a multi-wire detector and anticoincidence guard background suppression, for the continuous monitoring of ^{3}H in air [9.52]. Detection limits of 3700 Bq/m^3 were reported for 1 min count times.

Conversion of ^{3}H to methane and blending with argon to produce the counting gas offers the best detection sensitivity. The method has been used by the Shallow Underground Laboratory of Pacific Northwest National Laboratory (USA) to measure ^{3}H in landfill methane, whereby the gas sample was purified using gas absorption to remove CO_2 and then passed over a desiccant to remove water. The purified methane was then blended with argon to produce P10 gas suitable for proportional counting. Detection efficiencies of 66% were calculated using Monte Carlo modelling with the radiation transport code GEANT4 with the proportional counter operating at 7.0 atm (0.71 MPa) [9.35]. Subsequent studies on the measurement of ^{3}H in water following conversion to CH_4, calibrating against ^{241}Am, demonstrated efficiencies of 77.3% and background count rates of 51.5 counts per day, resulting in measurement limits of detection of 90 TU for a 12 d count [9.35].

A more detailed overview of the principles of gas flow proportional counting systems and applications is given in Ref. [9.53].

9.5. MEASUREMENT OF ^{3}H AND ^{3}He — NOBLE GAS MASS SPECTROMETRY

9.5.1. History

The production, precipitation and dispersal of tritium associated with atmospheric nuclear weapon testing provided oceanographers and geochemists with the perfect hydrological tracer/dye experiment in the world ocean [9.54]. However, it was quickly realized that to effectively measure the large numbers of samples being collected would require an ultrasensitive and efficient analytical method for tritium. The well established technology of LSC, a radiometric method, was incapable of providing the analytical speed, precision or sensitivity required to detect the ultralow concentrations expected in large numbers of ocean water samples. With LSC measurements, the detection limit is mainly limited by counting time and instrument noise. A significantly more advantageous and sensitive method is to count the atoms of the accumulated stable gaseous decay product of tritium, ^{3}He (termed tritiogenic or tritiugenic ^{3}He). This technique is commonly known as the ^{3}He ingrowth method or the ^{3}He accumulation method.

By the late 1960s, NGMS instruments were adapted and refined to measure the extremely small concentrations of He isotopes in water samples [9.55]. Later, Clarke et al. [9.56] published experimental details about the ^{3}He ingrowth method for tritium analysis. With the beginning of WOCE [9.57] in 1990, a large analytical throughput was needed to enable thousands of tritium samples to be measured with a desired detection limit of 0.01 TU. Sophisticated high resolution noble gas mass spectrometers with Nier type ion sources from Vacuum Generators (VG-5400) and Mass Analyser Products (MAP 215-50) were implemented in some laboratories in the late 1980s to perform this task. Subsequently, the method also became available for groundwater studies [9.58, 9.59]. The initially large anthropogenic tritium contribution from the atmospheric nuclear weapon tests in precipitation had almost vanished; however, this method could still detect the very small tritium concentration in surface water and groundwater in the following decades. The current state of the art mass spectrometers used for this purpose are produced by several manufacturers such as Thermo Fisher Scientific (Helix SFT), Isotopx (NGX) and Nu Instruments (Noblesse HR).

The fundamental operational steps in noble gas mass spectrometry (NGMS) for ^{3}He ingrowth measurements are described by Clarke et al. [9.56], while other researchers report their own variants of the method [9.60–9.68]. A comparison of the ^{3}He ingrowth method and ultralow level liquid scintillation spectrometry is shown in Table 9.3.

TABLE 9.3. COMPARISON BETWEEN ^{3}He INGROWTH MASS SPECTROMETRY AND ULTRALOW LEVEL LIQUID SCINTILLATION COUNTING

Tritium	^{3}He ingrowth method	Ultralow level LSC with electrolytic enrichment or special counting equipment[a]	Typical LSC
Detection limit (TU)	0.01		
Detection limit (Bq/L)		0.4[b,c,d]	2

TABLE 9.3. COMPARISON BETWEEN ^{3}He INGROWTH MASS SPECTROMETRY AND ULTRALOW LEVEL LIQUID SCINTILLATION COUNTING (cont.)

Tritium	^{3}He ingrowth method	Ultralow level LSC with electrolytic enrichment or special counting equipment[a]	Typical LSC
Standard error for concentration >5 TU	3%	10%	—[e]
Sample throughput per year at 10× detection limit of method	>1 000	>300	>1 000
Typical queue time for a busy laboratory	6 months	1–3 months	1 month
Samples per day for tritium preparation	16	6 samples per 6-worktube Pyrolyser type system (slower if electrolytic enrichment is also involved)	6 samples per 6-worktube Pyrolyser type system
Daily number of analyses including gas standards, blanks calibration	50	5	10
Cost of laboratory instrument (€)	1 000 000	120 000	80 000
Typical laboratory staffing needed (persons)	3	2	2
Typical sample type	Water and solids	Water, biota (possibly solids)	Water, biota (possibly solids)

TABLE 9.3. COMPARISON BETWEEN ^{3}He INGROWTH MASS SPECTROMETRY AND ULTRALOW LEVEL LIQUID SCINTILLATION COUNTING (cont.)

Tritium	^{3}He ingrowth method	Ultralow level LSC with electrolytic enrichment or special counting equipment[a]	Typical LSC
Applications	Groundwater dating, precipitation, ocean circulation	Groundwater dating, precipitation, radionuclide labelling for biological processes, waste characterization of decommissioning	Radionuclide labelling for biological processes, waste characterization of decommissioning samples

[a] Instrument names shown are examples and do not imply endorsement by the IAEA.
[b] Hitachi LSC-LB7 ALOKA (background count rate 3.5 cpm, sample volume 40 mL, counting time 24 h).
[c] HIDEX 600 SL Super Low Level (background count rate ~3 cpm).
[d] HIDEX ULLA (ultralow level analyser; background count rate <1 cpm).
[e] —: no data.

9.5.2. Basic principles

In NGMS samples, both liquids and solids can be processed to enable measurement of tritium via the ingrown ^{3}He. The main challenges of the process involve the following:

— A complete removal of all ^{3}He in the sample before storage;
— A He-tight gas system for storage of the sample for months;
— A high resolution, high precision gas source mass spectrometer for helium isotope analysis.

When a sample in which the number N_T of tritium atoms is placed in a closed system for time *t*, ^3He generated by tritium decay is quantitatively confined in the system. If the ^3He is separated and purified, and the number N_{He} of ^3He atoms is measured with a mass spectrometer, the number of tritium atoms can be obtained from the following equation:

$$N_T = \frac{N_{He}}{1 - e^{-\lambda t}}$$

where λ is the ^3H radioactive decay rate (s^{-1}) and *t* is the storage time (s).

9.5.3. Example procedure for water analysis by He ingrowth technique (University of Bremen approach)

9.5.3.1. Sample preparation

Elimination of initial ^3He in water samples prior to analysis is essential to reduce measurement errors. For example, a water sample of 500 g may contain as many as 10^9 ^3He atoms, which must be reduced by about six orders of magnitude prior to sample storage. This is achieved by vigorously agitating the water sample to dissipate the dissolved or entrained helium into the sample container headspace, which is continuously pumped using a roughing vacuum pump with a flow-limiting capillary. The pumped water vapour acts as the carrier gas. The turbulence applied to the water samples accelerates the extraction of free gases into the evacuated headspace, so that the outgassing of He into the headspace is completed in approximately 30 minutes. For samples of other substances, such as organic materials, it may be more difficult to extract all He from the samples. Organic material may inhibit turbulence and the gas extraction can be a longer process. Moreover, for solid material, the sample needs to be transformed into fluid form to reduce low diffusivities. For water samples, any change of the sample mass due to water vapour loss needs to be accounted for.

9.5.3.2. Storage

Containers must be He-gas-tight and free from He to prevent loss or contamination from the container material into the sample for up to one year. The overall acceptable He leak rate has to be less than 10^{-14} mol/a. Full metal containers cannot achieve these low rates, and reuse may be unreliable because of memory effects. Crimped copper tubes can serve as good storage containers, but elaborate handling is required and the volumes are restricted. In most laboratories, borosilicate glass bulbs are widely used, as they are non-permeable, have volumes of 100–1000 mL and can be reused after refitting seals and connection parts [9.69]. Each laboratory evaluates their experimental approaches to ensure resilience. Low permeability glass bulbs must be He-degassed by heating the glass to 400°C or more in a vacuum oven for about 48 h or flushing the glass in a ceramic oven at high temperatures with He-free pure nitrogen. Careful handling of the glass bulbs is needed and the size of the bulbs has to be restricted. Often, glass bulbs filled with sample water are stored in freezers to reduce He diffusion.

9.5.3.3. Analytical measurement

A high resolution static gas source sector field mass spectrometer (GS-SFMS) is needed to measure the minute amounts of ^{3}He accumulated from tritium decay and the ^{4}He. A schematic of a He isotope analysis instrument is shown in Fig. 9.11. The system operating at the University of Bremen is shown in Fig. 9.12, and Fig. 9.13 contains a close-up view of the gas inlet system with glass bulbs and an example of a mass spectrum with peaks from ^{3}He and HD/H_3.

The complete transfer of all accumulated ^{3}He from the sample to the GS-SFMS is essential. The glass bulbs are normally half-filled with water, and the low solubility of He (~0.8%) ensures that 99% of the He accumulates in the

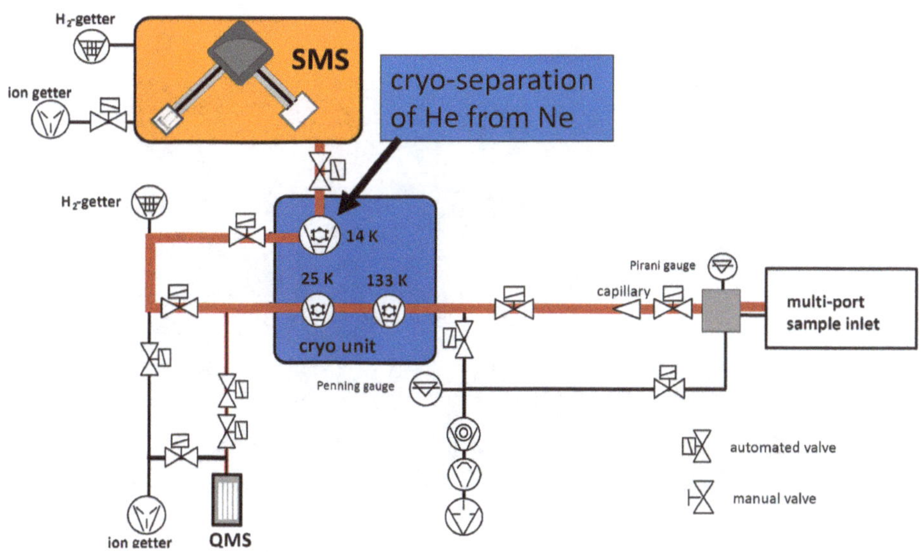

FIG. 9.11. Schematic of a He isotope instrument. The red line depicts the path for sample gases (courtesy of J. Sültenfuß, University of Bremen).

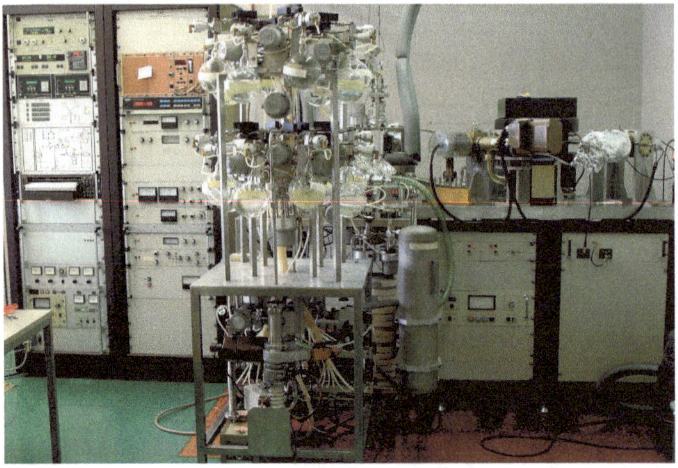

FIG. 9.12. The He isotope mass spectrometry system at the University of Bremen. The sample gas inlet unit with glass bulbs for tritium is in the foreground (courtesy of J. Sültenfuß, University of Bremen).

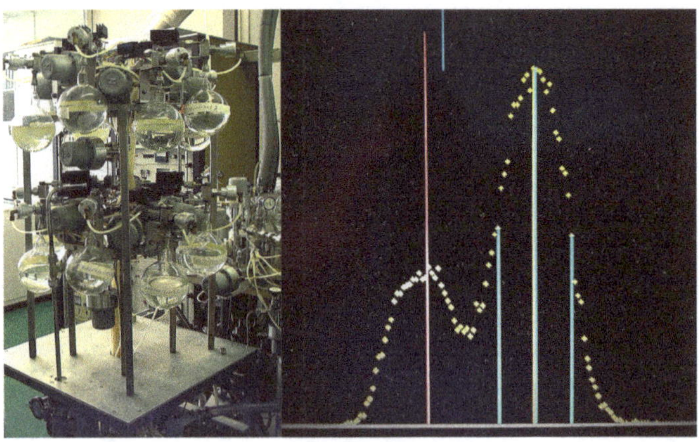

(a) (b)

FIG. 9.13. (a) Gas inlet system with 1000 mL glass bulbs for tritium analysis. (b) Mass spectrum with ^{3}He (left) and HD/^{3}H (right) peaks. (Courtesy of J. Sültenfuß, University of Bremen.)

headspace. A punching tool opens the tip of the glass bulb, and a continuous water vapour flow carries the ^{3}He (and other sample gases) from the headspace to a cryotrapping system. A set of capillaries prevents any backflow. Minimization of contamination is achieved using an all-metal sample inlet unit coupled to an integrated ultrahigh vacuum pumping system.

The following provides general insight into the procedural sequence of an NGMS tritium (via ^{3}He) analysis of a water sample:

— An integrated cryotrapping system isolates water vapour and other gases and allows He to pass to a charcoal trap kept at 14 K. This trap is later heated up to 45 K to release only He into the NGMS. A separate quadrupole mass spectrometer is used to check each sample for large He contamination, so that the main NGMS can be protected from inadvertent gas intrusion. The trap system is warmed up to room temperature once a week to remove condensed water. Turbo pumps provide the desired vacuum conditions.
— Maintaining the ultrahigh vacuum in the NGMS is achieved by ion getter pumps, and a static hydrogen getter reduces the hydrogen background. The NGMS can be maintained in a clean condition by baking at up to 300°C to reduce any background after opening the NGMS for routine maintenance.
— Special attention is always paid to the high mass resolution needed. The acceleration voltage is kept constant, and the magnetic field is precisely controlled. Adjacent to the ^{3}He peak (left) in the spectrum (Fig. 9.13) are the peaks of molecular hydrogen ions with deuterium (HD^{+}) and HHH^{+} ions, which interfere with the peak of ^{3}He. The mass difference is only 0.006 u, and a mass resolution of 600 u or better is needed to separate the peaks. To achieve good separation and stable ion source optics, very precise control of the magnetic field and a low H_2 background are crucial prerequisites. As H_2 is constantly released (but with a declining rate) from the inner metal walls of the GS-SFMS, pumping and regular baking are required. Weeks or months may be needed to achieve the required low H_2 background.
— Detection of ^{3}He ions is made using a channeltron multiplier (Galileo type 4860) in single-ion counting mode. The ^{4}He ion current is recorded as a voltage across a 10 GΩ resistor with a Faraday cup. Ion source optics and the magnet position have a large impact on the peak shape, and establishing the proper ion source parameters requires experimentation. After the ion beam is inspected using the ^{4}He peak shape, a slit in front of the ^{3}He detector is adjusted to gain maximum resolution. Additionally, a 90° electronic filter further reduces other scattered ions.
— To count the miniscule number of ^{3}He ions, the power supply for the channeltron multiplier must be adjusted to minimize white noise and achieve maximum sensitivity. Off-peak counting can be as low as 0.1 counts per

second (cps). It is crucial for the laboratory temperature to be stabilized to keep all electronic devices stable.

- To check the hermetic tightness of the system, continuous blank measurements are conducted. The net count rate is required to be less than 0.2 cps. For routine measurements, the counting time is set to a period of about 1000 s. During that time, about 1% of all atoms are recorded by the detectors.
- Data recording times are split into clusters of 10 s to allow outliers to be easily identified and rejected. In principle, a net counting rate of 0.2 cps from a low concentration tritium sample can be identified. This is equivalent to the ^{3}He accumulated from a 500 g water sample stored for six months with a concentration of 0.01 TU. Larger sample mass and storage time can further reduce the detection limit.
- The achievable analytical uncertainty is limited by counting statistics. The ^{4}He signal is used to identify non-tritiugenic ^{3}He. Even if not all non-tritiugenic ^{3}He sources can be traced, experience shows that it is advisable to use the atmospheric ^{3}He/^{4}He ratio within an error of 10% for correction of various contamination sources. It is notable that this results in larger uncertainty limits for tritium measurements with significant ^{4}He readings.
- For the calibration of the readings from the detectors, gas standards of atmospheric air are analysed continuously. An atmospheric He concentration of 5.24 ppm with a ^{3}He/^{4}He ratio of 1.384×10^{-6} is typically accepted by expert ^{3}He ingrowth laboratories. The atmospheric air is obtained from an undisturbed outside area and stored in a large steel container with a well defined volume. The temperature, air pressure and humidity are controlled during filling, as pipettes with accurate volumes are used to admit air standards to the system. In addition, linearity measurements are needed to check the sensitivity change of the ion source. Best performance and stability are achieved if the measurement system runs continuously for several weeks to months.

9.5.4. Examples of tritium distribution in the ocean and groundwater

The very high sensitivity of ^{3}He ingrowth mass spectrometry measurements for water samples allows detailed insights to be gained for low tritium concentration in groundwater. The precision and sensitivity of measurement that can be achieved is shown by the distribution of tritium concentrations in groundwater in some Central European countries (see Fig. 9.14).

Application of the high precision ^{3}He ingrowth method enables quantification of tritium content in oceanic water, including deep water layers with very low concentration of tritium. For example, Fig. 9.15 clearly presents

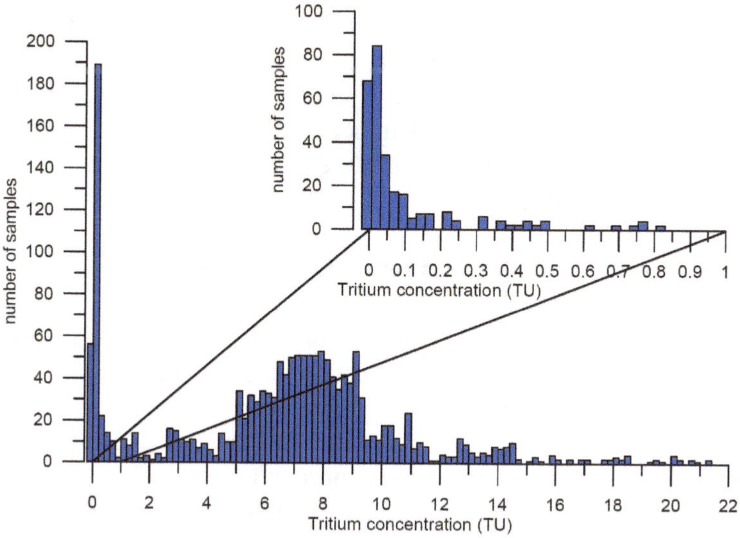

FIG. 9.14. Tritium distribution of several hundred groundwater samples collected in 2005–2012 in Denmark, The Netherlands and Germany (Data: Helium Isotope Lab, University of Bremen) (courtesy of J. Sültenfuß, University of Bremen).

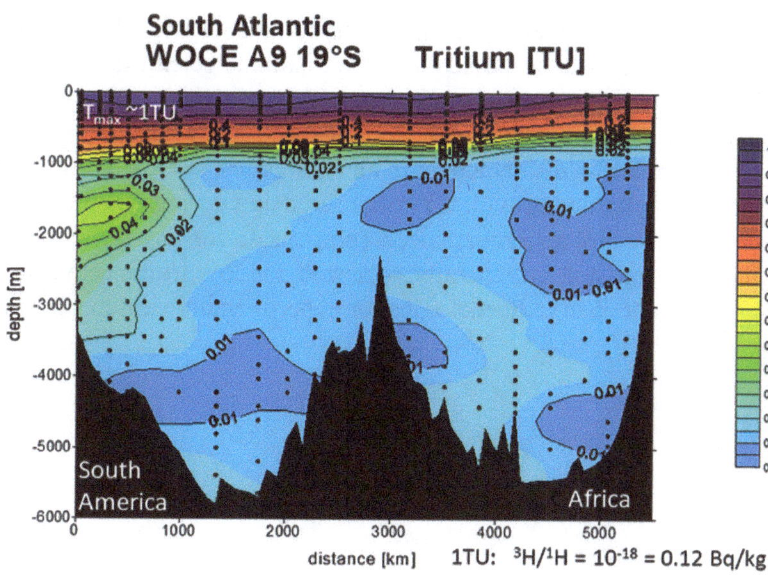

FIG. 9.15. Tritium concentration in South Atlantic in 1995, observed during the South Atlantic expedition of the WOCE programme. (Data: Helium Isotope Lab, University of Bremen; figure courtesy of J. Sültenfuß, University of Bremen.)

the systematic variations in tritium content with depth across the South Atlantic in 1995, which were obtained in an expedition within the framework of WOCE[6] [9.57]. The distribution clearly indicates the ventilated waters in the upper 800–1000 m. At the South American margin, the North Atlantic Deep Water can be identified at depths of 1500–2000 m, marked by elevated tritium concentrations of approximately 0.05 TU.

9.5.5. Analysis of organically bound tritium in solids

Measurement of OBT in a dry solid sample can be performed using the ^{3}He ingrowth method (an exemplary procedure described by Tamari et al. [9.70] is outlined in Box 9.1). The sample is placed in a glass storage container and outgassed to 10^{-4} Pa or lower, and subsequently the storage container is sealed and left for 2 months or longer to allow ingrowth of ^{3}He. The ^{3}He produced is measured by NGMS. The lower limit of detection of ^{3}He is about 5000 atoms, which approximately represents the decay of tritium for 3 months from an organic matter sample of 50 g. This corresponds to a detection limit of 0.02 Bq/L in the OBT combustion water. By comparison, a low background LSC method using a Hitachi LB-7 (Aloka) instrument provides an OBT detection limit of about 0.3 Bq/L.

The significantly lower detection limits obtainable with the He ingrowth (NGMS) method makes its use essential for demanding scientific investigations. Its significant cost and the relative slowness of delivering results are enormously outweighed by the improved sensitivity. The need for a He ingrowth period might be considered as a disadvantage in an emergency, but LSC can be used to co-investigate samples and provide an initial screening during ^{3}He ingrowth. The method's sensitivity and analytical speed compensate for any negative aspects. The He ingrowth method is generally preferred for low tritium environmental samples or applications in radioecological or experimental research studies.

---

6 http://woceatlas.ucsd.edu/

BOX 9.1. EXAMPLE OF ANALYTICAL PROCEDURE TO MEASURE CONCENTRATION OF OBT IN VEGETATION USING HIGH SENSITIVITY ^{3}He INGROWTH MASS SPECTROMETRY [9.70]

(1) Measurements are carried out for the processed sample (Portion A) and an unprocessed sample (Portion B).

(2) Free water is extracted from Portion A using freeze-drying.

(3) Low He permeability aluminosilicate glass container/ampoules are baked at ~400°C for several hours in a helium free atmosphere to remove any dissolved helium.

(4) A portion of the dried sample (5–20 g) is placed in an outgassed ampoule, which is then pumped to 10^{-4} Pa vacuum.

(5) After the evacuation to 10^{-4} Pa, the container is flame-sealed to ensure that no external He can diffuse into the evacuated sample over the long storage period.

(6) The glass containers are loaded on a suitable mass spectrometer to measure ^{3}He and ^{4}He. For dry samples, not all the He is necessarily admitted to the mass spectrometer, and controlled input of the available gas from the cryotrapping system is required.

(7) Possible contamination in the sample by atmospheric ^{3}He during sample storage is corrected for, assuming an atmospheric composition of ^{3}He/^{4}He = 1.384×10^{-6}. The determined tritium concentration is reported on a dry basis.

9.6. VALIDATION OF ANALYTICAL DATA

It is advisable to apply international standards for the measurement of ^{3}H. A number of standards have been issued that mainly focus on the measurement of ^{3}H in waters (e.g. ISO9698 [9.25]). Where standard methods are not available, laboratory developed procedures are typically implemented, which require robust validation to demonstrate their performance.

Validation of analytical measurement procedures is routinely achieved through use of reference materials, participation in proficiency testing programmes and laboratory intercomparison exercises. Proficiency testing programmes, typically organized by international organizations such as the IAEA (e.g. tritium intercomparisons TRIC2008 [9.71] and TRIC2018 [9.72]) or by national metrology institutes typically focus on the measurement of standardized ^{3}H aqueous solutions with the ^{3}H in the form of HTO. Although effective at assessing the accuracy of the measurement technique, such proficiency testing programmes do not permit a more thorough assessment of methodologies for non-aqueous ^{3}H measurement or measurement of ^{3}H in solid matrices.

Intercomparison exercises focussing on OBT are less common. The CANDU Owners Group and International OBT Working Group have organized a series of such intercomparisons covering a range of matrices including fish, Swiss chard, potato [9.44, 9.73], wheat [9.74] and sediments [9.75].

9.7. CONCLUSIONS

The radioactive isotope tritium and its radiogenic progeny ^{3}He are exceptionally powerful tracers of transient and steady state environmental processes. The acquisition over recent decades of a significant amount of high precision data on tritium concentration in environmental matrices has substantially contributed to advances in hydrological, atmospheric and terrestrial system studies.

Environmental tritium, originating from natural or anthropogenic sources, easily moves between various environmental compartments (atmosphere, hydrosphere, geosphere, biosphere). Because of this, and the often very low concentrations of tritium, a range of procedural approaches have been developed for its accurate and precise analysis. The quality and representativity of the measured data depend on many factors. These relate to the variability and magnitude of tritium source terms, its speciation in the environment, types of sample matrix as well as sampling and sample handling procedures, existing ranges of tritium concentration, possible interferences, sample contamination and the proper choice of analytical technology.

Tritium can replace stable hydrogen in various inorganic and organic molecules. Tritium may initially enter the environment as simple inorganic molecules (HT, HTO or CH_3T). In waters and soil, its prevalent form is HTO. Tritium atoms may also substitute for hydrogen in organic molecules, where the tritium may be present either in an exchangeable or a non-exchangeable form. Other distinct species, so far identified in some estuarine and river sediments, are persistent anthropogenic tritium organic compounds. These persistent tritiated organic compounds originated from specialized industrial sites where tritium was introduced into synthesized organic molecules.

Selection of an appropriate analytical procedure and its optimization depend on the type of sample being analysed and on whether ^{3}H activity concentration is sufficient for measurement or additional information on the ^{3}H distribution and speciation is required. The choice of an optimal measurement procedure is determined in part by practicalities such as available technologies but, more importantly, is guided by critical analytical sensitivity and reproducibility requirements. The two major measurement options are radiometric and mass spectrometric techniques. The long established and cost effective radiometric method of LSC typically allows the determination of tritium activity concentration in water above 1 Bq/L. The sensitivity of LSC can be enhanced by using electrolytic enrichment to preconcentrate tritium. An alternative radiometric method, gas flow proportional counting, can achieve detection limits of about 4 Bq/L.

The LSC method, with its many advantages, is not suited to the sensitivity demands of measuring the ultralow concentrations typical of ocean water and precipitation. With the best LSC instruments, the detection limit is mainly determined by instrument background and measurement time. Significantly stronger advantages are offered by noble gas source mass spectrometry (NGMS), which allows the measurement of tritium and its stable gaseous ^{3}He progeny. He ingrowth mass spectrometry can readily and precisely measure tritium activity concentrations down to 0.001 Bq/L. This method is typically utilized in specialized international laboratories that study variations in hydrological and atmospheric systems (ocean waters, precipitation, aquifers).

9.8. ACKNOWLEDGEMENT

The technical artwork support provided by K. Davis (UK) is gratefully acknowledged.

REFERENCES TO CHAPTER 9

[9.1] CROUDACE, I.W., WARWICK, P.E., MORRIS, J.E., Evidence for the preservation of technogenic tritiated organic compounds in an estuarine sedimentary environment, Environ. Sci. Technol. **46** 11 (2012) 5704–5712, https://doi.org/10.1021/es204247f

[9.2] KIM, S.B., BAGLAN, N., DAVIS, P.A., Current understanding of organically bound tritium (OBT) in the environment, J. Environ. Radioact. **126** (2013) 83–91, https://doi.org/10.1016/j.jenvrad.2013.07.011

[9.3] JEAN-BAPTISTE, P., et al., ^{3}He mass spectrometry for very low-level measurement of organic tritium in environmental samples, J. Environ. Radioact. **101** 2 (2010) 185–190,
https://doi.org/10.1016/j.jenvrad.2009.10.005

[9.4] KIM, S., KOROLEVYCH, V., Quantification of exchangeable and non-exchangeable organically bound tritium (OBT) in vegetation, J. Environ. Radioact. **118** (2013) 9–14,
https://doi.org/10.1016/j.jenvrad.2012.11.006

[9.5] POINTURIER, F., BAGLAN, N., ALANIC, G., A method for the determination of low-level organic-bound tritium activities in environmental samples, Appl. Radiat. Isot. **61** 2–3 (2004) 293–298,
https://doi.org/10.1016/j.apradiso.2004.03.024

[9.6] LE GOFF, P., et al., Measurement of tritium in the free water of milk: Spotting and quantifying some biases and proposing ways of improvement, J. Environ. Radioact. **127** (2014) 1–10,
https://doi.org/10.1016/j.jenvrad.2013.09.006

[9.7] MORGENSTERN, U., TAYLOR, C.B., Ultra low-level tritium measurement using electrolytic enrichment and LSC, Isot. Environ. Health Stud. **45** 2 (2009) 96–117,
https://doi.org/10.1080/10256010902931194

[9.8] NIVESSE, A., BAGLAN, N., MONTAVON, G., PÉRON, O., New insights into the accessibility of native cellulose to environmental contaminants toward tritium behavior prediction, J. Hazard. Mater. **420** (2021) 126619,
https://doi.org/10.1016/j.jhazmat.2021.126619

[9.9] WICKENDEN, D., Determination of tritiated water in biological matrices, Sci. Total Environ. **130** (1993) 121–128,
https://doi.org/10.1016/0048-9697(93)90065-E

[9.10] STROM, D., BOHNER, K., An inexpensive and practical ambient-temperature vacuum still for bioassays, Health Phys. **56** 3 (1989) 355–359.

[9.11] AKATA, N., et al., A new pretreatment technique for environmental tritium analysis with microwave heating method, Plasma Fusion Res. **11** (2016) 2405017,
https://doi.org/10.1585/pfr.11.2405017

[9.12] McCUBBIN, D., LEONARD, K.S., BAILEY, T.A., WILLIAMS, J., TOSSELL, P., Incorporation of organic tritium (^{3}H) by marine organisms and sediment in the Severn Estuary/Bristol Channel (UK), Mar. Pollut. Bull. **42** 10 (2001) 852–863,
https://doi.org/10.1016/S0025-326X(01)00039-X

[9.13] HUANG, Y.-J., JIANG, J., GUO, G.-Y., ZENG, F., LIU, X.-H., A wet-oxidation procedure of radioactive waste resin and waste concentrated liquid for ^{3}H and ^{14}C analysis, J. Radioanal. Nucl. Chem. **326** 1 (2020) 765–771,
https://doi.org/10.1007/s10967-020-07354-8

[9.14] AHN, H., et al., Application of a wet oxidation method for the quantification of ^{3}H and ^{14}C in low-level radwastes, Appl. Radiat. Isot. **81** (2013) 62–66,
https://doi.org/10.1016/j.apradiso.2013.03.059

[9.15] LOCKYER, J.F., LALLY, A.E., The determination of tritium, ^{14}C and ^{35}S in milk and crop samples using a combustion technique, Sci. Total Environ. **130** (1993) 337–344,
https://doi.org/10.1016/0048-9697(93)90088-N

[9.16] POINTURIER, F., BAGLAN, N., ALANIC, G., CHIAPPINI, R., Determination of organically bound tritium background level in biological samples from a wide area in the south-west of France, J. Environ. Radioact. **68** 2 (2003) 171–189,
https://doi.org/10.1016/S0265-931X(03)00053-5

[9.17] WARWICK, P., KIM, D., CROUDACE, I., OH, J., Effective desorption of tritium from diverse solid matrices and its application to routine analysis of decommissioning materials, Anal. Chim. Acta **676** 1–2 (2010) 93–102,
https://doi.org/10.1016/j.aca.2010.07.017

[9.18] DU, L., et al., An improved combustion apparatus for the determination of organically bound tritium in environmental samples, Appl. Radiat. Isot. **110** (2016) 218–223,
https://doi.org/10.1016/j.apradiso.2015.12.031

[9.19] NAYAK, S.R., et al., Organically bound tritium: Optimization of measurements in environmental matrices by combustion method and liquid scintillation spectrometry, J. Radioanal. Nucl. Chem. **319** 3 (2019) 917–926,
https://doi.org/10.1007/s10967-018-6395-y

[9.20] LIN, F., YU, W., GUO, J., LIAO, H., WANG, Y., A method for the determination of organically bound tritium in marine biota based on an improved tubular-combustion system, J. Environ. Radioact. **211** (2020) 106084,
https://doi.org/10.1016/j.jenvrad.2019.106084

[9.21] TRIVEDI, A., DUONG, T., LEON, J.W., LINAUSKAS, S.H., Measurement of Organically Bound Tritium in Urine and Feces, Report No. AECL-10925, Atomic Energy of Canada Limited, Chalk River (1993).

[9.22] CROUDACE, I.W., WARWICK, P., MARSH, R., A suite of robust radioanalytical techniques for the determination of tritium and other volatile radionuclides in decommissioning wastes and environmental matrices, Fusion Sci. Technol. **71** 3 (2017) 290–295,
https://doi.org/10.1080/15361055.2017.1293450

[9.23] MARSH, R.I., CROUDACE, I.W., WARWICK, P.E., COOPER, N., ST-AMANT, N., A new bomb-combustion system for tritium extraction, J. Radioanal. Nucl. Chem. **314** 2 (2017) 651–658,
https://doi.org/10.1007/s10967-017-5446-0

[9.24] WARWICK, P., CROUDACE, I., HOWARD, A.G., Improved technique for the routine determination of tritiated water in aqueous samples, Anal. Chim. Acta **382** 1–2 (1999) 225–231,
https://doi.org/10.1016/S0003-2670(98)00757-0

[9.25] INTERNATIONAL ORGANIZATION FOR STANDARDIZATION, Water Quality—Tritium—Test Method Using Liquid Scintillation Counting, ISO 9698:2019, ISO, Geneva (2019).

[9.26] AMERICAN PUBLIC HEALTH ASSOCIATION, "7500–^{3}H B liquid scintillation spectrometric method", Standard Methods for the Examination of Water and Wastewater, 19th edn, APHA Press, Washington, DC (1995), https://doi.org/10.2105/SMWW.2882.147

[9.27] GRÖNING, M., et al., Increasing the performance of tritium analysis by electrolytic enrichment, Isot. Environ. Health Stud. **45** 2 (2009) 118–125, https://doi.org/10.1080/10256010902872042

[9.28] LI, G., et al., RO film-based pretreatment method for tritium determination by LSC, Appl. Radiat. Isot. **166** (2020) 109343, https://doi.org/10.1016/j.apradiso.2020.109343

[9.29] ANAND, J., LAL, D., Synthesis of methane from water for tritium measurement, Nature **201** (1964) 775–777, https://doi.org/10.1038/201775a0

[9.30] POVINEC, P., Preparation of methane gas for proportional ^{3}H and ^{14}C counters, Radiochem. Radioanal. Lett. **9** (1972) 127–135.

[9.31] MORAN, J., et al., Improvements to sample processing and measurement to enable more widespread environmental application of tritium, Appl. Radiat. Isot. **126** (2017) 61–65, https://doi.org/10.1016/j.apradiso.2017.01.033

[9.32] GRABCZAK, J., KUC, T., ROZANSKI, K., FIEDOROWICZ, J., Routine tritium determination in water samples by gas proportional counting, Freib. Forschungshefte C **388** (1983) 179–193.

[9.33] BROWN, R.M., OGRAM, G.L., SPENCER, F.S., Oxidation and dispersion of HT in the environment: The August 1986 Field Experiment at Chalk River, Health Phys. **58** 2 (1990) 171–181, https://doi.org/10.1097/00004032-199002000-00004.

[9.34] UDA, T., SUGIYAMA, T., TANAKA, M., MUNAKATA, K., MOMOSHIMA, N., Developments of gaseous water, hydrogen and methane sampling system for environmental tritium monitoring, Fusion Eng. Des. **81** 8 (2006) 1385–1390, https://doi.org/10.1016/j.fusengdes.2005.08.077

[9.35] MACE, E.K., et al., First results of a simultaneous measurement of tritium and ^{14}C in an ultra-low-background proportional counter for environmental sources of methane, J. Environ. Radioact. **155** (2016) 122–129, https://doi.org/10.1016/j.jenvrad.2016.02.001

[9.36] EYROLLE, F., et al., An updated review on tritium in the environment, J. Environ. Radioact. **181** (2018) 128–137, https://doi.org/10.1016/j.jenvrad.2017.11.001

[9.37] KAUFMAN, S., LIBBY, W.F., The natural distribution of tritium, Phys. Rev. **93** 6 (1954) 1337–1344, https://doi.org/10.1103/PhysRev.93.1337

[9.38] KUMAR, B., et al., A compact tritium enrichment unit for large sample volumes with automated re-filling and higher enrichment factor, Appl. Radiat. Isot. **118** (2016) 80–86, https://doi.org/10.1016/j.apradiso.2016.07.018

[9.39] ALLEN, R., SMITH, D., OTLET, R., RAWSON, D., Low-level tritium measurements in water, Nucl. Instrum. Methods **45** 1 (1966) 61–71, https://doi.org/10.1016/0029-554X(66)90409-5

[9.40] METSON, P., A routine method for the determination of tritium by electrolytic concentration before liquid scintillation counting, Analyst, **1125** (1969) 1122–1129.

[9.41] WASSENAAR, L.I., KUMAR, B., DOUENCE, C., BELACHEW, D.L., AGGARWAL, P.K., Measurement of extremely ^{2}H-enriched water samples by laser spectrometry: Application to batch electrolytic concentration of environmental tritium samples, Rapid Commun. Mass Spectrom. **30** 3 (2016) 415–422, https://doi.org/10.1002/rcm.7459

[9.42] BELACHEW, D.L., et al., A laboratory information management system for the analysis of tritium (^{3}H) in environmental waters, Appl. Radiat. Isot. **137** (2018) 139–146, https://doi.org/10.1016/j.apradiso.2018.03.001

[9.43] MOMOSHIMA, N., NAGAO, Y., TOYOSHIMA, T., Electrolytic enrichment of tritium with solid polymer electrolyte for application to environmental measurements, Fusion Sci. Technol. **48** 1 (2005) 520–523, https://doi.org/10.13182/FST05-A980

[9.44] BAGLAN, N., et al., Organically bound tritium analysis in environmental samples, Fusion Sci. Technol. **67** 2 (2015) 250–253, https://doi.org/10.13182/FST14-T3

[9.45] POCHWALSKI, K., RADOSZEWSKI, T., Disintegration Rate Determination by Liquid Scintillation Counting Using the Triple to Double Coincidence Ratio (TDCR) Method, 1848/OPIDI/E/A, Institute of Nuclear Research, Warsaw (1979).

[9.46] BRODA, R., A review of the triple-to-double coincidence ratio (TDCR) method for standardizing radionuclides, Appl. Radiat. Isot. **58** 5 (2003) 585–594, https://doi.org/10.1016/S0969-8043(03)00056-3

[9.47] PRIYA, S., GOPALAKRISHNAN, R., GOSWAMI, A., TDCR measurements of ^{3}H, ^{63}Ni and ^{55}Fe using Hidex 300SL LSC device, J. Radioanal. Nucl. Chem. **302** 1 (2014) 353–359, https://doi.org/10.1007/s10967-014-3190-2

[9.48] JAKONIĆ, I., et al., Optimization of low-level LS counter Quantulus 1220 for tritium determination in water samples, Radiat. Phys. Chem. **98** (2014) 69–76, https://doi.org/10.1016/j.radphyschem.2014.01.012

[9.49] VARLAM, C., STEFANESCU, I., DULIU, O., FAURESCU, I., POPESCU, I., Applying direct liquid scintillation counting to low level tritium measurement, Appl. Radiat. Isot. **67** 5 (2009) 812–816, https://doi.org/10.1016/j.apradiso.2009.01.023

[9.50] GACA, P., WARWICK, P.E., CROUDACE, I.W., Liquid scintillation counters calibration stability over long timescales, J. Radioanal. Nucl. Chem. **314** 2 (2017) 753–760, https://doi.org/10.1007/s10967-017-5388-6

[9.51] AOYAMA, T., WATANABE, T., Windowless air proportional counters for detecting tritium β-rays, Nucl. Instrum. Methods Phys. Res. **205** 1–2 (1983) 311–317, https://doi.org/10.1016/0167-5087(83)90202-8

[9.52] AOYAMA, T., WATANABE, T., A new type of tritium-in-air monitor for fusion reactors, Fusion Eng. Des. **10** (1989) 423–427, https://doi.org/10.1016/0920-3796(89)90087-2

[9.53] SUCKOW, A., "Chapter 9: Analysis of radionuclides", Radioactivity in the Environment, Vol. 16 (FROEHLICH, K., Ed), Elsevier, Amsterdam (2009) 363–406.

[9.54] JENKINS, W.J., et al., A comprehensive global oceanic dataset of helium isotope and tritium measurements, Earth Syst. Sci. Data **11** 2 (2019) 441–454, https://doi.org/10.5194/essd-11-441-2019

[9.55] CLARKE, W.B., BEG, M., CRAIG, H., Excess ^{3}He in the sea: Evidence for terrestrial primodal helium, Earth Planet. Sci. Lett. **6** 3 (1969) 213–220, https://doi.org/10.1016/0012-821X(69)90093-4

[9.56] CLARKE, W.B., JENKINS, W., TOP, Z., Determination of tritium by mass spectrometric measurement of ^{3}He, Int. J. Appl. Radiat. Isot. **27** 9 (1976) 515–522, https://doi.org/10.1016/0020-708X(76)90082-X

[9.57] CHAPMAN, P., The world ocean circulation experiment (WOCE), Mar. Technol. Soc. J. **32** 3 (1998) 23–36.

[9.58] SCHLOSSER, P., STUTE, M., DÖRR, H., SONNTAG, C., MÜNNICH, K.O., Tritium/^{3}He dating of shallow groundwater, Earth Planet. Sci. Lett. **89** 3–4 (1988) 353–362, https://doi.org/10.1016/0012-821X(88)90122-7

[9.59] SCHLOSSER, P., STUTE, M., SONNTAG, C., MÜNNICH, K.O., Tritiogenic ^{3}He in shallow groundwater, Earth Planet. Sci. Lett. **94** 3–4 (1989) 245–256, https://doi.org/10.1016/0012-821X(89)90144-1

[9.60] BAYER, R., et al., "'Ultra low level' tritium measurements", Performance and Blank Components of a Mass Spectrometric System for Routine Measurement of Helium Isotopes and Tritium by the ^{3}He Ingrowth Method, Springer, Berlin (1989) 39–40, https://doi.org/10.1007/978-3-642-48373-8_6

[9.61] JEAN-BAPTISTE, P., MANTISI, F., DAPOIGNY, A., STIEVENARD, M., Design and performance of a mass spectrometric facility for measuring helium isotopes in natural waters and for low-level tritium determination by the ^{3}He ingrowth method, Int. J. Rad. Appl. Instrum. A **43** 7 (1992) 881–891, https://doi.org/10.1016/0883-2889(92)90150-D

[9.62] LUDIN, A., WEPPERNIG, R., BÖNISCH, G., SCHLOSSER, P., Mass Spectrometric Measurement of Helium Isotopes and Tritium in Water Samples, Technical Report, Lamont-Doherty Earth Observatory, Palisades, NY (1997).

[9.63] BEYERLE, U., et al., A mass spectrometric system for the analysis of noble gases and tritium from water samples, Environ. Sci. Technol. **34** 10 (2000) 2042–2050, https://doi.org/10.1021/es990840h

[9.64] SÜLTENFUSS, J., ROETHER, W., RHEIN, M., The Bremen mass spectrometric facility for the measurement of helium isotopes, neon, and tritium in water, Isot. Environ. Health Stud. **45** 2 (2009) 83–95, https://doi.org/10.1080/10256010902871929

[9.65] STANLEY, R.H., BASCHEK, B., LOTT, D.E., III, JENKINS, W.J., A new automated method for measuring noble gases and their isotopic ratios in water samples, Geochem. Geophys. Geosystems **10** 5 (2009) Q05008, https://doi.org/10.1029/2009GC002429

[9.66] PAPP, L., PALCSU, L., MAJOR, Z., RINYU, L., TÓTH, I., A mass spectrometric line for tritium analysis of water and noble gas measurements from different water amounts in the range of microlitres and millilitres, Isot. Environ. Health Stud. **48** 4 (2012) 494–511, https://doi.org/10.1080/10256016.2012.679935

[9.67] HUNT, A.G., U.S. Geological Survey Noble Gas Laboratory's Standard Operating Procedures for the Measurement of Dissolved Gas in Water Samples, U.S. Geological Survey Techniques and Methods 5–A11, USGS, Reston, VA (2015), https://doi.org/10.3133/tm5A11

[9.68] MABRY, J., et al., Development of the ^{3}He mass spectrometric low-level tritium analytical facility at the IAEA, J. Anal. At. Spectrom. **37** 12 (2022) 2502–2509, https://doi.org/10.1039/D2JA00173J

[9.69] SUCKOW, A., SCHLOSSER, P., RUPP, H., BAYER, R., Diffusion and permeation constants of helium in various glasses, Glass Technol. **31** 4 (1990) 160–164.

[9.70] TAMARI, T., et al., OBT measurement of vegetation by mass spectrometry and radiometry, Fusion Sci. Technol. **60** 4 (2011) 1252–1255, https://doi.org/10.13182/FST11-A12657

[9.71] GRÖNING, M., TATZBER, H., TRINKL, A., KLAUS, P., VAN DUREN, M., Eighth IAEA Interlaboratory Comparison on the Determination of Low-Level Tritium Activities in Water (TRIC2008), IAEA, Vienna (2009).

[9.72] COPIA, L., et al., Proficiency testing of 78 international laboratories measuring tritium in environmental waters by decay counting and mass spectrometry for age dating and water resources assessment, Rapid Commun. Mass Spectrom. **34** 17 (2020) e8832, https://doi.org/10.1002/rcm.8832

[9.73] KIM, S., et al., Canadian inter-laboratory organically bound tritium (OBT) analysis exercise, J. Environ. Radioact. **150** (2015) 236–241, https://doi.org/10.1016/j.jenvrad.2015.08.021

[9.74] BAGLAN, N., et al., Feedback of the third interlaboratory exercise organised on wheat in the framework of the OBT working group, J. Environ. Radioact. **181** (2018) 52–61, https://doi.org/10.1016/j.jenvrad.2017.09.009

[9.75] WARWICK, P.E., CROUDACE, I.W., MARSH, R.M., BAGLAN, N., KIM, S.B., A new reference material for tritium organic molecules in sediment: Results of an international intercomparison exercise, Geostand. Geoanal. Res. **42** 2 (2018) 253–262,
https://doi.org/10.1111/ggr.12207

LIST OF ABBREVIATIONS

ALPS	Advanced Liquid Processing System
AWE	Atomic Weapons Establishment
BWR	boiling light water cooled and moderated reactor
CANDU	Canadian deuterium–uranium reactor
CEA	French Atomic Energy and Alternative Energies Commission
CLIVAR	Climate and Ocean: Variability, Predictability and Change
CNL	Canadian Nuclear Laboratories Ltd
CR	concentration ratio
CRL	Chalk River Laboratories
DCRL	Derived Consideration Reference Level
DOE	US Department of Energy
EMRAS	Environmental Modelling for Radiation Safety
E-OBT	exchangeable organically bound tritium
FBR	fast breeder reactor
GCR	galactic cosmic ray
GEOSECS	Geochemical Ocean Sections Survey
GNIP	Global Network of Isotopes in Precipitation
GO-SHIP	Global Ocean Ship-based Hydrographic Investigations Program
GS-SFMS	gas source sector field mass spectrometer
HWR	heavy water reactor
ICRP	International Commission on Radiological Protection
JET	Joint European Torus
LET	linear energy transfer
LLNL	Lawrence Livermore National Laboratory
LSC	liquid scintillation counting
LWGR	light-water cooled, graphite moderated reactor
LWR	light water reactor
NE-OBT	non-exchangeable organically bound tritium
NGMS	noble gas mass spectrometry
NPP	nuclear power plant
NRU	National Research Universal
OBT	organically bound tritium
PHWR	pressurized heavy water reactor
PMT	photomultiplier tube
POM	particulate organic matter
PWR	pressurized light water moderated and cooled reactor

RBE	relative biological effectiveness
SAVE	South Atlantic Ventilation Experiment
SD	standard deviation
SFRP	spent fuel reprocessing plant
SPE	solid polymer electrolyte
SRBT	SRB Technologies, Inc.
SRNL	Savannah River National Laboratory
SRS	Savannah River site
STS	Semipalatinsk test site
TEPCO	Tokyo Electric Power Company Holding, Inc.
TFWT	tissue-free water tritium
TTO	Transient Tracers in the Ocean
T-TOM	technogenic tritiated organic molecules
TU	tritium unit
VNIIEF	All-Russian Research Institute of Experimental Physics
VNIITF	All-Russian Research Institute of Technical Physics
WMO	World Meteorological Organization
WOCE	World Ocean Circulation Experiment
WOMARS	Worldwide Marine Radioactivity Studies

CONTRIBUTORS TO DRAFTING AND REVIEW

Baglan, N.	Autorité de Sûreté Nucléaire (ASN), France
Balonov, M.I.	Institute of Radiation Hygiene, Russian Federation
Connan, O.	Institut de Radioprotection et de Sûreté Nucléaire (IRSN), France
Croudace, I.W.	University of Southampton, United Kingdom
Eyrolle, F.	Institut de Radioprotection et de Sûreté Nucléaire (IRSN), France
Finashov, L.	Southern Urals Biophysics Institute (SUBI), Russian Federation
Gröning, M.	International Atomic Energy Agency
Kakiuchi, H.	Institute for Environmental Sciences, Japan
Kim, S.B.	Canadian Nuclear Laboratories, Canada
Lukashenko, S.N.	Russian Institute of Radiology and Agroecology, Russian Federation
Ramzaev, V.P.	Institute of Radiation Hygiene, Russian Federation
Sültenfuß, J.	University of Bremen, Germany
Terzer-Wassmuth, S.	International Atomic Energy Agency
Ulanowski, A.	International Atomic Energy Agency
Vostrotin, V.	Southern Urals Biophysics Institute (SUBI), Russian Federation
Warwick, P.E.	University of Southampton, United Kingdom
Yankovich, T.	International Atomic Energy Agency
Yanov, A.	Southern Urals Biophysics Institute (SUBI), Russian Federation

Consultants Meetings

Vienna, Austria: 5–9 November 2018, 14–18 October 2019, 14–18 June 2021

IAEA
International Atomic Energy Agency

CONTACT IAEA PUBLISHING

Feedback on IAEA publications may be given via the on-line form available at:
www.iaea.org/publications/feedback

This form may also be used to report safety issues or environmental queries concerning IAEA publications.

Alternatively, contact IAEA Publishing:

Publishing Section
International Atomic Energy Agency
Vienna International Centre, PO Box 100, 1400 Vienna, Austria
Telephone: +43 1 2600 22529 or 22530
Email: sales.publications@iaea.org
www.iaea.org/publications

Priced and unpriced IAEA publications may be ordered directly from the IAEA.

ORDERING LOCALLY

Priced IAEA publications may be purchased from regional distributors and from major local booksellers.

24-03571E-T

Kolibri 360 GmbH
Ferdinand-Jühlke-Straße 7
99095 Erfurt, Deutschland
produktsicherheit@kolibri360.de